Recent Advances in Information-Centric Networks (ICNs)

Recent Advances in Information-Centric Networks (ICNs)

Editors

José Carlos López Ardao
Miguel Rodríguez Pérez
Sergio Herrería Alonso

MDPI

Basel • Beijing • Wuhan • Barcelona • Belgrade • Novi Sad • Cluj • Manchester

Editors
José Carlos López Ardao
Universidade de Vigo
Vigo
Spain

Miguel Rodríguez Pérez
Universidade de Vigo
Vigo
Spain

Sergio Herrería Alonso
Universidade de Vigo
Vigo
Spain

Editorial Office
MDPI
St. Alban-Anlage 66
4052 Basel, Switzerland

This is a reprint of articles from the Special Issue published online in the open access journal *Future Internet* (ISSN 1999-5903) (available at: https://www.mdpi.com/journal/futureinternet/special_issues/ICN_NDN).

For citation purposes, cite each article independently as indicated on the article page online and as indicated below:

Lastname, A.A.; Lastname, B.B. Article Title. *Journal Name* **Year**, *Volume Number*, Page Range.

ISBN 978-3-0365-9857-4 (Hbk)
ISBN 978-3-0365-9858-1 (PDF)
doi.org/10.3390/books978-3-0365-9858-1

Contents

About the Editors . . . . **vii**

José Carlos López-Ardao, Miguel Rodríguez-Pérez and Sergio Herrería-Alonso
Recent Advances in Information-Centric Networks (ICNs)
Reprinted from: *Future Internet* **2023**, *15*, 392, doi:10.3390/fi15120392 . . . . **1**

Salahadin Seid Musa, Marco Zennaro, Mulugeta Libsie and Ermanno Pietrosemoli
Convergence of Information-Centric Networks and Edge Intelligence for IoV: Challenges and Future Directions
Reprinted from: *Future Internet* **2022**, *14*, 192, doi:10.3390/fi14070192 . . . . **7**

Shoujiang Dang and Rui Han
An In-Network Cooperative Storage Schema Based on Neighbor Offloading in a Programmable Data Plane
Reprinted from: *Future Internet* **2022**, *14*, 18, doi:10.3390/fi14010018 . . . . **39**

Abdelkader Tayeb Herouala, Chaker Abdelaziz Kerrache, Benameur Ziani, Carlos T. Calafate, Nasreddine Lagraa and Abdou el Karim Tahari
Controlling the Trade-Off between Resource Efficiency and User Satisfaction in NDNs Based on Naïve Bayes Data Classification and Lagrange Method
Reprinted from: *Future Internet* **2022**, *14*, 48, doi:10.3390/fi14020048 . . . . **59**

Leanna Vidya Yovita, Nana Rachmana Syambas, Ian Joseph Matheus Edward
Weighted-CAPIC Caching Algorithm for Priority Traffic in Named Data Network
Reprinted from: *Future Internet* **2022**, *14*, 84, doi:10.3390/fi14030084 . . . . **73**

Luís Gameiro, Carlos Senna and Miguel Luís
Insights from the Experimentation of Named Data Networks in Mobile Wireless Environments
Reprinted from: *Future Internet* **2022**, *14*, 196, doi:10.3390/fi14070196 . . . . **89**

Pablo Iglesias-Sanuy, José Carlos López-Ardao, Miguel Rodríguez-Pérez, Sergio Herrería-Alonso, Andrés Suárez-González and Raúl F. Rodríguez-Rubio
An Efficient Location-Based Forwarding Strategy for Named Data Networking and LEO Satellite Communications
Reprinted from: *Future Internet* **2022**, *14*, 285, doi:10.3390/fi14100285 . . . . **107**

Yong Xu, Hong Ni and Xiaoyong Zhu
A Novel Multipath Transmission Scheme for Information-Centric Networking
Reprinted from: *Future Internet* **2023**, *15*, 80, doi:10.3390/fi15020080 . . . . **119**

Zhiyuan Wang, Hong Ni and Rui Han
A Replica-Selection Algorithm Based on Transmission Completion Time Estimation in ICN
Reprinted from: *Future Internet* **2023**, *15*, 120, doi:10.3390/fi15040120 . . . . **135**

Ahmed Benmoussa, Chaker Abdelaziz Kerrache, Carlos T. Calafate, and Nasreddine Lagraa
NDN-BDA: A Blockchain-Based Decentralized Data Authentication Mechanism for Vehicular Named Data Networking
Reprinted from: *Future Internet* **2023**, *15*, 167, doi:10.3390/fi15050167 . . . . **153**

Jiacheng Hou, Tianhao Tao, Haoye Lu and Amiya Nayak
Intelligent Caching with Graph Neural Network-Based Deep Reinforcement Learning on SDN-Based ICN
Reprinted from: *Future Internet* **2023**, *15*, 251, doi:10.3390/fi15080251 . . . . **171**

About the Editors

José Carlos López Ardao

José Carlos López Ardao was born in Vigo, Spain, in 1967. He received his M.Sc. and Ph.D. in telecommunication engineering from the University of Vigo, Spain, in 1990 and 1999, respectively. He is currently an Associate Professor with the Department of Telematics Engineering, Universidade de Vigo, where he is also an affiliated member of the co-located Networking Laboratory. He has co-authored more than 60 papers for conferences and journals. His main research interests include energy-efficient networking, IoT technologies, new protocols and architectures for the future Internet, and technology and innovation in education (social learning, learning analytics, gamification, and flipped learning).

Miguel Rodríguez Pérez

Miguel Rodríguez Pérez was born in Vigo, Spain, in 1978. He received his M.Sc. and Ph.D. in telecommunication engineering from the University of Vigo, Spain, in 2001 and 2006, respectively. He is currently an Associate Professor with the Department of Telematics Engineering, Universidade de Vigo, where he is also an affiliated member of the co-located Networking Laboratory. He has co-authored more than 60 papers for conferences and journals. His research interests, from performance evaluation and congestion control protocols to energy efficiency, focus primarily on computer networks. In recent years, he has explored how satellite networks can benefit from new network architectures.

Sergio Herrería Alonso

Sergio Herrería Alonso received his M.Sc. and Ph.D. in telecommunication engineering from the University of Vigo, Spain, in 2001 and 2006, respectively. Since 2010, he has been an Associate Professor with the Department of Telematics Engineering, University of Vigo, where he is also an affiliated member of the co-located Networking Laboratory. He has co-authored more than 50 papers in peer-reviewed international conferences and journals, most of them with a high impact factor (Q1 and Q2 quartiles) in the WOS Journal Citation Report. His main research interests include energy-efficient networking, IoT technologies, and new protocols and architectures for the future Internet.

future internet

Editorial

Recent Advances in Information-Centric Networks (ICNs)

José Carlos López-Ardao *, Miguel Rodríguez-Pérez and Sergio Herrería-Alonso

atlanTTic Research Center for Telecommunication Technologies, Universidade de Vigo, 36310 Vigo, Spain; miguel@det.uvigo.gal (M.R.-P.); sha@det.uvigo.es (S.H.-A.)
* Correspondence: jardao@det.uvigo.es

The great success of the Internet has been essentially based on the simplicity and versatility of its TCP/IP architecture, which imposes almost no restrictions on either the underlying network technology or on the data being transmitted. In essence, applications built on top of TCP/IP are provided with a point-to-point channel that can be used to transmit arbitrary information.

This approach has served quite satisfactorily for many years, but it no longer suits some of the new communication paradigms currently being used on the Internet [1]. Nowadays, most Internet applications are no longer interested in using a point-to-point channel to exchange traffic between two endpoints. In contrast, what they really want is to access information that is available in the other network nodes. Information-centric networks (ICNs) offer a service where users can directly obtain pieces of information, regardless of their location. This service is more suited to current network uses and to support the highly dynamic nature of mobile networks. Additionally, the legacy point-to-point communication model can be provided on top of it, if so desired.

Named data networking (NDN) is an implementation of ICN that has emerged as a promising candidate for future Internet architecture. In contrast to traditional networking protocols, NDN focuses on the content itself, rather than on its location. NDN enables name-based routing and location-independent data retrieval [2].

This Special Issue, titled "Recent Advances in Information-Centric Networks (ICNs)", aims to highlight and to present recent advances about different problems related to using information-centric network architecture. It contains nine high-quality research papers and one review paper. As usual, the *Future Internet* journal standards required that all the submitted manuscripts went through a rigorous peer-review process.

In the following, we will use the authors' own words to avoid misinterpretation and to provide a more accurate presentation of the contributions of each paper.

The work by Dang and Han (contribution 1), titled "An In-Network Cooperative Storage Schema Based on Neighbor Offloading in a Programmable Data Plane", was the first published paper in this Special Issue. Numerous studies have been conducted on distributed storage in academia and industry, but most storage systems are constructed at the application layer. As a result of the recent development of programmable network technology, solutions have been developed that use in-network computing to offload part of the storage function into the network [3]. This paper proposes the realization of data storage in the forwarding process and the cooperative offloading of the data beyond the local storage capacity to neighboring storage nodes. The proposed solution consists of dynamically splitting traffic among selected neighbor nodes to sequentially amortize excess traffic. The neighbor selection mechanism is based on the Local Name Mapping and Resolution System, in which the node weights are computed through combining the link bandwidth and the node storage capability and determining whether to split traffic via comparing normalized weight values with thresholds. Evaluation shows that the proposed schema is more efficient compared with end-to-end transmission and ECMP in terms of bandwidth usage and transfer time.

In the second contribution, titled "Controlling the Trade-Off between Resource Efficiency and User Satisfaction in NDNs Based on Naïve Bayes Data Classification and

Citation: López-Ardao, J.C.; Rodríguez-Pérez, M.; Herrería-Alonso, S. Recent Advances in Information-Centric Networks (ICNs). *Future Internet* **2023**, *15*, 392. https://doi.org/10.3390/fi15120392

Received: 28 November 2023
Accepted: 30 November 2023
Published: 1 December 2023

Lagrange Method", Herouala et al. address the trade-offs between resource efficiency and user satisfaction in the limited environments of named data networks (NDNs). The authors propose a strategy, named RADC (resource allocation-based data classification), which aims to manage such trade-offs through controlling the fairness index of the system. As indicated in the title, the method uses a machine learning technique based on multinomial naïve Bayes to classify the received contents, together with an adaptive resource allocation strategy based on the Lagrange method. Simulation experiments confirm the potential usefulness of the strategy to manage the trade-off between efficiency and user satisfaction.

One of the most important components in NDN is the cache strategy, since it improves bandwidth usage, server load, and service time [4]. Yovita et al. address this issue in the third contribution, titled "Weighted-CAPIC Caching Algorithm for Priority Traffic in Named Data Network". They propose the Weighted-CAPIC caching algorithm to improve the priority class performance, through adjusting different weights for every content class in addition to considering the variation of the content in providing a cache portion in the NDN router, without having to coordinate with other routers. The work shows that the proposed caching algorithm provides a higher cache hit ratio for both the priority class and the whole system.

Another work related to cache issues is presented in the paper by Hou et al. titled "Intelligent Caching with Graph Neural Network-Based Deep Reinforcement Learning on SDN-Based ICN" (contribution 10), the last paper to be accepted and published. The work focuses on enhancing cache performance through maximizing the cache hit ratio in software-defined networking (SDN), that is, a context where the ICN offers flexibility, scalability, and efficient resource management [5]. The authors propose a statistical model that generates users' content preferences, incorporating key elements observed in real-world scenarios. Furthermore, they introduce a graph neural network–double deep Q-network (GNN-DDQN) agent to make caching decisions for each node based on the user request history. Simulation results demonstrate that this caching strategy achieves a cache hit ratio significantly higher than the state-of-the-art policy.

The Internet of Vehicles (IoV) is also a promising research area in the field of the Internet of Things [6]. Advances in ICN, edge computing (EC), and artificial intelligence (AI) will transform and help to realize the Intelligent Internet of Vehicles (IIoV). In this context, in contribution 4, Musa et al. provide us with a comprehensive review of utilizing intelligence in EC and ICN to address current challenges in their application to the IIoV. In particular, they focus on intelligent edge computing and networking, offloading, intelligent mobility-aware caching and forwarding, and overall network performance. Furthermore, they discuss potential solutions and highlight potential research directions.

Despite the large amount of works published in the literature about different implementations of NDN architecture, there are few NDN implementations in real networks [7]. Contribution 5 by Gameiro et al., titled "Insights from the Experimentation of Named Data Networks in Mobile Wireless Environment", evaluates the performance of an NDN-based implementation in a mobile wireless network, as part of a smart city infrastructure. This work is especially interesting since ICN has been considered an alternative for overcoming the drawbacks of host-centric architectures when applied to mobile Internet of Things (IoT) networks. Such drawbacks are the frequent topology changes and intermittent connectivity caused by the mobility of the network nodes. The implementation is evaluated in three scenarios, and the results show that the implementation works properly, also deriving insights about correct NDN parameterization.

Contribution 6 brings us another interesting area of application for NDN architecture: satellite communications using LEO constellations with inter-satellite links (ISLs). At present, it is not clear which will be the networking design of these large constellations of LEO satellites, but the idea of using NDN in them is gradually gaining attention since this architecture is especially convenient for highly dynamic network environments [8,9]. Its strongest advantage over IP is mobility management, which is one of the biggest challenges of LEO constellations, given the high orbital speed of these satellites.

Among other native facilities, such as inbuilt security, NDN readily supports the mobility of clients, thus helping to overcome one of the main problems raised in LEO satellite networks. In this work, Iglesias-Sanuy et al. propose a new location-based forwarding strategy for LEO satellite networks that takes advantage of the knowledge of the relative position of the satellites and the grid structure formed by the ISLs. Forwarding at each node is performed using only local information (node and destination locations), without the need of interchanging information between nodes, as is the case with conventional routing protocols. Simulation experiments show that the proposed forwarding strategy is a good candidate to promote the efficient and effective future use of NDN architecture in LEO satellite networks.

Contribution 7, by Xu et al., is closely related to the previous paper. In this work, a novel multipath transmission scheme for NDN is presented, designing a path management mechanism, including path selection and path switching, that can determine the initial path based on historical transmission information and switch to other optimal paths according to the congestion degree during transmission. The experimental results show that the proposed method can improve the average throughput and reduce the average flow completion time and the average chunk completion time.

As the routers in the ICN network can cache the content, there are multiple replicas of a given content in the network. So, when a user requests some content, it is very important to find and select a suitable replica node to improve the user download rate and network throughput, which has been studied by many researchers [10,11]. Contribution 8 by Wang et al. addresses this issue and proposes a replica-selection algorithm to reduce the transmission completion time. The algorithm estimates the transmission completion time of different replica nodes and selects the smallest one. When no replica is found, the nearest-replica algorithm will be used, and so the traffic in the core network will not increase. Experiments show that the algorithm effectively improves the user's download rate and edge node throughput, reduces download rate fluctuations and user download delay, and improves fairness.

Another important issue related to NDN is security. The security features of the NDN design are intrinsic and require signatures to be used in data packets [12]. Producers are responsible for digitally signing all data they create, which enables consumers to validate the authenticity of the retrieved information. Additionally, any NDN node can store data, and consumers can receive data packets from any source. The NDN architecture relies on several security components to ensure data security [13]. But this data verification process may cause significant delays, especially in mobile and vehicular networks. This aspect makes it unsuitable for time-critical and sensitive applications such as the sharing of safety messages. This problem is addressed in contribution 9, where Benmoussa et al. propose NDN-BDA, a blockchain-based decentralized mechanism that enables local data verification while providing a secure authentication mechanism for producer vehicles. By using blockchain, a novel proposal in this field, there is no need for centralized authorities, thereby reducing associated overhead. NDN-BDA was thoroughly evaluated for efficiency against various attacks, which makes it a promising solution to provide a faster and more efficient data authenticity mechanism for NDN-based vehicular networks.

List of Contributions

1. Dang, S.; Han, R. An In-Network Cooperative Storage Schema Based on Neighbor Offloading in a Programmable Data Plane. *Future Internet* **2022**, *14*, 18. https://doi.org/10.3390/fi14010018.
2. Herouala, A.T.; Kerrache, C.A.; Ziani, B.; Calafate, C.T.; Lagraa, N.; Tahari, A.e.K. Controlling the Trade-Off between Resource Efficiency and User Satisfaction in NDNs Based on Naïve Bayes Data Classification and Lagrange Method. *Future Internet* **2022**, *14*, 48. https://doi.org/10.3390/fi14020048.

3. Yovita, L.V.; Syambas, N.R.; Edward, I.J.M. Weighted-CAPIC Caching Algorithm for Priority Traffic in Named Data Network. *Future Internet* **2022**, *14*, 84. https://doi.org/10.3390/fi14030084.
4. Musa, S.S.; Zennaro, M.; Libsie, M.; Pietrosemoli, E. Convergence of Information-Centric Networks and Edge Intelligence for IoV: Challenges and Future Directions. *Future Internet* **2022**, *14*, 192. https://doi.org/10.3390/fi14070192.
5. Gameiro, L.; Senna, C.; Luís, M. Insights from the Experimentation of Named Data Networks in Mobile Wireless Environments. *Future Internet* **2022**, *14*, 196. https://doi.org/10.3390/fi14070196.
6. Iglesias-Sanuy, P.; López-Ardao, J.C.; Rodríguez-Pérez, M.; Herrería-Alonso, S.; Suárez-González, A.; Rodríguez-Rubio, R.F. An Efficient Location-Based Forwarding Strategy for Named Data Networking and LEO Satellite Communications. *Future Internet* **2022**, *14*, 285. https://doi.org/10.3390/fi14100285.
7. Xu, Y.; Ni, H.; Zhu, X. A Novel Multipath Transmission Scheme for Information-Centric Networking. *Future Internet* **2023**, *15*, 80. https://doi.org/10.3390/fi15020080.
8. Wang, Z.; Ni, H.; Han, R. A Replica-Selection Algorithm Based on Transmission Completion Time Estimation in ICN. *Future Internet* **2023**, *15*, 120. https://doi.org/10.3390/fi15040120.
9. Benmoussa, A.; Kerrache, C.A.; Calafate, C.T.; Lagraa, N. NDN-BDA: A Blockchain-Based Decentralized Data Authentication Mechanism for Vehicular Named Data Networking. *Future Internet* **2023**, *15*, 167. https://doi.org/10.3390/fi15050167.
10. Hou, J.; Tao, T.; Lu, H.; Nayak, A. Intelligent Caching with Graph Neural Network-Based Deep Reinforcement Learning on SDN-Based ICN. *Future Internet* **2023**, *15*, 251. https://doi.org/10.3390/fi15080251.

Funding: This work has received financial support from grant PID2020-113240RB-I00, financed by MCIN/AEI/10.13039/501100011033.

Conflicts of Interest: The authors declare no conflict of interest.

References

1. Zhang, L.; Afanasyev, A.; Burke, J.; Jacobson, V.; Claffy, K.; Crowley, P.; Papadopoulos, C.; Wang, L.; Zhang, B. Named data networking. *ACM SIGCOMM Comput. Commun. Rev.* **2014**, *44*, 66–73. [CrossRef]
2. Saxena, D.; Raychoudhury, V.; Suri, N.; Becker, C.; Cao, J. Named Data Networking: A survey. *Comput. Sci. Rev.* **2016**, *19*, 15–55. [CrossRef]
3. Liu, M.; Luo, L.; Nelson, J.; Ceze, L.; Krishnamurthy, A.; Atreya, K. IncBricks: Toward In-Network Computation with an In-Network Cache. In Proceedings of the Twenty-Second International Conference on Architectural Support for Programming Languages and Operating Systems, Xi'an, China, 8–12 April 2017; Volume 52, pp. 795–809. [CrossRef]
4. Jin, H.; Xu, D.; Zhao, C.; Liang, D. Information-centric mobile caching network frameworks and caching optimization: A survey. *Eurasip J. Wirel. Commun. Netw.* **2017**, *2017*, 33. [CrossRef]
5. Aldaoud, M.; Al-Abri, D.; Awadalla, M.; Kausar, F. Leveraging ICN and SDN for Future Internet Architecture: A Survey. *Electronics* **2023**, *12*, 1723. [CrossRef]
6. Kaiwartya, O.; Abdullah, A.H.; Cao, Y.; Altameem, A.; Prasad, M.; Lin, C.T.; Liu, X. Internet of Vehicles: Motivation, Layered Architecture, Network Model, Challenges, and Future Aspects. *IEEE Access* **2016**, *4*, 5356–5373. [CrossRef]
7. Aboodi, A.; Wan, T.C.; Sodhy, G.C. Survey on the Incorporation of NDN/CCN in IoT. *IEEE Access* **2019**, *7*, 71827–71858. [CrossRef]
8. Bhattacherjee, D.; Singla, A. Network topology design at 27,000 km/h. In Proceedings of the 15th International Conference on Emerging Networking Experiments and Technologies, Orlando, FL, USA, 9–12 December 2019; pp. 341–354. [CrossRef]
9. Pan, H.; Yao, H.; Mai, T.; Zhang, N.; Liu, Y. Scalable Traffic Control Using Programmable Data Planes in a Space Information Network. *IEEE Netw.* **2021**, *35*, 35–41. [CrossRef]
10. Ioannou, A.; Weber, S. A Survey of Caching Policies and Forwarding Mechanisms in Information-Centric Networking. *IEEE Commun. Surv. Tutor.* **2016**, *18*, 2847–2886. [CrossRef]
11. Song, Y.; Ni, H.; Zhu, X. An Enhanced Replica Selection Approach Based on Distance Constraint in ICN. *Electronics* **2021**, *10*, 490. [CrossRef]

12. Zhang, Z.; Yu, Y.; Zhang, H.; Newberry, E.; Mastorakis, S.; Li, Y.; Afanasyev, A.; Zhang, L. An overview of security support in named data networking. *IEEE Commun. Mag.* **2018**, *56*, 62–68. [CrossRef]
13. Tehrani, P.F.; Osterweil, E.; Schmidt, T.C.; Wählisch, M. SoK: Public key and namespace management in NDN. In Proceedings of the 9th ACM Conference on Information-Centric Networking, Osaka, Japan, 19–21 September 2022; pp. 67–79. [CrossRef]

future internet

Review

Convergence of Information-Centric Networks and Edge Intelligence for IoV: Challenges and Future Directions

Salahadin Seid Musa [1,2], Marco Zennaro [2,*], Mulugeta Libsie [1] and Ermanno Pietrosemoli [2]

1 Department of Computer Science, Addis Ababa University, Addis Ababa P.O. Box 1176, Ethiopia; salubinseid@gmail.com (S.S.M.); mulugeta.libsie@aau.edu.et (M.L.)
2 STI Unit, Abdus Salam International Centre for Theoretical Physics, 34151 Trieste, Italy; ermanno@ictp.it
* Correspondence: mzennaro@ictp.it

Abstract: Recently the Internet of Vehicles (IoV) has become a promising research area in the field of the Internet of Things (IoT), which enables vehicles to communicate and exchange real-time information with each other, as well as with infrastructure, people, and other sensors and actuators through various communication interfaces. The realization of IoV networks faces various communication and networking challenges to meet stringent requirements of low latency, dynamic topology, high data-rate connectivity, resource allocation, multiple access, and QoS. Advances in information-centric networks (ICN), edge computing (EC), and artificial intelligence (AI) will transform and help to realize the Intelligent Internet of Vehicles (IIoV). Information-centric networks have emerged as a paradigm promising to cope with the limitations of the current host-based network architecture (TCP/IP-based networks) by providing mobility support, efficient content distribution, scalability and security based on content names, regardless of their location. Edge computing (EC), on the other hand, is a key paradigm to provide computation, storage and other cloud services in close proximity to where they are requested, thus enabling the support of real-time services. It is promising for computation-intensive applications, such as autonomous and cooperative driving, and to alleviate storage burdens (by caching). AI has recently emerged as a powerful tool to break through obstacles in various research areas including that of intelligent transport systems (ITS). ITS are smart enough to make decisions based on the status of a great variety of inputs. The convergence of ICN and EC with AI empowerment will bring new opportunities while also raising not-yet-explored obstacles to realize Intelligent IoV. In this paper, we discuss the applicability of AI techniques in solving challenging vehicular problems and enhancing the learning capacity of edge devices and ICN networks. A comprehensive review is provided of utilizing intelligence in EC and ICN to address current challenges in their application to IIoV. In particular, we focus on intelligent edge computing and networking, offloading, intelligent mobility-aware caching and forwarding and overall network performance. Furthermore, we discuss potential solutions to the presented issues. Finally, we highlight potential research directions which may illuminate efforts to develop new intelligent IoV applications.

Keywords: Internet of Vehicles; AI; edge intelligence; information-centric network; edge computing

Citation: Musa, S.S.; Zennaro, M.; Libsie, M.; Pietrosemoli, E. Convergence of Information-Centric Networks and Edge Intelligence for IoV: Challenges and Future Directions. *Future Internet* **2022**, *14*, 192. https://doi.org/10.3390/fi14070192

Academic Editors: José Carlos Lopez-Ardao, Miguel Rodríguez Pérez and Sergio Herrería Alonso

Received: 1 May 2022
Accepted: 22 June 2022
Published: 25 June 2022

1. Introduction

Advancements in digital infrastructure, computing, and learning algorithms have enabled the emergence of massive connectivity and digital intelligence. In particular, the growth in the Internet of Things (IoT) has facilitated the connection of a number of smart devices, including smart vehicles, vehicles equipped with advanced computing, sensing, networking, and communication technologies, to the Internet in what is known as the Internet of Vehicles (IoV). The original intention of vehicular networking, a key driver for intelligent transport systems (ITS), was to bring connectivity to vehicles [1].

Beyond mere connectivity, however, IoV aims to significantly transform existing ITS by enabling seamless cooperative driving in autonomous vehicles, which reduces the risk of traffic accidents. It also enhances infotainment, real-time tracking, traffic information and parking management, thus reducing commuting time and CO_2 emissions [2]. Therefore, the future of IoV in the era of 5G/B5G [3] and artificial intelligence (AI) has attracted attention from both academia and industry as a means to realize the Intelligent Internet of Vehicles (IIoV).

The full scale realization of IoV is challenged by various issues, including the heterogeneity of networking and communications interfaces, the requirements of latency sensitive applications, the heterogeneity of vehicle resources, security concerns, reliability, handling and processing of generated data and the lack of an accepted framework that supports collaborative IoV networks. Moreover, these issues are further exacerbated by frequent changes in network topology as a result of vehicular mobility. Some of these challenges have been addressed using emerging technologies, such as edge computing, information centric networks, multi-access radio technology and AI.

Edge computing [4] is a relatively recent computing paradigm that aims to provide computational, storage, and other cloud services in close proximity to the devices using them, thus achieving low latency, high bandwidth, and location and mobility-aware contextual services. Related edge-computing approaches, such as fog computing [5], cloudlets [6], and mobile edge computing have been proposed. Intelligent caching, intelligent forwarding, and computational offloading are some of the future directions in edge intelligence.

ICN [7] has emerged as a paradigm promising to cope with the limitations of the current host-based network architecture by providing mobility support, efficient content distribution, scalability and security based on content names regardless of the location. The existing host-based network architecture faces significant challenges to meet demands in terms of scalability, mobility, security, and resource utilization [8]. Numerous investigations have confirmed the feasibility of applying ICN in different domains of IoV applications [9]. Ghasemi et al. [10] demonstrated video streaming over ICN/NDN on a global NDN (named data networking) test bed. Other studies [11,12] have investigated the potential of ICN using NDN with IEEE802.15.4 for constrained IoT devices to implement a real-world smart agriculture scenario.

Our paper aims to provide a comprehensive survey of the development and state-of-the-art of the convergence of edge-computing intelligence and ICN for realizing IIoV (Intelligent Internet of Vehicles) solutions. In the past few years, some survey articles have been presented in the context of ICN for IoT summarizing the applicability and potential of information centric networks for the Internet of Things [13–17]. Khelifi et al. [18] carried out a brief survey on named data networking in vehicular ad hoc networks, while Kerche et al. [19] analyzed IoV and artificial intelligence for information-centric networks [20]. Our survey differs from previous studies in that it examines intelligence computing, edge caching, and networking in the context of ICN-based IoV. Figure 1 shows the converged architecture for IIoV.

To the best of our knowledge, this is the only comprehensive study that discusses AI based solutions for the convergence of ICN-based IoV with edge computing as well as their potential benefits. In this regard, we summarize the major contributions of the paper as follows:

- Presenting a comprehensive review regarding the utilization of edge intelligence and ICN for IoV solutions, addressing existing challenges.
- Performing an in-depth review of the leading enablers of IoV, including wireless communications, mobile edge/fog computing, machine learning/AI, intelligent computing, and information-centric networking.
- Providing a detailed ICN architecture suitable to meet the requirements of IoV networks, such as naming, caching, mobility, and security.
- Highlighting challenges and issues faced by ICN-based IoV, indicating future research directions which may illuminate efforts to develop new intelligent IoV applications.

The rest of the paper is organized as follows: In Section 2, a review of IoV, ICN, EC and AI is presented. Section 3 introduces a general model for ICN and edge computing convergence for IIoV with detailed discussion of the key elements. Section 4 reviews existing research on the application of AI to wireless networks and communication to achieve intelligent networking, intelligent edge computing and caching. Challenges and open issues are identified and highlighted in Section 5. Finally, the conclusions of the paper are provided in Section 6 (Figure 2).

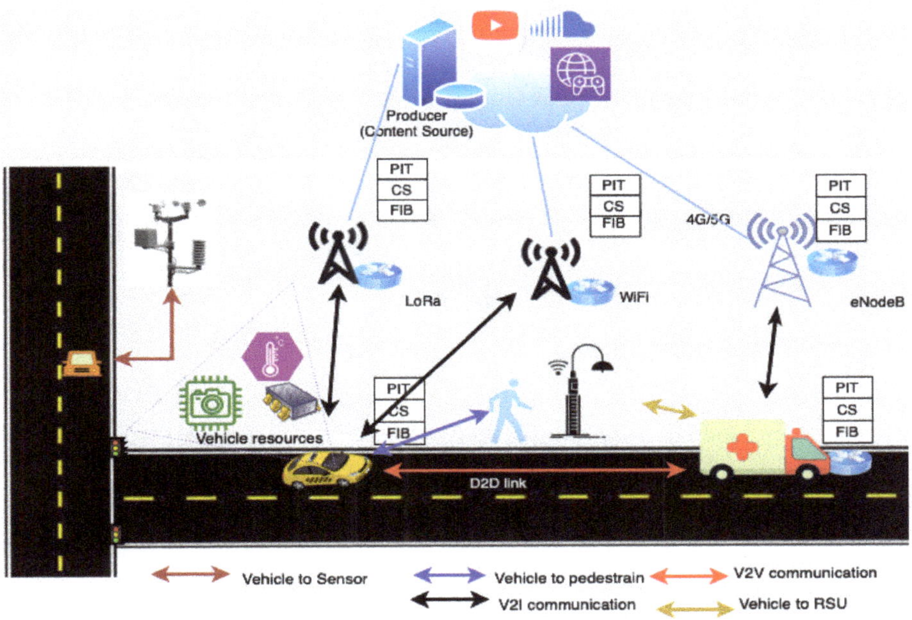

Figure 1. Converged Architecture for IIoV.

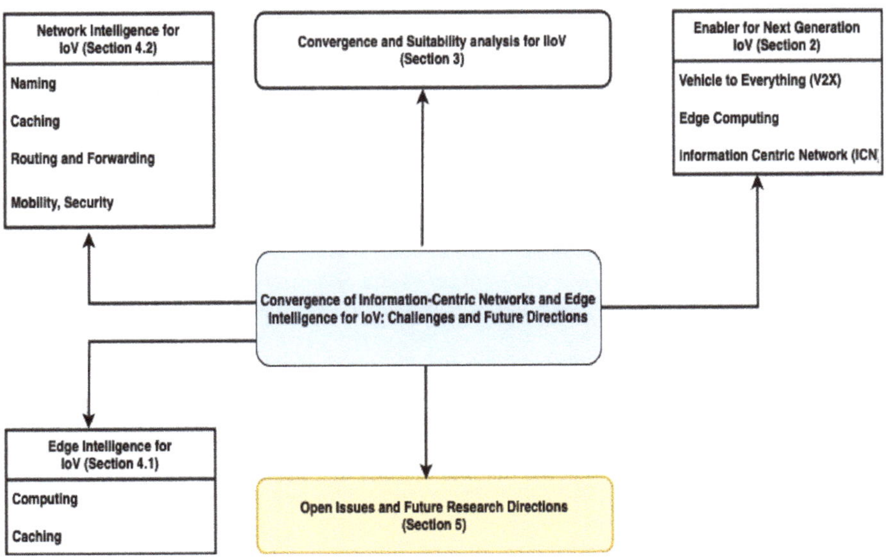

Figure 2. The organization of the remaining parts of the paper.

2. Internet of Vehicles (IoV)

IoV has attracted a great deal of attention as a useful Internet of Things (IoT) scenario for addressing the grand challenges of modern transportation problems. Connected and fully automated vehicles are expected to revolutionize the transportation systems of the future on a global scale, significantly enhancing road safety, alleviating traffic congestion, reducing fuel consumption, and enhancing driving experience and infotainment. IoV evolved from vehicular ad hoc networks (VANET) which mainly focused on providing road safety applications based on communication between vehicles and roadside units, constituting a special case of mobile ad hoc networks (MANET) [1].

In IoV, vehicles are equipped with computing and storage capabilities, a variety of sensors, including cameras, GPS, RADAR, LIDAR (light detection and ranging), and different communication wireless interfaces, leading to independent autonomous vehicles that can connect and cooperate with each other, with roadside infrastructure, with pedestrians, and with other entities through vehicle-to-everything (V2X) communications.

Several access technologies have already emerged to support V2X communication, such as wireless local area networks (WLAN), including IEEE 802.11p/ WAVE [21–23] and its evolution IEEE 802.11bd [24], wireless wide area networks, cellular technologies, such as LTE/5G [21,25,26], as well as LoRa [27–30] and Bluetooth. These wireless access technologies offer diverse communication options among different smart things. In IoV, each vehicle is envisioned as a smart object having sensing, computing and storage capabilities enabling it to connect with other entities through the vehicle-to-everything (V2X) umbrella, comprising vehicle-to-vehicle (V2V), vehicle-to-infrastructure (V2I), vehicle to sensor (V2S), vehicle to roadside unit (V2R), and vehicle to personal device (V2P).

2.1. *IoV Applications*

IoV opens up new opportunities and applications and offers many benefits to drivers, pedestrians, society and business. According to the World Health Organization (WHO), every year, approximately 1.3 million people die and 7 million are injured in traffic accidents. Moreover, people waste billions of hours because of traffic problems (e.g., through traffic jams and accidents) [31].

According to the 5G Automotive Association (5GAA), IoV applications are clustered into four categories.

- Safety-based: The primary goal is to reduce the severity and frequency of vehicle crashes. Connected vehicles exchange warning alerts to prevent possible collisions, and provide real-time crash information with location information to emergency teams.
- Traffic efficiency: Traffic efficiency covers a range of applications intended to optimize the flow of road traffic. For instance, scheduling traffic signals at intersections according to the volume of traffic to reduce waiting time.
- Cooperative driving: Cooperative adaptive cruise control (CACC) requires throughput of the order of 5 Mbps and latency of the order of 2–10 ms.
- Infotainment: Efficient provision of non-driving-related information and entertainment to drivers and passengers with stringent QoS is crucial for IoV. Many vehicular users are now interested in accessing HD video streaming and use various latency-sensitive mobile applications, such as real-time online gaming, virtual sports, video streaming, instant messaging among passengers, and geo-specific advertisements. Such services are characterized by relatively short latency.

Typical IoV applications and their requirement are listed and described in Table 1.

Table 1. IoV applications and characteristics.

Application	Examples	Communication Support	Bandwidth Req.	Latency Req.
Traffic Safety	Pre-sense Crash Warning	VRU, V2V, V2P	Low	20–100 ms
Traffic Efficiency	Navigation System, Traffic Light	V2R, V2I	High	100–500 ms
Cooperative Driving	CACC	V2V, V2I, V2S	High	2–10 ms
Infotainment	Video and Music Streaming, News	V2I, V2U	High	500–1000 ms

2.2. Radio Access Technologies

There is a need to combine short-range technologies, such as Bluetooth, with long-range technologies for seamless network availability. In this section, we describe their basic characteristics, summarized in Table 2.

1. ***IEEE 802.11p / WAVE***: The 802.11p (WAVE: wireless access in vehicular environments) protocol adapted the IEEE 802.11 PHY/MAC layer specification to the vehicular environment and termed it dedicated short-range communications (DSRC). It provides V2V and V2I communication using a 10 MHz wide channel at 5.9 GHz. Various regional bodies define spectrum utilization for IoV. In Europe, ITS-G5 allocated a total of 40 MHz in the frequency range of 5875–5915 MHz for safety-related road ITS. In addition, the frequency range 5915–5925 MHz was prioritized for urban rail ITS, but could also be used for road ITS once ETSI establishes polite protocols and proper co-channel sharing mechanisms between road ITS and urban rail ITS [32]. IEEE 802.11p uses binary convolutional coding (BCC) which is weaker than low-density parity-check coding (LDPC) or turbo-coding in higher modulation and coding schemes (MCS). This is addressed in the IEEE 802.11bd amendment which offers twice the performance, supporting speeds up to 500 km/h, and twice the communication range of 802.11p, due to its use of LDPC coding and 256 QAM modulation. 802.11bd is expected to be approved in 2022.
   Similarly, the US Federal Communication Commission had allocated 75 MHz (frequency range from 5850 MHz to 5925 MHz) for V2X applications. However, on 3 May 2021 it decided to reallocate 45 MHz of the 75 MHz spectrum to unlicensed devices, such as Wi-Fi. As a result, from this date on, only the 30 MHz between 5895 and 5925 will remain for safety applications using vehicle-to-everything (V2X) communication [33].
2. ***LTE Cellular V2X (C-V2X)***: LTE cellular vehicle-to-everything (C-V2X) is a 3GPP standard that uses the PC5 interface at 5.9 GHz. Employing high capacity, large cell infrastructure, it enables communication between vehicles, humans, and infrastructure. It consists of vehicle-to-vehicle (V2V), vehicle-to-infrastructure (V2I) and vehicle-to-person (V2) communication. It represents an alternative to the IEEE 802.11p standard, that supports direct vehicle-to-vehicle communication even without cellular base station involvement. C-V2X is continually evolving over 3GPP releases. It was introduced in 3GPP release 14, completed in 2017. In Release 15, further enhancements were introduced. In 5G, further performance improvements will be attained, replacing C-V2X with New Radio (NR). It should be noted that C-V2X and DSRC are not compatible with each other.
3. ***5G and Beyond 5G NR V2X***: Some IoV applications, such as self-driving vehicles, require that the transmission delay is less than one millisecond. The IoV of tomorrow will require higher throughput, lower latency, higher reliability and multicast capabilities, while supporting mobility up to 500 km/h and application-aware services. Thus, NR C-V2X in release 16 offers backward compatibility with Rel. 14 and Rel. 15 for safety, multicast messages, and advanced-use-case (transport profile) capabilities. Spectrum and energy efficiencies are improved by deploying massive multi-input multi-output (MIMO) and millimeter-wave technologies.

Figure 3 shows the spectrum allocation in Europe and the US. The frequencies from 5945 to 6425 MHz have been assigned to very low power wireless access systems (VLPWAS), including radio local area networks (RLANs) for public and private applications, regardless of the underlying network topology in Europe, with a maximum effective isotropic radiated power (EIRP) of 14 dBm [34]. In the US, FCC has allocated the same band under the control of an automated frequency control (AFC) system to protect incumbent users, at the standard power levels that are currently permitted in the 5 GHz band [35]. More details can be found in [32].

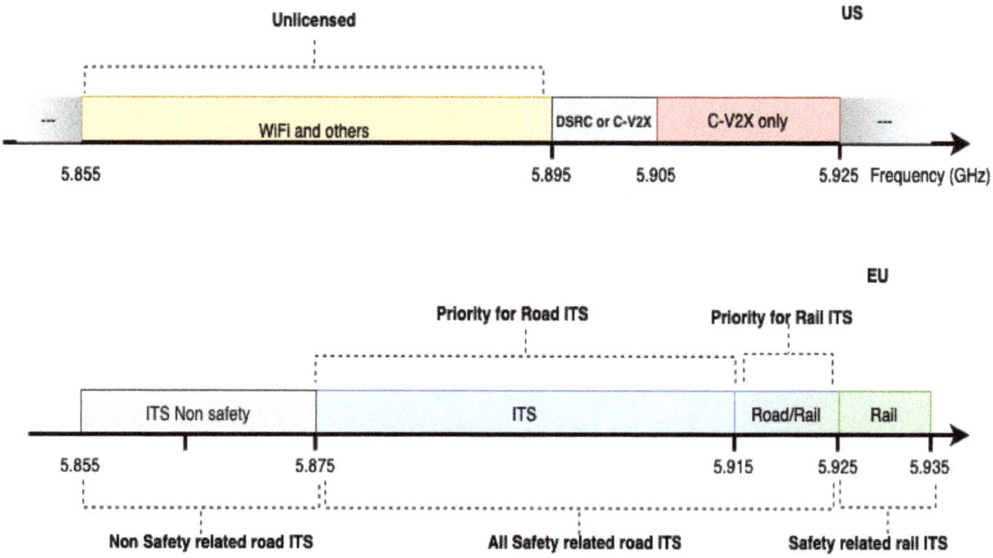

Figure 3. ITS Spectrum allocation in US and Europe.

4. *Heterogeneous Wireless Access*: Combining the strengths of both IEEE802.11p and cellular leverages while tempering their weaknesses. QoS-based service can be provided with cellular networks offering high-throughput entertainment services while DSRC handles vehicle safety.

Table 2. Comparison of IEEE 802.11p, IEEE 802.11bd, 4G and 5G technologies for IoV.

Factors	IEEE 802.11p	IEEE 802.11bd	LTE C-V2X	NR V2X
Base technology	IEEE 802.11a,n	IEEE 802.11n/ac	4G	5G
Radio band	5.9 GHz	5.9 GHz and 60 GHz	Cellular licensed and 5.9 GHz	5.9–52.6 GHz
Peak data rate	27 Mbps		28.8 Mbps	
Delay	5 ms	20 ms		<5 ms
Mobility	252 km/h	500 km/h	350 km/h	500 km/h
Physical layer	OFDM1	OFDM	SC-FDMA	OFDM
MAC layer		CSMA		Model1: gNodeB scheduling Mode 2: Flexible sub-modes
Channel-coding	BCC	LDPC	Data: Turbo-coding Control: Convolution coding	Data: LDPC Control: Polar-coding
Modes	Broadcast	Broadcast, groupcast	Direct C-V2X, broadcast	Broadcast, groupcast, unicast
Base station	AP	AP	Macro	Micro base station

2.3. Layered Architectures and Standards

Several IoV architectures have been devised by researchers [1,36–39]. Bonomi [40] proposed a four-layer architecture that includes embedded devices and sensors, multi-service edge, core, data center and cloud. Kaiwartya et al. [1] proposed a five-layer architecture (perception layer, coordination, artificial intelligence, application, business). A 13-layer scheme with three-dimensional system architecture (dimension of IoV, intelligent computing, and real-time big data analytics) has even been proposed [39]. Zhuang et al. [41] presented an SDN/NFV-enabled network architecture for IoV. Chen et al. [42] proposed a Cognitive Internet of Vehicles architecture.

Various researchers and industries are pursuing the development of novel network architectures and solutions that could facilitate IoV implementation [43]. The European Telecommunications Standards Institute (ETSI) in Europe and the Federal Communication Commission (FCC) in the US have proposed architectural standards for IoV. Approaches including cooperative awareness messages (CAMs) for transmitting geographically aware information with relevant data for other vehicles, decentralized environmental notification messages (DENMs), local dynamic maps (LDMs), a key technology for integrating static, temporary, and dynamic information in a geographical context, and geonetworking addressing for geolocation-based routing, have been added to the traditional TCP/IP-based model.

Despite ongoing efforts, there is no universal and comprehensive architecture for IoV. To fully implement IoV, different sectors (researchers, automotive manufacturers, standardization organizations, city planners, etc.) must collaborate.

3. Information Centric Networking (ICN)

Current Internet architecture is a host-centric design based on the TCP/IP protocol which was developed in the 1970s and adopted as the protocol standard for ARPANET (the predecessor of the Internet) in 1983. A great deal of research has been undertaken to address the limitations of TCP/IP networks in terms of scalability, transmission efficiency, mobility, and security.

ICN is an emerging paradigm for the future Internet that involves a radical change. It allows content access regardless of its location by leveraging forwarding, name-based routing, security, and caching. It is proposed to replace the current TCP/IP architecture offering a flexible network structure in the IoT environment.

Many ICN architectures have been proposed over recent years [44], including data-oriented network architecture (DONA) [45], content-centric network (CCN) [7] and named-data networking (NDN) [46]. NDN is one of the four NSF FIA projects aimed at designing the future Internet architectures with the objective of completely overhauling the Internet by replacing IP with content chunks as a universal element of transport. Figure 4 shows the comparison between TCP/IP and NDN.

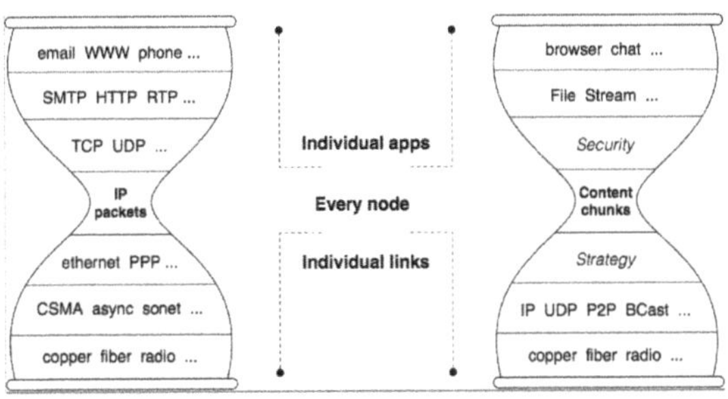

Figure 4. Comparison of host-centric and information-centric stack [46].

Traditional IP networks need to carry out two operations. The first is performed in the routing (or control) plane, in which the routers communicate routing updates and select the best path for constructing the forwarding information table (FIB). The second operation is performed in the forwarding (or data) plane, and involves following the directives contained in the FIB to perform the forwarding. The routing plane is stateful and can adapt to network changes, including link failure, router crashes, new routes, or alternative better paths. However, the IP forwarding is stateless and cannot adapt to anything without instruction from the routing control plane. This has often been referred to as "smart routing and dumb forwarding".

NDN replaces the IP dumb-forwarding by stateful forwarding. Instead of distributing IP prefixes, the routing protocols distribute the name prefixes. Stateful forwarding is a process that records incoming interest packets in the PIT and uses the recorded information to forward the retrieved data packets back to the consumers(s). The recorded information can also be used to measure data plane performance, e.g., to adjust interest forwarding strategy decisions. All communication in NDN is performed using two distinct types of packets: interest and data. As shown in Figure 5, each NDN router (vehicles, RSU, BS, MBS, etc.) maintains three major data structures: a content store (CS) for temporary caching of received data packets, a pending interest table (PIT), and a forwarding table (FIB) [47].

Instead of sending packets between a source and destination devices identified by their numeric IP addresses, NDN disseminates named data at the network level, forwarding names directly that carry application semantics. Moreover, each packet in NDN is secured at the time of production, allowing data replication anywhere in the network and preserving the security properties of the data over its lifetime [48]. The ICN architecture building blocks enabling the shift from a host-centric to a content-centric paradigm are elaborated in the following subsections.

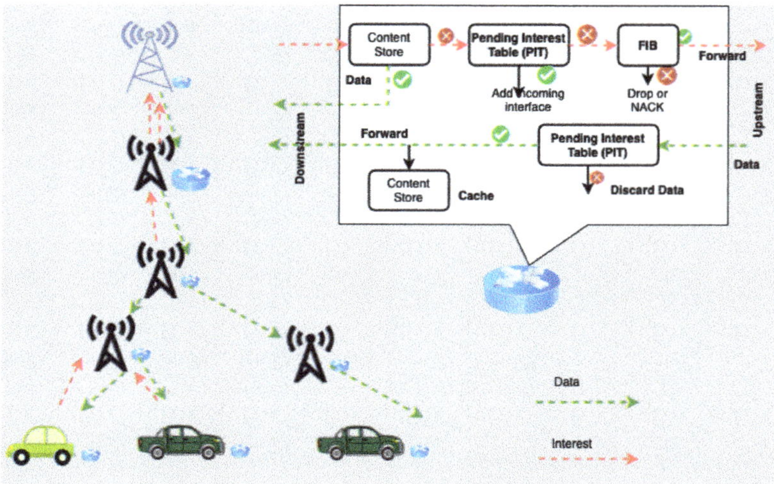

Figure 5. Typical interest and data communication in NDN communication.

3.1. Content Naming

Content-naming has a direct influence on the other aspects of ICN-based networks, such as routing and forwarding [49]. As ICN changes the focus of traditional networks from host-based to content-based, the data being requested is considered to be an object and given a name that helps users to identify it. Named data objects (NDOs) are the core of addressing and forwarding in the ICN network. NDOs are addressable data units in ICN networks that can represent a collection of bytes or a piece of information. Every NDO, including documents, videos, music, and photos must have a globally unique name.

Four types of naming techniques are used in the ICN literature: hierarchical, flat, attribute-based and hybrid.

- Hierarchical names are given in plain, human-readable components, separated by "/". CCN and NDN follow this naming convention.
- Flat names are self-certifying and used in ICN architectures such as MobilityFirst [50] and DONA [45]. The content name has two parts of the form $P : L$ and is associated with metadata.
- The attribute-based technique identifies content by attribute-value pairs (AVP). Contents are requested by applying constraints (also called predicates) to the attributes.
- Hybrid-naming combines the previous three techniques.

3.2. Routing and Forwarding

Routing is one of the most critical parts of the ICN/NDN network since it is significantly different from the current Internet architecture. As opposed to routing by the address of a host, ICN routes by the names of the requested data. Two packet types are used in NDN to identify content: interest packets and content packets. Each router maintains three data structures: content store (CS), pending interest table (PIT), and forwarding interest base (FIB). CS stores or caches data, PIT records interest entries and FIB is used to map the outgoing interface for each Interest request. Currently, two approaches are used for routing: name-based and name resolution-based. In name-based routing, the Interest (request for specific content) is routed from the consumer to the producer of the content or cache node using the name of the requested data to resolve the next hop in the network. Name resolution-based routing uses a name-resolution service employing two steps: the consumer node sends messages to the name-resolution server (NRS) with the name of the desired content, and the NRS server resolves the name with a single, or set of, addresses of content producers or caches that have the requested content. NRS provides the name resolution function for translating an object name into some other information, such as a locator, another name, metadata, etc., that is used for forwarding the object requested. More specifically, it is a service that is provided by the ICN infrastructure to help a consumer to reach a specific piece of information. The consumer provides a persistent name and the NRS returns a name or locator that can reach a current instance of the requested object [51].

3.3. Caching

Content caching is one of the fundamental architectural building blocks of ICN networks, enabling fast, reliable, and scalable content delivery. Each NDN node in the network has the capability to cache named content and respond to requests for it. Reacting to a content request, an NDN node either responds directly if it has the content in its local cache, or sends a request message to other nodes. Content is cached once the request is fulfilled. ICN caching democratizes content delivery since caching is implemented by all network routers involved instead of a few specialized cache nodes. Several caching schemes [52] have been proposed for optimal caching and efficient resource utilization.

3.4. Mobility

Mobility refers to the changes in speed and direction of nodes over time. Mobility support at the network layer is a strength of ICN that addresses one of the limitations of IP networks. Consumer mobility is naturally supported in ICN architecture, so a change in the physical location of the consumer node does not affect the delivery of the content. However, producer mobility is more difficult to support as the name resolution system or routing tables in named-based routing need to be updated. To support producer mobility in ICN, different techniques are suggested in the literature [53]. Zafar [54] presented an adaptive content dissemination solution for VANET networks that includes a hierarchical context-aware content naming scheme that enables safety-critical and non-safety applications. Fog-computing-assisted mobility support with information-centric IoV is addressed in [55].

3.5. Security

Data protection and privacy are considered to be fundamental requirements for IoV services. IoV is fundamentally a multi-domain environment with a large number of heterogeneous nodes and services. When designing IoV systems, security and privacy should be carefully considered and integrated. ICN has dramatically changed the traditional security perspective from securing the communication path to securing the content itself. Protection and trust are implemented at the packet level rather than at the communication channel level. As a result, data trust is decoupled from how and where data are stored and no secure connection setup is necessary [56].

4. Edge Computing

Cloud computing began as a solution for the storage and processing of large-scale data. However, in recent years, the massive growth in smart and connected devices, coupled with technologies such as IoT, V2X, augmented reality (AR), etc., has presented challenges to centralized cloud computing. The requirements for ultra-low latency and real-time response, high bandwidth, mobility support, and context-awareness services are the focus of researchers in academia and industry.

According to Gartner, it is estimated that 75% of data in 2025 will be processed outside the traditional data center [57]. Edge computing aims to bring cloud computing services closer to the end-user for a real-time, context-aware response. It comprises a continually expanding set of connected devices and systems that gather and analyze data close to the users. According to Intel predictions, an autonomous vehicle will generate about 4 TB of data in an hour (1 GB of data is generated every second). In the following, we review edge computing approaches including fog computing [5], cloudlets [6], and mobile edge computing (MEC) [58]. A detailed comparison of fog computing, cloudlets, and MEC is presented in [59]. Table 3 shows a comparison of different computing strategies. Various vendors, including Microsoft Azure [60], Cisco Edge Computing [61], and IBM [62] are already employing edge computing. Some approaches to edge computing described in the literature are:

- Fog computing: Fog computing is proposed to address the issue of the large latency between mobile devices and the cloud. Bonomi et al. [5] presented its purpose and key characteristics. It is a highly virtual platform to provide computing, storage, and networking services closer to end devices than the Cloud. In general, it offers the following benefits: quick response to delay-sensitive applications, such as emergency and multimedia applications, data aggregation, data protection and security, context-aware and location-aware service provisioning [63]. Details on fog computing can be found in [64].
- Cloudlets: Cloudlets are a project developed by CMU (Carnegie Mellon Unversity) [6]. They provide the middle tier in a three-tier hierarchy: mobile device–cloudlet–cloud. Their main goal is to merge mobile computing and cloud computing by introducing a middle-tier forming a multi-hierarchical structure. They can be viewed as "data centers in a box", whose goal is to "bring the cloud closer" by means of low end-to-end latency and high bandwidth. Mobile end devices can offload computational tasks to a cloudlet.
- Mobile edge computing (MEC). Mobile edge computing is a European Telecommunications Standards Institute (ETSI) initiative to provide IT and cloud-computing capabilities within the RAN (radio area network) in close proximity to mobile subscribers. The RAN edge offers a service environment with ultra-low latency and high bandwidth, as well as direct access to real-time RAN information (subscriber location, cell load, channel load, etc.) and is well-suited to provide context-related services.

Table 3. Comparison of Cloud Computing, Fog Computing, Mobile Edge Computing and Cloudlets.

Features	Cloud Computing	Fog Computing	Mobile Edge Computing	Cloudlets
Latency	High	Low	Low	Low
Node devices	Servers	Routers, switches, AP, ...	Server on base station	Data center in box
Node location	Centralized (Remote)	Between end to cloud	Macro BS	Local / out installation
Software architecture	SOA (Service Oriented Architecture) and EDA (Event Driven Architecture)	Fog abstraction-layer-based	Mobile-orchestrate-based	Cloudlet agent-based
Context awareness	Low	Medium	High	Low
Mobility support	No	Yes	Yes	Yes
Proximity	Global	One or multiple hops	One hop	One hop
Access mechanism	Fixed and wireless	WiFi, mobile networks	mobile networks	WiFi
Inter-node communication	Not supported	Supported	Partial	Partial
Application	Computational intensive and delay tolerant	Latency sensitive	Latency sensitive	Latency sensitive
Server density	Low	High	Low	High
Backhaul usage	Frequent	Infrequent	Infrequent	Infrequent

5. Machine Learning, Deep Learning and Artificial Intelligence

AI is a discipline inspired by human intelligence used to predict, automate, and optimize tasks that humans have historically performed, such as speech and vision recognition, and decision-making. Application areas include computer vision and natural language processing (NLP). Machine learning, deep learning, generative adversarial networks (GAN) [65] and federated learning are some of the AI methods used.

Machine learning (ML) allows machines to become more accurate at prediction or classification without explicitly being programmed to do so. Machine learning algorithms use training data to build a model. There are three broad classes of ML: supervised learning, unsupervised learning and reinforcement learning. It leverages tagged data sets to build a model that is then applied to a specific inference task.

Deep learning is a subset of machine learning that uses several layers of artificial neural networks (ANN), such as convolutional neural networks (CNN), recurrent neural networks (RNN) and deep reinforcement learning (DRL). In recent years it has advanced significantly in the performance of classification and prediction tasks.

5.1. *Convolutional Neural Networks*

The CNN-based model led to a breakthrough in object detection and recognition. This development made CNNs attractive for entirely novel application domains in IIoV. Intelligent vehicles rely heavily on object detectors for perception, pathfinding and other decision-making. CNNs are mainly applied to analyze image-based applications. Since 2012, several CNN algorithms and architectures have been proposed, such as R-CNN and its variants, as well as YOLO. R-CNN is a region-based CNN, proposed in [66], which combines region-based algorithms with CNN. The purpose of R-CNN is to extract 2000 regions through selective search, then, instead of working on the whole image, classification is performed on the selected regions. YOLO was introduced in 2016 using an approach known as, "You only look once" [67]. In contrast to region-based approaches, YOLO passes the n-by-n image only once in a fully convolutional neural network (FCNN), which makes it quite fast. It splits the image into grids of dimension m-by-m and generates bounding boxes and their class probabilities. Figure 6 shows a typical CNN learning process and object detection (e.g., vehicle, pedestrian, RSU).

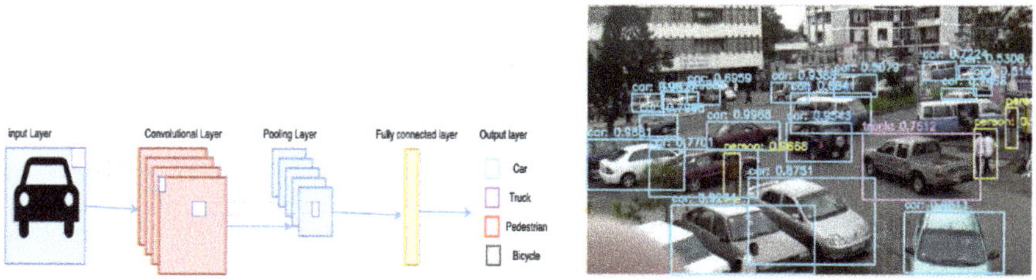

(a) (b)

Figure 6. CNN and example of object-detection application of autonomous vehicle in IoV (**a**) Typical CNN learning process. (**b**) Object detection in IoV context.

5.2. Recurrent Neural Networks

Recurrent neural networks (RNN) employ sequential or time-series data and are distinguished by their "memory" elements, which aid in the storage of earlier knowledge and influence the present input and output of the network. Unlike traditional deep neural networks, in which inputs are independent of outputs, RNN output is dependent on the sequence's preceding parts. Variant RNN architectures proposed include vanilla recurrent neural networks, long short-term memory (LSTM), and gated recurrent units (GRU). RNN in IoV can be used to predict the future position and velocity of each of the dynamic objects detected in the scene (for example, vehicles and pedestrians) and thus provide essential input information, such as autonomous cruise control and autonomous emergency braking. Figure 7 shows a typical computational graph of a basic RNN and its application in vehicle positioning and speed prediction.

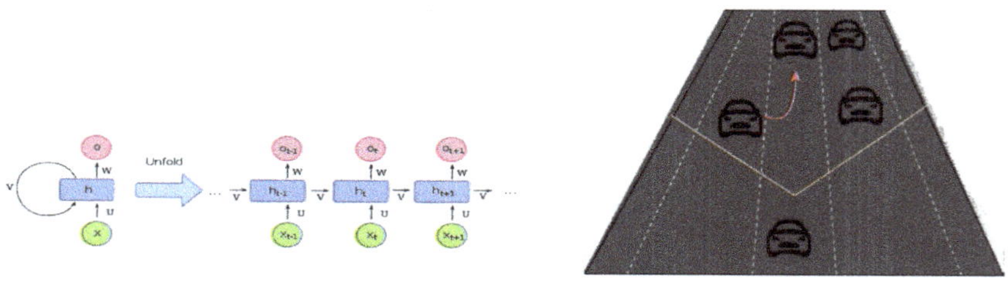

(a) (b)

Figure 7. RNN and example of position prediction application of autonomous vehicle in IoV [68]. (**a**) Typical RNN. (**b**) Position prediction.

5.3. Reinforcement Learning

Reinforcement learning is a special type of machine learning in which the learner (agent) interacts with the environment to learn optimal policies required to map states to actions. The agent senses the current status of the environment and takes the corresponding action, after which it earns a reward, while the environment changes to a new state. The goal of RL is to create an intelligent agent that can perform efficient policies to maximize the rewards. Figure 8 demonstrates the general RL framework.

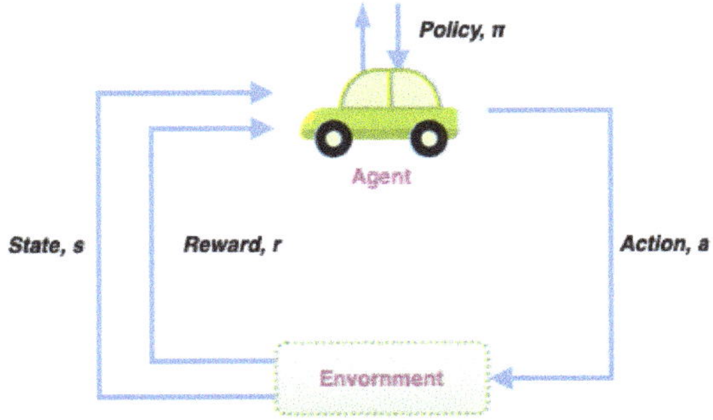

Figure 8. Typical reinforcement learning process.

5.4. Federated Learning

Federated learning, also known as collaborative learning, trains an algorithm across multiple decentralized devices without sharing training data samples. Federated learning allows numerous players to develop a common, strong machine learning model addressing crucial issues such as data privacy, security, data access rights, and access to heterogeneous data. Figure 9 shows a typical federating learning IoV scenario; in Figure 9a, the RSU enabled the aggregation of federated learning, and in Figure 9b, the selected vehicle enabled the model aggregation.

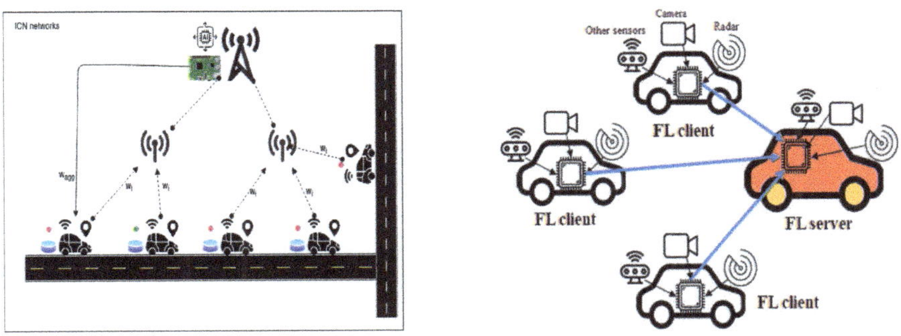

(**a**) (**b**)

Figure 9. IoV FL process [69]. (**a**) Federated learning performed on RSU (V2I). (**b**) Federated learning among vehicles (V2V).

5.5. TinyML

TinyML is a growing field of machine learning technologies and applications, comprising hardware, algorithms and software, which are capable of processing sensor data at low power, thus enabling a variety of applications. TinyML enables application of deep learning models to micro controllers, allowing machine learning models to run on embedded systems that have limited computing and storage resources [70]. TinyML has enormous potential in the context of IoV. For example, the authors of [71] utilized TinyML for IoV in anomaly detection. Similarly, [72] proposed autonomous driving for mini-vehicles considering limited onboard storage and computing capabilities. Figure 10 shows TinyML applied in an IOV environment.

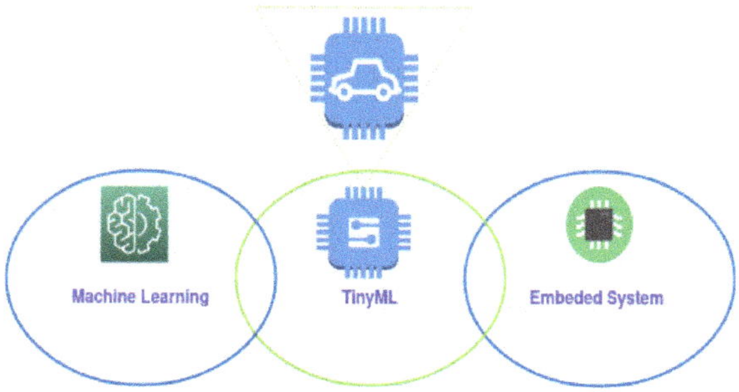

Figure 10. Typical IoV TinyML.

6. ICN and Edge Computing Convergence Suitability Analysis for IIoV

In this section, we present a convergence and suitability analysis of ICN, edge computing and AI for existing and future IoV applications that require efficient, reliable, secure and scalable communication infrastructure and architecture for seamless integration, and cooperation among entities in IoV networks.

6.1. Converged IoV Infrastructure

The benefits of ICN for IoV stem from the fact that a requested content might be located in a server far away from the current vehicle position, but it also might have been cached in a closer repository, easily accessed by its named content, thus reducing the latency and backhaul traffic. This is particularly relevant due to the dynamic aspect of IoV; the content might even be cached in a nearby vehicle. ICN promises efficient content delivery, and supports mobility, scalability, and security. It allows Internet applications to be developed based on named requests, caching, and any-casting. Because of its various advantages, ICN offers an interesting solution for IoV communications and seems to fit perfectly with many IoV applications [18,19]. Its in-network content caching and forwarding methods enable the dissemination of information between different vehicles in the network. In the ICN context, vehicles can be consumers, producers, and data forwarders at the same time; RSUs, BSs, and core network elements can provide NDN forwarding and caching capability. For an efficient ICN-based IoV network, all ICN core elements described in Section II should be optimized to meet the frequent topology changes of IoV networks. Intelligent and adaptive naming, routing/forwarding, security/privacy, and mobility-aware design are necessary to provide efficient IoV applications and services, as described in the following section.

Edge computing is another enabling technology that meets the time-sensitive requirements of IoV to offer pervasive services for massive numbers of connected vehicles with low latency and very large bandwidth. Traditionally, cloud computing offered computational, storage, and on-demand services to a far away user. However, as many IoV applications require low latency and high bandwidth, edge computing provides services in close proximity to end-users. Future Internet applications require low latency, high security, mobility, and scalable content distribution. Edge computing for IoV is an ideal solution to remove the burden on backhaul networks. However, existing TCP/IP-based edge computing solutions have some weaknesses when applied to support cloud services at the edge of the network. Edge computing offers many benefits associated with latency challenges, as it provides high bandwidth close to the user. In addition, advanced computing enables the processing of real-time tasks closer to the source or even on it.

In recent years, the marriage of edge computing and AI has spawned a new research area called edge intelligence [70]. Similarly, the marriage of ICN and AI has spawned intelligent networking. Recent advances in deep learning have accelerated the realization of autonomous vehicles. Edge computing, which is one of the strategies of 5G, provides an optimal service for real-time access and high bandwidth. ICN and edge computing can be combined, resulting in synergistic benefits for IoV. Several AI approaches have been used for solving problems in IoV and autonomous vehicles. Table 4 shows the mutual benefits of the synergy of ICN and edge computing. Figure 11 shows the convergence of ICN, edge computing and AI for IIoV.

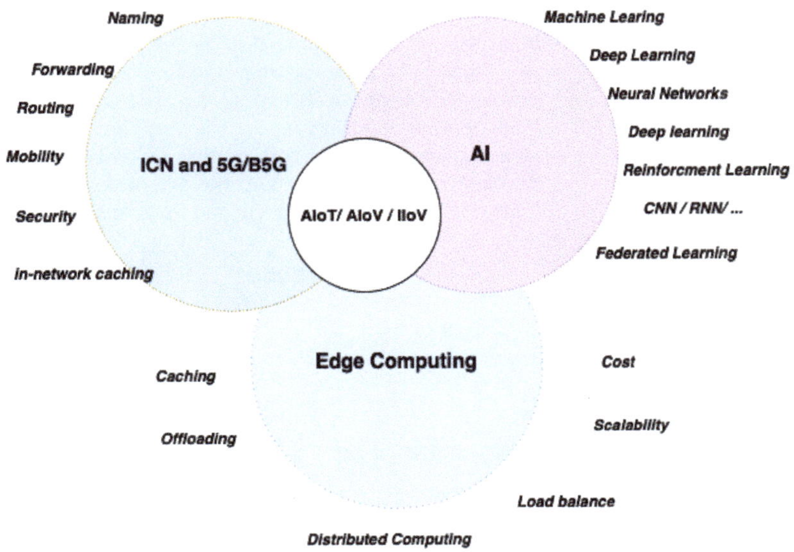

Figure 11. The convergence of ICN-Edge Computing-AI for IIoV.

Table 4. Role of AI in mutual benefits of ICN and edge computing for IIoV.

Domain	Key Area	Mutual Benefit for IoV	AI Role
ICN	Naming	Name-based routing, naming and latency, naming caching, naming caching, naming mobility	AI helps analyze and aggregate content name for efficient forwarding
	Routing and forwarding	Edge node support by running routing and forwarding algorithm	Learning algorithm can be used to predict delay and dynamic selection of forwarding path
	Mobility	Edge computing to support producer mobility	Mobility prediction
	In network caching	Edge node support caching of content which is useful for high bandwidth and low-latency edge computing applications, such as IoV and multimedia streaming	Efficient caching algorithm
Edge	Local processing	Locally processed data can be easily contextualized to different users	AI can be used to continuously analyze large quantities of data to determine what exactly is happening on the network, to make predictions and to respond to events
	Storage	Enable ICN caching services at the edge server to provide mobility and delay aware requirement	Optimal cache management
	Load balancing	ICN provides load balancing	Intelligent learning algorithm helps in optimal task distribution

ICN and edge computing are essential for achieving this goal. By leveraging the ICN perspective, vehicular users can obtain multimedia chunks via the vehicle-edge-cloud approach to improve the delivery quality. ICN with edge computing is an excellent fit and allows adaptive video delivery optimization. This enables traffic optimization on the edge, in addition to backhaul and core networks using in-network caching. Caching multimedia content, based on popularity, request pattern, and quality awareness (e.g., user device capability and type, available bandwidth, etc.), is possible through ICN and edge computing. Moreover, with the prevalence of content caching in edge computing, it is possible to have multiple active links and load balancing on dynamics and mobile conditions.

Figure 12 shows the proposed IIoV layered protocol stack. It consists of different layers, such as physical, access, edge, ICN, and application, as well as security, AI, and intelligence dimensions. The ICN layer is an intermediate layer that provides core NDN aspects (naming, forwarding/routing, mobility, caching). The physical layer encompasses the various wireless communication interfaces (Wi-Fi, 4G/5G) employed by the ICN node for better availability of content and its distribution.

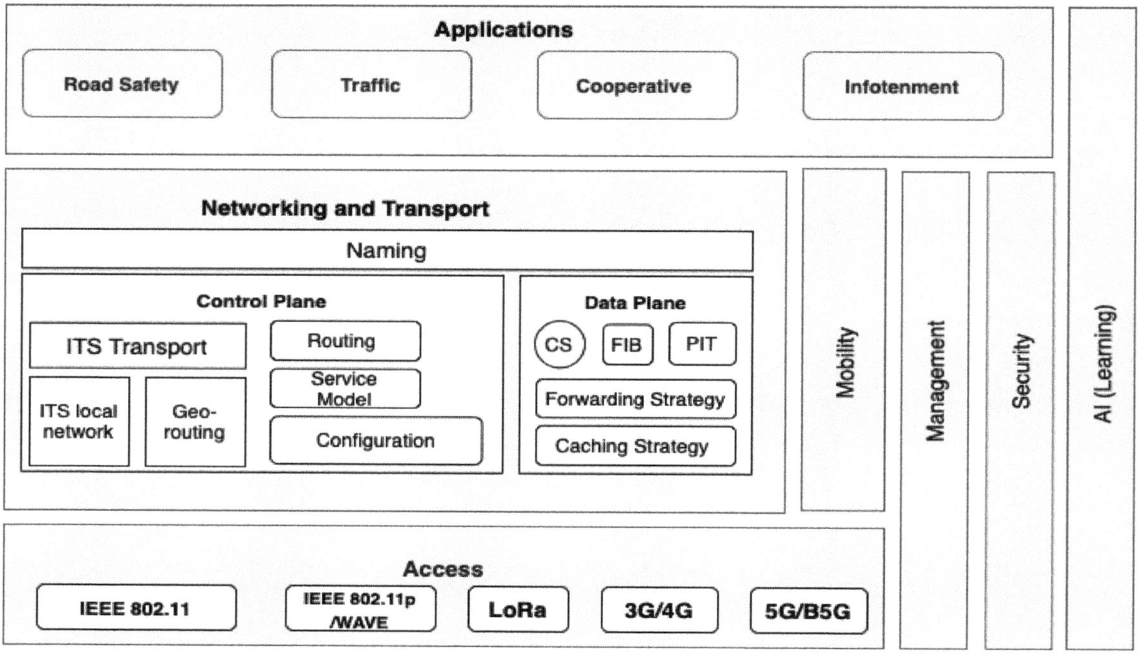

Figure 12. Protocol stack of converged architecture for ICN IIoV.

6.2. Collaborative Perception and Driving: Use Case

To demonstrate the potential of ICN and EC convergence in the realization of IIoV, we conducted a suitability analysis by exploring enabling technologies and considering cooperative perception and driving as a use case. In the cooperative driving use case, large vehicles, such as trucks and minivans, share camera views of road conditions ahead of them with lagging vehicles. A driver in front might warn the lagging driver intending to pass about an oncoming vehicle in the passing lane with the goal of preventing catastrophic head-on collisions during overtaking. Consider the scenario shown in Figure 13, where two connected autonomous vehicles are moving one behind the other, in the same lane, while engaged in cooperative collaborative perception to improve their individual capabilities and exchange safety information through V2X communications. The driver of vehicle A in front spots a pedestrian ahead on the adjacent lane, which is not in the field of view of vehicle B in the rear, as vehicle A has obstructed the view. Vehicle B is planning to make a passing/overtaking maneuver, but it is unaware of the pedestrian. The knowledge about the pedestrian becomes crucial for vehicle B, as it would not have enough time to avoid colliding with the pedestrian after the lane change. The remedy would be to avoid performing the lane change at this time. Accurate perception of the surrounding environment is a vital requirement for safety. In cooperative driving, vehicles exchange safety information in a variety of ways—vehicles could share raw sensor data; however it would require significant communication bandwidth which cannot always be guaranteed. Another option is to exchange information about the detected objects including their position, speed, etc., as cooperative awareness messages (CAMs). Thus, cooperative and autonomous driving is one application of IoV that should benefit from the synergy of ICN and edge computing. The content distribution and caching capability of ICN networks can be leveraged to share cooperative awareness messages (CAMs) among vehicles. On the other hand, edge computing enables services to be deployed and provides local computing and storage. For instance, by caching and reusing computation results, other consumers can avoid repeating computations. Interests aggregation can also be performed on edge nodes thus avoiding duplicates of interests for the same content from multiple consumers. Therefore, ICN and edge computing share complementary properties, local storage, and the ability to reduce latency, enhance mobility, and ensure scalability.

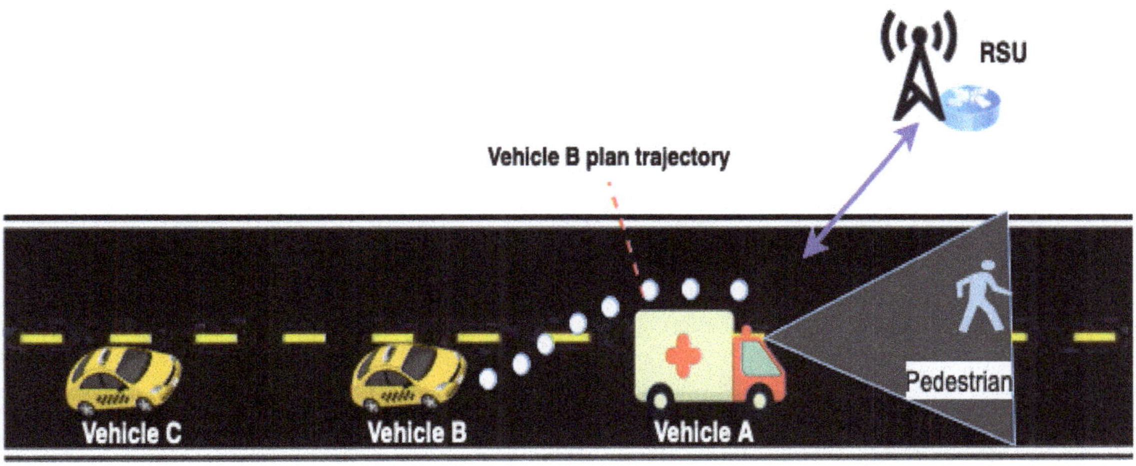

Figure 13. Collaborative driving—see through.

7. Edge Intelligence and Networking for IoV

With the goal of providing intelligent AI applications, a new computing and networking paradigm named Edge AI has been introduced. In the following, we present recent research on intelligent edge networking, edge computing, and caching in the context of the IoV domain. Table 5 summarizes the area where AI/ML/DL techniques solve bottleneck problems and realize converged IIoV.

7.1. Edge Computing Intelligence for IoV

Intelligent IoV should use a variety of intelligent computing platforms, such as cloud computing intelligence, edge computing, and on-vehicle intelligence. Providing intelligence at each layer of the IoV network architecture helps to meet the different requirements of IoV applications. IIoV will facilitate urban traffic management, vehicle insurance, road infrastructure building and repair, logistics and transit, in addition to vehicle driving and safety. IoV must be interconnected with other systems, such as Smart City, as a special case of the Internet of Things (IoT). Cloud computing and intelligence are mostly employed in this scenario for training and inference of AI models with massive datasets, as well as aggregating non-real-time data for wide-area networks deployed on the cloud.

To overcome excessive latency in cloud computing, edge computing and intelligence can support IoV for processing delay-sensitive computational tasks, such as road-side image processing, aggregating, and storing local information. Edge computing devices can be deployed on different edge nodes, such as on the RSU, BS, and vehicles having powerful onboard processing capabilities. Edge nodes provide computational services. The vehicle can offload tasks using either binary offloading or partial offloading schemes. Binary offloading is when the device determines whether the task is to be executed locally or to be offloaded to the edge server, depending on the workload of the server and the QoS requirements. In partial offloading the computational task is partitioned and executed on both the local device and an edge server. In [73], an edge learning framework is proposed and is demonstrated to reduce network traffic by 80% and running time by 69% over state-of-the-art cloud-based solutions. The author of [74] proposes edge-based data analysis and task allocation. Edge computing resources are utilized and task distribution for edge nodes based on game theory is applied. In [75], a novel real-time video analytics process is demonstrated using low-cost IoT devices and LoRaWAN networks to realize new services and applications that include traffic management through IoT edge computing.

Edge computing and intelligence also support privacy-aware services employing federated learning at the edge. Edge devices provide a communication efficient and privacy-preserving federated learning framework for enhancing the performance of IoV, whereby in-vehicle learning models are trained by exchanging inputs, outputs, and their learning parameters locally, as described in [22]. Vehicles can transform into mobile data centers, perform federated learning to enhance transmission control protocol (TCP) performance over WiFi, and react in a timely way to the Internet to meet vehicles' needs.

An energy-efficient computational offloading scheme employing deep reinforcement learning for the Intelligent Internet of Vehicles is discussed in [76].

Addressing the issue of running DNN, which require tremendous computational resources, high energy consumption and significant latency, the authors of [70] propose a collaborative framework for DNN co-inference between edge and end devices. The effectiveness of the framework was tested on a Raspberry Pi.

Newer vehicles are equipped with a powerful onboard unit that can provide computing capabilities for other sensors in the vehicle or cooperate with other vehicles. On-vehicle intelligence is an approach for processing high delay-sensitive tasks, such as real-time decision-making for vehicle control. Intelligent sensors utilize advanced signal processing techniques, data fusion, intelligent algorithms, and AI to understand sensor data, for better integration of sensors and feature extraction, leading to measures that can be used in smart sensing applications. In [77], a vehicle vibration sensor is used to classify and monitor road conditions. TinyML is a novel paradigm that enables low-end client devices to leverage machine learning techniques by deploying lean models previously developed on freely available cloud platforms [78]. It enables the performance of predictive maintenance by monitoring certain parameters and drawing conclusions based on ML to predict future faults.

7.2. Intelligent Edge Caching for IoV

The contents of IoV applications, either safety or non-safety, are likely to be transient, characterized by a limited lifespan and to become invalid after a certain period of time. Information, such as traffic conditions of certain road segments, will change after a few minutes after being reported by the source (producer) [79]. Thus, these transient contents have an impact on in-networking caching and forwarding decisions for NDN-based networks.

A comprehensive literature review on caching transient contents in vehicular NDN is provided in [80].

Caching improves the distribution and availability of contents and services in ICN [14]. It enhances the network performance by storing contents or services that are frequently used, to improve retrieval time and throughput, thus guaranteeing the quality of user experience (QoE). Cache management is one of the essential elements of NDN. Routing and forwarding are supported since nodes can cache content and interests. Content-caching issues include cache placement and cache replacement. Cache placement refers to which content and where to cache, while cache replacement refers to how long to cache in order to maximize the long-term average hit ratio. Traditional cache replacements include FIFO, LRU, and LFU. Content caching is challenging because of the dynamic and high mobility nature of IoV networks, content growth and sporadic demand of content, delay-sensitive applications, and limited cache capacity.

In IoV networks, caching can be performed on vehicles, on RSUs, on BSs, or jointly. For example, RSU caching minimizes the latency of retrieving contents and increases the cache hit ratio. Quick response time and storage efficiency are the two main factors to be considered for the cache management scheme. Caching popular content near the user improves efficiency.

Several cache management methods have been proposed to improve the efficiency of NDN. In the vehicular context they include: mobility-aware caching, delay-aware caching, proactive caching, probabilistic caching, and cooperative caching. For example, The authors of [81] proposed a limited domain cache (LdC), which simultaneously considers cache hit rate and the popularity of contents factors.

Recently, learning-based approaches have been used to design and optimize edge caching. In the following, we discuss representative studies that employ intelligent caching for IoV. The authors of [82] propose a cooperative edge caching framework which aims to exploit cooperation among BSs, RSUs, and connected vehicles. In [83], joint task caching and computational offloading for vehicular edge computing are proposed.

Performance evaluation and applicability analysis of different machine learning models for caching and routing have been addressed in [84]. Content popularity and request stream correlation are factors that affect cache performance and routing.

In [85], a deep-learning-based content caching framework, DeepCache, is proposed. The popularity of the content is predicted using an LSTM-based model. Similarly, [86] uses content popularity and user location to create an intelligent cache for content-centric networks. The approach is that of an optimization problem to determine which type of content should be cached in which router at a given time, thus classifying it as either normal or popular.

Content replacement is also an important issue in content caching. In [87], a dueling deep-q-network-based delay-aware cache update policy for fog radio access is employed and compared with the traditional caching replacement policies, i.e, FIFO, LRU, and LFU. The paper considers user mobility and time-varying channel states to improve the performance of the caching strategy.

7.3. *Edge Network Intelligence for IoV*

An edge network, in the context of IoV, can be defined as an autonomous network consisting of smart vehicles, a set of RSUs and BSs offering services without any prior configuration. The nodes in the network should have computation and caching capabilities. Each node in the network could have a multi-wireless interface to access services (e.g., 4G/5G, Wi-Fi) single-hop or multi-hop.

Network intelligence should allow autonomous devices to join edge networks, discover the services offered, gain seamless access to the services, and be coordinated with other entities in the network. Thus, edge networks need to have built-in embedded intelligence to allow better agility, resiliency, faster customization, and security. Indeed, embedding intelligence into the network will provide a much higher level of automation, enabling faster deployment, dynamic provisioning adapted to the nature of network functions, and end-to-end orchestration for coherent deployment of network infrastructures and service chains. It will also result in enhanced resilience and improved availability of future networks and services. In the next subsection, we review some of the latest research on this topic.

7.3.1. ICN-Based Naming and Resource Discovery for IoV Networks

Naming is one of the most significant networking aspects in NDN networks, used to identify the contents, as well as other network functionalities, including forwarding, caching, mobility, security, etc. [88]. ICN decouples content from its location and enables it to be easily retrieved, not only from the original server (producer), but also from the in-network caching nodes in the subsequent content request. Furthermore, it also provides interest aggregation for the requests from different consumers for common content, which, in turn, reduces interest packet traffic. ICN naming brings significant help to IoV communications by allowing vehicles to cache content, and also to forward interest, based on the content name instead of the address of the host, as the IoV environment is characterized by rapid changes and short connectivity duration.

However, NDN-based naming for IoV is challenging and most existing ICN frameworks assume that consumers are ex ante aware of the service and resource name. The consumer node sends out an interest packet that carries a name that identifies the desired content. However, vehicular users connecting to the network may not be aware of the name of the content or service of interest. A robust content naming design is needed to allow producers to describe precisely what it has and consumers to express which information they need. Existing research attempts to design content names for the IoV are based on specific use cases.

In [54], a hierarchical context-aware content naming scheme for NDN-based VANET, that enables naming of both safety and non-safety application contents, is proposed. It exploits the coding scheme to represent most of the content names. It follows the following naming scheme: content type (1 bit, safety or non-safety), content scope (1 bit, local or global), content format (2 bits, 00 for text, 11 for video, 01 for picture, and 10 for audio), application types (4 bits for main type + 4 bits subtype), to represent 32 different applications, timestamps, GPS locations, content sizes, and content file types.

In [89], a naming scheme for a vehicular content-centric network is proposed with the purpose of uniquely identifying the content generated by vehicles using a hierarchical and hash-based naming scheme. The hierarchical naming scheme is partly used for content name prefix aggregation and as an aid for a routing and hash-based naming scheme to help maintain content integrity.

In [90], the authors argue that the existing conventional hierarchical or hybrid NDN naming schema are inefficient to extract contents that are transient in nature and present instead a named-based query and command mechanism (NINQ) framework for ICN-based IoT. The authors of [54] present an adaptive content dissemination solution for VANET networks that includes hierarchical context-aware content naming schemes that enable both safety-critical and non-safety applications. To guarantee that the forwarding and processing rules of the packets are correct, [91] proposes a name space analysis framework to model and analyze NDN data planes using header space analysis.

Another challenge with NDN naming is the issue of name lookup in a large-scale namespace in the forwarding plane of the content router. Managing FIBs at an NDN router is challenging having long content names or name prefixes. In [92], a method is proposed to compress FIB with the goals of reducing the required router storage space and meeting the demands of routing efficiently

The authors of [93] propose a neural network-based solution named Pyramid-NN, which builds an index with the aim of reducing memory consumption and false positive rates while increasing the lookup speed.

Thus, an adaptive naming mechanism that exploits AI techniques is required to help vehicular nodes to identify available resources and to provide seamless service and resource discovery. For example, a naming technique based on natural language processing (NLP) is proposed in [94]. In this study, an AI-based application module assigns names for the content based on the event captured by cameras and sensors that describe the types of events and their geographical locations.

7.3.2. Intelligent Forwarding and Routing for IIoV

The named-data link state routing (NLSR) protocol [95] is one of the routing protocols for ICN networks that allows the router to return the content if it exists in the cache or to forward the interest packets to the nearest content router without predicting the probable location of the content. The same protocol was evaluated in [96] to test its real-world performance in a test-bed. Multipath forwarding and in-network caching (MUCA), proposed in [97], is an intra-domain name-based routing protocol that utilizes multi-path forwarding, link state database synchronization and in-network caching. The paper claims that MUCA outperforms the existing named-data link-state routing protocol (NLSR) [95]; however, according to [98] it is obsolete. The authors of [99] propose instead a multi-criteria decision-making (MCDM) approach based on interest forwarding and cooperative data caching.

Recently some research has been undertaken in designing an intelligent forwarding strategy. For example, [100] used a support vector machine (SVM) to forecast the success of interest packets in predicting content cached locations with the aim of minimizing the content search time and to maximize throughput. As an NDN-forwarding plane enables choosing the next forwarding hop independently, without relying on routing, an intelligent forwarding strategy can be designed for adaptive and intelligent data forwarding [101]. Similarly, [101] proposed FS-RL, an intelligent forwarding strategy based on RL in named-data networking. Moreover, the authors of [102] considered the feasibility of using reinforcement learning, specifically Q-learning for efficient and adaptive forwarding in NDN networks. They achieved higher interest satisfaction and shorter delay compared with the existing BestRoute and Flooding strategies. In [103], a random neural-network-based reinforcement learning interest forwarding scheme was proposed. The forwarding scheme takes advantage of the in-network caching capability of NDN architecture.

A routing protocol for vehicular named data networking was proposed in [104] to address the problem of reverse path partitioning (RPP) during consumer mobility. The mechanism, auxiliary forwarding set (AFS), considers mobility factors to identify and select forwarding nodes. The authors of [105] proposed a new method using Bloom filters which combine routing of network connectivity with a building and forwarding engine. A fuzzy interest forwarding approach that exploits semantic similarities between the names carried in interest packets and the names of potentially matching data in CS and entries in FIB is proposed in [106]. These are promising initiatives but they do not evaluate large-scale, highly dynamic, and frequently changing topologies which are a primary characteristic of IoV networks.

7.3.3. Mobility Support for IIoV

Mobility is an integral part of IoV. Thus, mobility management is an important network function, which still presents a challenge. Mobility management should support continuous user communication while on the move without interruption. Different mobility types and scenarios should be considered when developing mobility support solutions for IIoV [107]. In an ICN/NDN-based IoV environment, each vehicle might operate simultaneously as an information producer and consumer. Vehicular consumers send their requests directly to the next NDN router mentioning the content name, but they do not know the location of the content in which they are interested. The NDN router is then responsible for forwarding the consumer's interest to the node where the content is stored. Experimental-based NDN solutions for VANET supporting various communications categories (V2V, V2I, and V2R) have been produced and have demonstrated their potential to improve the QoS of VANETs [108]. Furthermore, the authors of [109] examined the promising potential of NDN for connected vehicle applications. However, the support of fast mobility and a large number of vehicle-aware solutions and frameworks needs to be improved. Most ICN architecture designs claim to inherently support consumer mobility; however, the challenge is when the producer node also moves. Recent studies, such as [110], argue that the existing schemes to address consumer mobility have shortcomings as they waste bandwidth and mobile consumers may miss real-time content during handover. The paper proposes a softwarized and MEC-enabled protocol architecture supporting consumer mobility in ICN and claims improvement in the communication overhead by 99.93%.

Producer mobility is more complicated since the NDN naming scheme does not include location information. A number of studies, such as those of [111–115], address producer mobility. Four producer mobility approaches have been presented in the literature: location resolution, triangular, locator/identifier separation, and routing-based. They can be divided into two categories: anchor-based and anchorless schemes. Most approaches utilize anchor nodes to hide the producer node movement. These use an anchor node or rendezvous server to track the current locations of contents. The anchor node holds the binding information for content and its location. It always interacts with the mobile producer to maintain updated binding information. This is a similar approach to that of home agent mobility in IP networks. However, such an approach causes suboptimal routing problems and signaling overhead. In the anchor-less scheme [112], mobile producers send special announcements to the previously contacted NDN access router (NAR) whenever it changes the point of attachment.

Recently, intelligent mechanisms have been proposed to address producer mobility in the ICN context. In [116] a proxy-based mobility support approach called PMNDN is proposed. The proxy entity is responsible for maintaining the reachability of the user. The authors of [113] present producer mobility solutions using NDN in an IoT context. Seamless service-driven mobility support, as a service that caters to both host-driven and network driven mobility, is considered in [117]. In [118], anchor-less producer mobility management in named data networking for real-time multimedia is proposed. Similarly, in [119], the authors proposed supporting producer mobility via named data networking in an integrated space-terrestrial scheme. The optimal producer mobility support solution (OPMSS) [111] is proposed to minimize unnecessary interest packet losses and delay, while providing an optimization path for data when the mobile producer relocates. In [115], the authors propose a lightweight machine-learning-based strategy to predict the position of the producer in advance and to determine whether or not handover occurs. It uses an echo state network (ESN), which is a type of fast converging recurrent neural network that uses simple linear regression.

7.3.4. Security and Privacy for IIoV

Security and privacy concerns are still challenging for NDN-based IoV networks [2]. In a non-trustable distributed environment, blockchain technology is being used to establish a creditworthy ecosystem among independent participants. IoV networks can leverage the potential of blockchain technology in IoV environments. Researchers discuss different solutions to utilize blockchain technology. For example, [120] proposes blockchain-based security architecture for NDN-based IoV networks using consensus algorithms and performance evaluations of its effectiveness in key management, cache positioning defense, and access control. In [121], the authors also propose a protocol for secure and privacy-friendly service provisioning at the mobile edge that provides a security service between the consumers and producers based on the NDN paradigm. In [122], the researchers apply communication sequential processes (CSP) modeling and verification tools for NDN-based IoV networks to verify vital properties, including deadlock freedom, data availability, PIT deletion faking, and CS caching pollution. The model fails to ensure the security of data in the presence of an intruder, so the authors utilized blockchain to guarantee the security of NDN-based IoV. Recently, machine-learning-based security and privacy solutions have been introduced for intrusion detection in IoT networks to reduce smart vehicle accidents and to detect malicious attacks in vehicular networks. RNN-based solutions might help to detect malicious attacks in IoV. For example, [123] proposes a hybrid deep learning (DL) model based on LSTM and GRU-based solutions for cyber-attack detection in IoV.

Table 5. Summary of AI techniques in different aspects of ICN and edge computing.

ICN	Naming and lookup	[89]	Hierarchical and hash-based naming with efficient Compact Trie name management scheme for vehicular content-centric networks (VCCN)	
		[94]	NLP based naming	NLP
		[93]	Smart name lookup for NDN	Neural network
	Routing and forwarding	[102]	Reinforcement learning-based algorithm for efficient and adaptive forwarding in named data networking	RL
		[101]	Forwarding plane enables each router to select the next forwarding hop independently without relying on routing	Intelligent forwarding strategy based on RL (IFS-RL)
		[124]	Select suitable forwarding interface based on the current network status	Deep reinforcement learning
		[125]	Adaptive forwarding strategy	RL DQN
		[103]	Adaptive forwarding strategy	RL with random NN
	Producer Mobility	[115]	Predict the location of producer node using its past movements	Echo state network (ESN) type of RNN
	Security	[122]	Guarantee the security of NDN-based IoV	Blockchain-based mechanism
		[120]	Mechanism for cache poisoning protection and access control in NDN vehicular edge computing networks	Blockchain based solution
Edge	Caching	[85]	DeepCache	LSTM
		[86]	Caching scheme based on content popularity and user location	Optimization
		[87]	Delay-aware cache update policy	Dueling DQN
	Computing	[74]	Efficient route planning for automated vehicles through big data acquisition and analysis architecture in ICN	Game theory
		[76]	Three-layer offloading framework in intelligent IoV to minimize the overall energy consumption while satisfying users' delay constraints	deep reinforcement learning

8. Issues and Recommendations for Future Research

In spite of the developments discussed, there are still open issues and challenges that need to be addressed to realize the advantages of IIoV. The main purpose of this section is to stimulate the interest of the ICN, Edge, IoV, and AI research communities in the requirements of future intelligent IoV networks. The challenges still to be addressed include: how to build an efficient network framework with edge computing and ICN in IoVs; how and where to place edge servers to service a large and dynamic number of vehicular users; and, how to design efficient caching, offloading, and communication strategies for IoV to provide intelligent services.

8.1. Intelligent Connected Platform

Following the success of cloud-based AI services, products such as Google's Cloud IoT Edge, Amazon Edge, and Azure IoT Edge, are being provided which extend data processing, analysis, and storage close to endpoints and provide edge computing services. Thus, 5GAA was established with the aim of bringing the automotive and telecommunication industries together to develop end-to-end solutions for future mobility and transportation services. Further partnering has occurred with standardization organizations, such as ETSI. Thus, edge intelligence can become a pervasive paradigm allowing edge devices to develop powerful edge AI functionalities. IIoV as an intelligent connected platform relies on the support of high-tech technologies, such as mobile internet, edge computing, cloud clouting, big data, and AI. The platform should support flexible service deployment, splitting and deploying independently as needed to meet the needs of specific vendors. To fully harness the potential of edge intelligence in IoV, several key challenges must be overcome. Intelligent connected platforms should be compatible with heterogeneous systems. The market potential of this technology will attract many vehicle manufacturers and IT vendors in the future. Thus, a common open standard should be set such that IoV stakeholders can enjoy seamless and smooth services across heterogeneous intelligent connected platforms anytime and anywhere.

8.2. Architecture and Network Heterogeneity Issues

Designing a universal network architecture that encompasses the heterogeneity of IoV is a challenging task. Mobile cloud services can be integrated with the ICN-based IoV to provide access to the information. However, this creates several challenges. Edge nodes execute a considerable amount of computational tasks, including content caching and networking. Using edge nodes, the processing burden and latency are significantly decreased. Furthermore, future smart vehicles will be equipped with various wireless interfaces that should be supported, including WAVE, LTE, 5G, and even satellite-based ones. For efficient content retrieval and communication, the network stack should exploit these multiple wireless interfaces. Interoperability between different variants of ICN, and with existing TCP/IP architecture, must be maintained as IP is currently the dominant protocol on the Internet. Supporting heterogeneity and interoperability is the major challenge.

8.3. Intelligent Distributed Computing and Autonomous Networking

An important aspect of intelligent IoV is the training and deployment of deep learning models. It may be too difficult to model and run computationally intensive deep learning models directly on vehicles. Directly moving data to a central server for training, e.g., at a cloud server, will introduce prohibitive communication overheads. Offloading them to the edge server may cause impairments due to time-varying wireless fading channels and lead to excessive latency when the offloaded data size is large. Besides enjoying the general benefits of edge intelligence, various challenges are presented, such as meeting the QoS requirements of applications, including real time requirements, such as 20–100 ms latency in safety application, high level accuracy and reliability for autonomous vehicles in 3D object detection and inference. Various techniques have been proposed to improve the performance of edge intelligence to enable inferences in IoV networks. Model compression is one enabling technique to alleviate the issue of resource-intensive deep learning models and resource-constrained edge devices. Model partition is another technique to alleviate the burden of executing deep learning models on vehicles, offloading the computational intensive tasks to the edge node (RSU) or other vehicles to obtain models to perform inference. Model partitions can be between vehicle and edge nodes or among vehicles for vehicle-edge node collaboration, according to the current system environment factors, such as network bandwidth, device resource and edge-server load.

Autonomous networking design is important to achieve efficient edge intelligence service provisioning under dynamic heterogeneous wireless network coexistence (e.g., LTE/5G/Wi-Fi/Bluetooth), allowing newly added edge devices to self-configure in a plug-and-play manner. Furthermore, ultra-reliable low-latency communication (URLLC) has been defined in 5G new radio for mission-critical application scenarios that demand low delay and high reliability. Thus, it will be valuable to integrate 5G URLLC capability with edge computing to provide ultra-reliable and low-latency edge services.

8.4. Smart Naming and Discovery

As IoV networks are distributed, intelligent IoV applications will be available in multiple vehicles and edge devices. Vehicles and edge nodes may run different IoV services and generate content. Content- and service-naming enable addressing and identification of content and services. Thus, it is critical to design efficient service- and content-naming schemes and also to service discovery protocols so that consumers can identify and locate the relevant content/service in a timely manner. However, naming in a dynamic environment is challenging due to intermittent connectivity caused by rapid mobility and topology changes. Solutions such as intelligent attribute-value-based naming can help. An adaptive encoding technique would be a viable option to tackle the issue of associating names. However, much more research is needed on this topic.

8.5. Mobility and Context Awareness Issues

In IoV, a thorough knowledge of vehicle mobility is a crucial aspect of continuous IoV services and network management. As vehicles move, the link condition between them and the RSU can change quickly and become intermittent. More significantly, an urban environment can be highly dynamic and complex with many unpredictable factors, such as time-varying traffic densities, buildings, and other structures affecting the V2V and V2I link performance. ICN was considered an alternative to bypass the limitations imposed by traditional IP networks, especially those relating to mobility. Even so, several research challenges remain unresolved, including frequent handovers that may degrade QoS and QoE. ICN supports consumer mobility natively. However, it becomes more complex if the producer also moves. A potential solution could be to locate mobile producers by flooding them with interest packets for path discovery. However, this would increase network traffic and the likelihood of network congestion. Vehicle contextual information, including position, speed, direction, acceleration, road traffic, road condition, and weather information can be effectively obtained from onboard sensors and shared among vehicles to design intelligent context-aware IoV protocols.

8.6. Security and Privacy Preservation

As a result of IoV's open deployment nature, it could be intercepted by potential attackers resulting in degradation of IoV services. We need to revise security and privacy at different layers (e.g., physical, link, network, session, and application). Blockchain-based solutions are used to provide security and are a suitable fit due to their decentralized nature, as mentioned in Section 7.3.4. However, in specific use cases, such as in connected and autonomous vehicles, vehicles perform collaborative perception and collaborate using collective perception messages (CPMs) to improve road safety and efficiency. Attackers may target CPMs and send false information that jeopardizes safety. Current standardization efforts do not yet include vehicle misbehavior specifications for advanced V2X services, such as collaborative perception (CP) [126].

8.7. Deployment and Orchestration

The IoV network entities, including vehicles, RSUs, and BSs, have heterogeneous storage, computing, routing, and forwarding capabilities. The ideal edge intelligence deployment technique to support as many vehicles as possible to meet the coverage, connectivity, and support of various intelligent services is still a research problem. Thus, the placement of edge intelligence is of paramount importance to improve communication, computing, and caching performance in terms of optimizing the resource allocation with minimal deployment costs and message exchange overheads. Given the large scale of physical IoV networks, we need to determine the minimum number of edge nodes, their location placement, network topology, and their association with BS and MBS to achieve an optimal allocation of resources. An intelligent optimization solution that considers the types, amount, and update frequency of information and available underlying network elements is needed. A clean-slate deployment requires all the network entities to support name-based routing and intelligent forwarding strategies.

9. Conclusions

Due to the fast growth of IoT and IoV, the conventional host-based networking paradigm faces a number of issues, including those related to scalability, mobility, addressing and security. Furthermore, recent applications demand real-time data which necessitates the rethinking of network models. The Internet of Vehicles (IoV) is different from the traditional Internet and other IoT networks, such as weather monitoring, due to the highly dynamic nature, mobility and high density of vehicles, and thus requires special attention. Security is a major concern in vehicle communication. In this paper, we have presented detailed reviews of the convergence of various technologies supporting IIoV, including ICN, edge computing, and AI. Edge intelligence accelerates V2X communication, vehicular edge computing, and intelligent networking, We also analyzed different techniques reported in the literature that use ICN-based communications for IoV networks. Lastly, we presented the requirements and challenges of building an intelligent communication networks architecture and intelligent computation for IoV. We hope that this work will serve as a useful reference in the area of intelligent IoV and inspire new topics of research.

Author Contributions: Conceptualization, S.S.M. and M.Z.; methodology, M.Z.; investigation, S.S.M., M.Z., M.L. and E.P.; resources, S.S.M.; writing—original draft preparation, S.S.M. and E.P.; writing—review and editing, E.P.; visualization, S.S.M.; supervision, M.Z. and M.L.; project administration, M.Z.; funding acquisition, M.Z. All authors have read and agreed to the published version of the manuscript.

Funding: This research received no external funding.

Institutional Review Board Statement: Not applicable.

Informed Consent Statement: Not applicable.

Data Availability Statement: Not applicable.

Acknowledgments: The authors would like to thank the STEP and TRIL programmes of the Abdus Salam International Centre for Theoretical Physics (ICTP).

Conflicts of Interest: The authors declare no conflict of interest.

Abbreviations

The following abbreviations are used in this manuscript:

AI	Artificial Intelligence	LRU	Least Recently Used
ANN	Artificial Neural Network	LSTM	Long Short-Term Memory
APU	AI Processing Unit	MANET	Mobile Ad Hoc Networks
BS	Base Station	MDP	Markov Decision Process
C-RAN	Cloud-Radio Access Networks	MEC	Mobile (Multi-access) Edge Computing
CDN	Content Delivery Network	ML	Machine Learning
CNN	Convolutional Neural Network	MLP	Multi-Layer Perceptron
CAM	Cooperative Awareness Message	NDN	Named Data Networking
CS	Content Store	NFV	Network Function Virtualization
CV	Computer Vision	NLP	Natural Language Processing
D2D	Device to Device	NPU	Neural Processing Unit
DNN	Deep Neural Network	PIT	Pending Interest Table
DQL	Deep Q-Networks	QoE	Quality of Experience
DRL	Deep Reinforcement Learning	QoS	Quality of Service
EC	Edge Computing	RL	Reinforcement Learning
ETSI	European Telecommunication Standardization Institute	RSU	Roadside Unit
FCNN	Fully Connected Neural Network	SDN	Software Defined Network
FIB	Forwarding Interest Table	SGD	Stochastic Gradient Descent
FIFO	First In First Out	TL	Transfer Learning
FL	Federated Learning	V2I	Vehicle to Infrastructure
GAN	Generative Adversarial Network	V2V	Vehicle to Vehicle
GNNs	Graph Neural Networks	V2X	Vehicle to Everything
ICN	Information Centric Networks	VANET	Vehicular Ad Hoc Networks
IIoV	Intelligent Internet of Vehicles	VR	Virtual Reality
IoV	Internet of Vehicle	WAVE	Wireless Access Vehicular Environment
ITS	Intelligent Transport Systems		

References

1. Kaiwartya, O.; Abdullah, A.H.; Cao, Y.; Altameem, A.; Prasad, M.; Lin, C.T.; Liu, X. Internet of Vehicles: Motivation, Layered Architecture, Network Model, Challenges, and Future Aspects. *IEEE Access* **2016**, *4*, 5356–5373. [CrossRef]
2. Contreras-Castillo, J.; Zeadally, S.; Guerrero-Ibanez, J.A. Internet of Vehicles: Architecture, Protocols, and Security. *IEEE Internet Things J.* **2018**, *5*, 3701–3709. [CrossRef]
3. Storck, C.R.; Duarte-Figueiredo, F. A 5G V2X ecosystem providing internet of vehicles. *Sensors* **2019**, *19*, 550. [CrossRef] [PubMed]
4. Shi, W.; Cao, J.; Zhang, Q.; Li, Y.; Xu, L. Edge Computing: Vision and Challenges. *IEEE Internet Things J.* **2016**, *3*, 637–646. [CrossRef]
5. Bonomi, F.; Milito, R.; Zhu, J.; Addepalli, S. Fog computing and its role in the internet of things. In Proceedings of the 1st ACM Mobile Cloud Computing Workshop, Helsinki, Finland, 17 August 2012; pp. 13–15. [CrossRef]
6. Satyanarayanan, M.; Chen, Z.; Ha, K.; Hu, W.; Richter, W.; Pillai, P. Cloudlets: At the leading edge of mobile-cloud convergence. In Proceedings of the 6th International Conference on Mobile Computing, Applications and Services, Austin, TX, USA, 6–7 November 2014; pp. 1–9. [CrossRef]
7. Jacobson, V.; Smetters, D.K.; Briggs, N.H.; Thornton, J.D.; Plass, M.F.; Braynard, R.L. Networking Named Data. In Proceedings of the ACM CoNEXT, Rome, Italy, 1–4 December 2009; pp. 1–12.
8. Sabir, Z.; Amine, A. NDN vs TCP/IP Which One Is the Best Suitable for Connected Vehicles. In *Recent Advances in Mathematics and Technology*; Springer: Berlin/Heidelberg, Germany, 2020.
9. Afanasyev, A.; Burke, J.; Refaei, T.; Wang, L.; Zhang, B.; Zhang, L. A Brief Introduction to Named Data Networking. In Proceedings of the IEEE Military Communications Conference MILCOM, Los Angeles, CA, USA, 29–31 October 2018; pp. 605–611. [CrossRef]
10. Ghasemi, C.; Yousefi, H.; Zhang, B. Internet-Scale Video Streaming over NDN. *IEEE Netw.* **2021**, *35*, 174–180 . [CrossRef]
11. Abane, A.; Daoui, M.; Muhlethaler, P.; Afifi, H. A down-to-earth integration of Named Data Networking in the real-world IoT. In Proceedings of the 2018 6th International Conference on Future Internet of Things and Cloud Workshops (FiCloudW), Barcelona, Spain, 6–8 August 2018.
12. Abane, A.; Daoui, M.; Bouzefrane, S.; Banerjee, S.; Muhlethaler, P. A realistic deployment of named data networking in the internet of things. *J. Cyber Secur. Mobil.* **2020**, *9*, 1–27 [CrossRef]
13. Arshad, S.; Azam, M.A.; Rehmani, M.H.; Loo, J. Recent advances in information-centric networking-based internet of things (ICN-IoT). *IEEE Internet Things J.* **2019**, *6*, 2128–2158. [CrossRef]
14. Din, I.U.; Asmat, H.; Guizani, M. A review of information centric network-based internet of things: Communication architectures, design issues, and research opportunities. *Multimed. Tools Appl.* **2018**, 30241–30256. [CrossRef]
15. Mars, D.; Gammar, S.M.; Lahmadi, A.; Azouz, L.; Mars, D.; Gammar, S.M.; Lahmadi, A.; Azouz, L.; Using, S. Using Information Centric Networking in Internet of Things: A Survey **2019**, *10*, 87–103. [CrossRef]
16. Nour, B.; Sharif, K.; Li, F.; Biswas, S.; Moungla, H.; Guizani, M.; Wang, Y. A survey of Internet of Things communication using ICN: A use case perspective. *Comput. Commun.* **2019**, *142–143*, 95–123. [CrossRef]
17. Hail, M.A. IoT-NDN: An IoT architecture via named data netwoking (NDN). In Proceedings of the 2019 IEEE International Conference on Industry 4.0, Artificial Intelligence, and Communications Technology, IAICT 2019, Bali, Indonesia, 1–3 July 2019; pp. 74–80. [CrossRef]
18. Khelifi, H.; Luo, S.; Nour, B.; Moungla, H.; Faheem, Y.; Hussain, R.; Ksentini, A. Named Data Networking in Vehicular Ad Hoc Networks: State-of-the-Art and Challenges. *IEEE Commun. Surv. Tutor.* **2020**, *22*, 320–351. [CrossRef]
19. Kerrche, C.A.; Ahmad, F.; Elhoseny, M.; Adnane, A.; Ahmad, Z.; Nour, B. *Internet of Vehicles Over Named Data Networking: Current Status and Future Challenges*; Springer International Publishing: Berlin/Heidelberg, Germany, 2020; Volume 242, pp. 83–99. [CrossRef]
20. Yao, H.; Li, M.; Du, J.; Zhang, P.; Jiang, C.; Han, Z. Artificial Intelligence for Information-Centric Networks. *IEEE Commun. Mag.* **2019**, *57*, 47–53. [CrossRef]
21. Hameed Mir, Z.; Filali, F. LTE and IEEE 802.11p for vehicular networking: A performance evaluation. *Eurasip J. Wirel. Commun. Netw.* **2014**, *2014*. [CrossRef]
22. Pokhrel, S.R.; Choi, J. Improving TCP Performance over WiFi for Internet of Vehicles: A Federated Learning Approach. *IEEE Trans. Veh. Technol.* **2020**, *69*, 6798–6802. [CrossRef]
23. Bazzi, A.; Masini, B.M.; Zanella, A.; Thibault, I. On the performance of IEEE 802.11p and LTE-V2V for the cooperative awareness of connected vehicles. *IEEE Trans. Veh. Technol.* **2017**, *66*, 10419–10432. [CrossRef]
24. Naik, G.; Choudhury, B.; Park, J.M. IEEE 802.11bd amp; 5G NR V2X: Evolution of Radio Access Technologies for V2X Communications. *IEEE Access* **2019**, *7*, 70169–70184. [CrossRef]
25. Singh, A.; Singh, B. A Study of the IEEE802.11p (WAVE) and LTE-V2V Technologies for Vehicular Communication. In Proceedings of the International Conference on Computation, Automation and Knowledge Management, ICCAKM 2020, Dubai, United Arab Emirates, 9–10 January 2020; pp. 157–160. [CrossRef]
26. Amadeo, M.; Campolo, C.; Molinaro, A.; Harri, J.; Rothenberg, C.E.; Vinel, A. Enhancing the 3GPP V2X architecture with information-centric networking. *Future Internet* **2019**, *11*, 199. [CrossRef]
27. Murillo, F.J.; Yoshioka, J.S.Q.; López, A.D.V.; Salazar-Cabrera, R.; de la Cruz, Á.P.; Molina, J.M.M. Experimental evaluation of lora in transit vehicle tracking service based on intelligent transportation systems and IoT. *Electronics* **2020**, *9*, 1950. [CrossRef]

28. Arena, F.; Pau, G.; Severino, A. A review on IEEE 802.11p for intelligent transportation systems. *J. Sens. Actuator Netw.* **2020**, *9*, 22. [CrossRef]
29. Ferrari, P.; Sisinni, E.; Carvalho, D.F.; Depari, A.; Signoretti, G.; Silva, M.; Silva, I.; Silva, D. On the use of LoRaWAN for the Internet of Intelligent Vehicles in Smart City scenarios. In Proceedings of the 2020 IEEE Sensors Applications Symposium, SAS, Kuala Lumpur, Malaysia, 9–11 March 2020; pp. 6–11. [CrossRef]
30. Haque, K.F.; Abdelgawad, A.; Yanambaka, V.P.; Yelamarthi, K. Lora architecture for v2x communication: An experimental evaluation with vehicles on the move. *Sensors* **2020**, *20*, 6876. [CrossRef]
31. World Health Organization (WHO). *Road Traffic Injuries: The Facts*; World Health Organization: Geneva, Switzerland, 2018; pp. 1–32.
32. ETSI TR 103 667 Intelligent Transport Systems (ITS); Study on Spectrum Sharing between ITS-G5 and LTE-V2X Technologies in the 5 855 MHz–5 925 MHz Band. Technical Report, 2021. Available online: https://www.etsi.org/deliver/etsi_tr/103600_103699/103667/01.01.01_60/tr_103667v010101p.pdf (accessed on 19 March 2022).
33. Use of the 5.850-5.925 GHz Band. Technical Report, 2021. Available online: https://www.federalregister.gov/documents/2021/05/03/2021-08802/use-of-the-5850-5925-ghz-band (accessed on 19 March 2022).
34. Official Journal of the European Union, L232. Technical Report, 2021. Available online: https://eur-lex.europa.eu/legal-content/EN/TXT/PDF/?uri=OJ:L:2021:232:FULL&from=EN (accessed on 20 March 2022).
35. FCC Adopts New Rules For The 6 GHz Band, Unleashing 1,200 Megahertz Of Spectrum For Unlicensed Use. Technical Report, 2020. Available online: https://www.fcc.gov/document/fcc-opens-6-ghz-band-wi-fi-and-other-unlicensed-uses (accessed on 20 March 2022).
36. Wan, J.; Zhang, D.; Zhao, S.; Yang, L.T.; Lloret, J. Context-aware vehicular cyber-physical systems with cloud support: Architecture, challenges, and solutions. *IEEE Commun. Mag.* **2014**, *52*, 106–113. [CrossRef]
37. Raza, S.; Wang, S.; Ahmed, M.; Anwar, M.R. A survey on vehicular edge computing: Architecture, applications, technical issues, and future directions. *Wirel. Commun. Mob. Comput.* **2019**, *2019*, *3159762*. [CrossRef]
38. Contreras-Castillo, J.; Zeadally, S.; Ibáñez, J.A.G. A seven-layered model architecture for internet of vehicles. *J. Inf. Telecommun.* **2017**, *1*, 4–22. [CrossRef]
39. Darwish, T.S.; Abu Bakar, K. Fog Based Intelligent Transportation Big Data Analytics in The Internet of Vehicles Environment: Motivations, Architecture, Challenges, and Critical Issues. *IEEE Access* **2018**, *6*, 15679–15701. [CrossRef]
40. Bonomi, F. *The Smart and Connected Vehicle and the Internet of Things*; WSTS: San Jose, CA, USA, 2013.
41. Zhuang, W.; Ye, Q.; Lyu, F.; Cheng, N.; Ren, J. SDN/NFV-Empowered Future IoV With Enhanced Communication, Computing, and Caching. *Proc. IEEE* **2020**, *108*, 274–291. [CrossRef]
42. Chen, M.; Tian, Y.; Fortino, G.; Zhang, J.; Humar, I. Cognitive Internet of Vehicles. *Comput. Commun.* **2018**, *120*, 58–70. [CrossRef]
43. Tuyisenge, L.; Ayaida, M.; Tohme, S.; Afilal, L.E. Network Architectures in Internet of Vehicles (IoV): Review, Protocols Analysis, Challenges and Issues. In Proceedings of the 5th International Conference, IOV 2018, Paris, France, 20–22 November 2018; pp. 3–13. [CrossRef]
44. Yu, K.; Eum, S.; Kurita, T.; Hua, Q.; Sato, T.; Nakazato, H.; Asami, T.; Kafle, V.P. Information-Centric Networking: Research and Standardization Status. *IEEE Access* **2019**, *7*, 126164–126176. [CrossRef]
45. Koponen, T.; Chawla, M.; Chun, B.G.; Ermolinskiy, A.; Kim, K.H.; Shenker, S.; Stoica, I. A data-oriented (and Beyond) network architecture. *Comput. Commun. Rev.* **2007**, *37*, 181–192. [CrossRef]
46. Zhang, L.; Afanasyev, A.; Burke, J.; Jacobson, V.; Claffy, K.; Crowley, P.; Papadopoulos, C.; Wang, L.; Zhang, B. Named Data Networking. *SIGCOMM Comput. Commun. Rev.* **2014**, *44*, 66–73. [CrossRef]
47. Yi, C.; Afanasyev, A.; Moiseenko, I.; Wang, L.; Zhang, B.; Zhang, L. A case for stateful forwarding plane. *Comput. Commun.* **2013**, *36*, 779–791. [CrossRef]
48. Singh, V.P.; Ujjwal, R.L. A walkthrough of name data networking: Architecture, functionalities, operations and open issues. *Sustain. Comput. Inform. Syst.* **2020**, *28*, 100419. [CrossRef]
49. Bari, M.F.; Chowdhury, S.R.; Ahmed, R.; Boutaba, R.; Mathieu, B. A survey of naming and routing in information-centric networks. *IEEE Commun. Mag.* **2012**, *50*, 44–53. [CrossRef]
50. Seskar, I.; Nagaraja, K.; Sam, N.; Raychaudhuri, D. MobilityFirst Future Internet Architecture project. In Proceedings of the Asian Internet Engineeering Conference, AINTEC 2011, Bangkok Thailand, 9–11 November 2011; pp. 1–3. [CrossRef]
51. Hong, J. Design Guidelines for Name Resolution Service in ICN. Available online: https://tools.ietf.org/id/draft-irtf-icnrg-nrs-requirements-03.html (accessed on 19 March 2022).
52. Yu, M.; Li, R.; Liu, Y.; Li, Y. A caching strategy based on content popularity and router level for NDN. In Proceedings of the 2017 IEEE 7th International Conference on Electronics Information and Emergency Communication, ICEIEC 2017, Macau, China, 21–23 July 2017; pp. 195–198. [CrossRef]
53. Tyson, G.; Sastry, N.; Rimac, I.; Cuevas, R.; Mauthe, A. A survey of mobility in information-centric networks: Challenges and research directions. In Proceedings of the International Symposium on Mobile Ad Hoc Networking and Computing (MobiHoc), Hilton Head, SC, USA, 11 June 2012; pp. 1–6. [CrossRef]
54. Zafar, W.U.I.; Rehman, M.A.U.; Jabeen, F.; Kim, B.S.; Rehman, Z. Context-aware naming and forwarding in ndn-based vanets. *Sensors* **2021**, *21*, 4629. [CrossRef] [PubMed]

55. Wang, M.; Wu, J.; Li, G.; Li, J.; Li, Q.; Wang, S. Toward mobility support for information-centric IoV in smart city using fog computing. In Proceedings of the 2017 5th IEEE International Conference on Smart Energy Grid Engineering, SEGE 2017, Oshawa, ON, Canada, 14–17 August 2017; pp. 357–361. [CrossRef]
56. Yu, Y.; Li, Y.; Du, X.; Chen, R.; Yang, B. Content Protection in Named Data Networking: Challenges and Potential Solutions. *IEEE Commun. Mag.* **2018**, *56*, 82–87. [CrossRef]
57. What Edge Computing Means for Infrastructure and Operations Leaders. Available online: https://www.gartner.com/smarterwithgartner/what-edge-computing-means-for-infrastructure-and-operations-leaders (accessed on 28 March 2022).
58. Patel, M.; Naughton, B.; Chan, C.; Sprecher, N.; Abeta, S.; Neal, A. Mobile-edge computing introductory technical white paper. *White Pap.-Mob.-Edge Comput. (Mec) Ind. Initiat.* **2014**, *29*, 854–864.
59. Muniswamaiah, M.; Agerwala, T.; Tappert, C.C. A Survey on Cloudlets, Mobile Edge, and Fog Computing. In Proceedings of the 2021 8th IEEE International Conference on Cyber Security and Cloud Computing (CSCloud)/2021 7th IEEE International Conference on Edge Computing and Scalable Cloud (EdgeCom), Washington, DC, USA, 26–28 June 2021; pp. 139–142. [CrossRef]
60. Azure IoT Edge. Available online: https://azure.microsoft.com/it-it/services/iot-edge/ (accessed on 28 March 2022).
61. Edge Computing Solutions. Available online: https://www.cisco.com/c/en/us/solutions/service-provider/edge-computing.html. (accessed on 28 March 2022).
62. IBM Solutions for 5G and Edge Computing. Available online: https://www.ibm.com/cloud/edge-computing. (accessed on 28 March 2022).
63. Aazam, M.; Zeadally, S.; Harras, K.A. Fog Computing Architecture, Evaluation, and Future Research Directions. *IEEE Commun. Mag.* **2018**, *56*, 46–52. [CrossRef]
64. Mukherjee, M.; Shu, L.; Wang, D. Survey of fog computing: Fundamental, network applications, and research challenges. *IEEE Commun. Surv. Tutor.* **2018**, *20*, 1826–1857. [CrossRef]
65. Goodfellow, I.J.; Pouget-Abadie, J.; Mirza, M.; Xu, B.; Warde-Farley, D.; Ozair, S.; Courville, A.; Bengio, Y. Generative Adversarial Networks. *arXiv* **2014**, arXiv:1406.2661.
66. Girshick, R.; Donahue, J.; Darrell, T.; Malik, J. Rich Feature Hierarchies for Accurate Object Detection and Semantic Segmentation. In Proceedings of the 2014 IEEE Conference on Computer Vision and Pattern Recognition, Columbus, OH, USA, 23–28 June 2014; pp. 580–587. [CrossRef]
67. Redmon, J.; Divvala, S.; Girshick, R.; Farhadi, A. You Only Look Once: Unified, Real-Time Object Detection. In Proceedings of the 2016 IEEE Conference on Computer Vision and Pattern Recognition (CVPR), Las Vegas, NV, USA, 27–30 June 2016; pp. 779–788. [CrossRef]
68. Kim, B.; Kang, C.M.; Lee, S.H.; Chae, H.; Kim, J.; Chung, C.C.; Choi, J.W. Probabilistic Vehicle Trajectory Prediction over Occupancy Grid Map via Recurrent Neural Network. In Proceedings of the 2017 IEEE 20th International Conference on Intelligent Transportation Systems (ITSC), Yokohama, Japan, 16–19 October 2017.
69. Du, Z.; Wu, C.; Yoshinaga, T.; Yau, K.; Ji, Y.; Li, J. Federated Learning for Vehicular Internet of Things: Recent Advances and Open Issues. *IEEE Open J. Comput. Soc.* **2020**, *1*, 45–61. [CrossRef]
70. Li, E.; Zhou, Z.; Chen, X. Edge Intelligence: On-Demand Deep Learning Model Co-Inference with Device-Edge Synergy. In Proceedings of the 2018 Workshop on Mobile Edge Communications, Budapest, Hungary, 20 August 2018; Association for Computing Machinery: New York, NY, USA, 2018; pp. 31–36. [CrossRef]
71. Andrade, P.; Silva, I.; Signoretti, G.; Silva, M.; Dias, J.; Marques, L.; Costa, D.G. An Unsupervised TinyML Approach Applied for Pavement Anomalies Detection Under the Internet of Intelligent Vehicles. In Proceedings of the 2021 IEEE International Workshop on Metrology for Industry 4.0 IoT (MetroInd4.0 IoT), Rome, Italy, 7–9 June 2021; pp. 642–647. [CrossRef]
72. de Prado, M.; Rusci, M.; Capotondi, A.; Donze, R.; Benini, L.; Pazos, N. Robustifying the Deployment of tinyML Models for Autonomous Mini-Vehicles. *Sensors* **2021**, *21*, 1339. [CrossRef]
73. Huang, Y.; Ma, X.; Fan, X.; Liu, J.; Gong, W. When deep learning meets edge computing. In Proceedings of the 2017 IEEE 25th International Conference on Network Protocols (ICNP), Toronto, ON, Canada, 10–13 October 2017; pp. 1–2. [CrossRef]
74. Zhao, C.; Dong, M.; Ota, K.; Li, J.; Wu, J. Edge-MapReduce-Based Intelligent Information-Centric IoV: Cognitive Route Planning. *IEEE Access* **2019**, *7*, 50549–50560. [CrossRef]
75. Seid, S.; Zennaro, M.; Libsie, M.; Pietrosemoli, E.; Manzoni, P. A Low Cost Edge Computing and LoRaWAN Real Time Video Analytics for Road Traffic Monitoring. In Proceedings of the 2020 16th International Conference on Mobility, Sensing and Networking (MSN), Tokyo, Japan, 17–19 December 2020; pp. 762–767. [CrossRef]
76. Ning, Z.; Dong, P.; Wang, X.; Guo, L.; Rodrigues, J.J.P.C.; Kong, X.; Huang, J.; Kwok, R.Y.K. Deep Reinforcement Learning for Intelligent Internet of Vehicles: An Energy-Efficient Computational Offloading Scheme. *IEEE Trans. Cogn. Commun. Netw.* **2019**, *5*, 1060–1072. [CrossRef]
77. Seid, S.; Zennaro, M.; Libse, M.; Pietrosemoli, E. Mobile Crowdsensing Based Road Surface Monitoring Using Smartphone Vibration Sensor and Lorawan. In Proceedings of the 1st Workshop on Experiences with the Design and Implementation of Frugal Smart Objects, London, UK, 21 September 2020; Association for Computing Machinery: New York, NY, USA, 2020; pp. 36–41. [CrossRef]
78. Ray, P.P. A review on TinyML: State-of-the-art and prospects. *J. King Saud Univ.-Comput. Inf. Sci.* **2021**, *34*, 1595–1623. [CrossRef]
79. Zhang, J.; Letaief, K.B. Mobile Edge Intelligence and Computing for the Internet of Vehicles. *Proc. IEEE* **2020**, *108*, 246–261. [CrossRef]

80. Amadeo, M. A Literature Review on Caching Transient Contents in Vehicular Named Data Networking. *Telecom* **2021**, *2*, 75–92. [CrossRef]
81. Kim, D.Y.; Lee, J. An NDN cache management for MEC. *Appl. Sci.* **2020**, *10*, 896. [CrossRef]
82. Qiao, G.; Leng, S.; Maharjan, S.; Zhang, Y.; Ansari, N. Deep Reinforcement Learning for Cooperative Content Caching in Vehicular Edge Computing and Networks. *IEEE Internet Things J.* **2020**, *7*, 247–257. [CrossRef]
83. Tang, C.; Wu, H. Joint optimization of task caching and computation offloading in vehicular edge computing. *Peer-to-Peer Netw. Appl.* **2021**, *15*, 854–869. [CrossRef]
84. Kulkarni, A.; Seetharam, A. Model and Machine Learning based Caching and Routing Algorithms for Cache-enabled Networks *arXiv* **2020**, arXiv:2004.06787. [CrossRef]
85. Ndikumana, A.; Ullah, S.; Kim, D.H.; Hong, C.S. Deepauc: Joint deep learning and auction for congestion-aware caching in Named Data Networking. *PLoS ONE* **2019**, *14*, e0220813. [CrossRef]
86. Wu, H.T.; Cho, H.H.; Wang, S.J.; Tseng, F.H. Intelligent data cache based on content popularity and user location for Content Centric Networks. *Hum.-Centric Comput. Inf. Sci.* **2019**, *9*, 44. [CrossRef]
87. Guo, B.; Zhang, X.; Sheng, Q.; Yang, H. Dueling deep-q-network based delay-aware cache update policy for mobile users in fog radio access networks. *IEEE Access* **2020**, *8*, 7131–7141. [CrossRef]
88. Serhane, O.; Yahyaoui, K.; Nour, B.; Moungla, H. A Survey of ICN Content Naming and In-Network Caching in 5G and beyond Networks. *IEEE Internet Things J.* **2021**, *8*, 4081–4104. [CrossRef]
89. Bouk, S.H.; Ahmed, S.H.; Kim, D. Hierarchical and hash based naming with Compact Trie name management scheme for Vehicular Content Centric Networks. *Comput. Commun.* **2015**, *71*, 73–83. [CrossRef]
90. Rehman, M.A.U.; Ullah, R.; Kim, B.S. NINQ: Name-integrated query framework for named-data networking of things. *Sensors* **2019**, *19*, 2906. [CrossRef]
91. Jahanian, M.; Ramakrishnan, K.K. Name space analysis: Verification of named data network data planes. In Proceedings of the ICN '19:2019 Conference on Information-Centric Networking, Macao, China, 24–26 September 2019; pp. 44–54. [CrossRef]
92. Karrakchou, O.; Samaan, N.; Karmouch, A. FCTree: A Space Efficient FIB Data Structure for NDN Routers. In Proceedings of the 2018 IEEE 43rd Conference on Local Computer Networks (LCN), Chicago, IL, USA, 1–4 October 2018; pp. 589–596. [CrossRef]
93. Li, Z.; Liu, J.; Yan, L.; Zhang, B.; Luo, P.; Liu, K. Smart Name Lookup for NDN Forwarding Plane via Neural Networks. In *IEEE/ACM Transactions on Networking*; IEEE: Piscataway, NJ, USA, 2021. [CrossRef]
94. Mochida, T.; Nozaki, D.; Okamoto, K.; Qi, X.; Wen, Z.; Sato, T.; Yu, K. Naming scheme using NLP machine learning method for network weather monitoring system based on ICN. In Proceedings of the 2017 20th International Symposium on Wireless Personal Multimedia Communications (WPMC), Bali, Indonesia, 17–20 December 2017; pp. 428–434. [CrossRef]
95. Hoque, A.K.M.M.; Amin, S.O.; Alyyan, A.; Zhang, B.; Zhang, L.; Wang, L. NLSR: Named-data Link State. In Proceedings of the 3rd ACM SIGCOMM workshop on Information-centric networking-ICN '13, New York, NY, USA, 12 August 2013; p. 15.
96. Friyanto, A.; Ariefianto, W.T.; Syambas, N.R. Analysis Operation NLSR with Ubuntu as NDN Router. In Proceeding of the 2019 5th International Conference on Wireless and Telematics, ICWT 2019, Yogyakarta, Indonesia, 25–26 July 2019; pp. 20–23. [CrossRef]
97. Ghasemi, C.; Yousefi, H.; Shin, K.G.; Zhang, B. MUCA: New Routing for Named Data Networking. In Proceedings of the 2018 IFIP Networking Conference IFIP Networking and Workshops, IFIP Networking 2018-Proceedings, Zurich, Switzerland, 14–16 May 2018; pp. 289–297. [CrossRef]
98. Wang, L.; Lehman, V.; Mahmudul Hoque, A.K.; Zhang, B.; Yu, Y.; Zhang, L. A Secure Link State Routing Protocol for NDN. *IEEE Access* **2018**, *6*, 10470–10482. [CrossRef]
99. Pu, C. Pro NDN: MCDM-Based Interest Forwarding and Cooperative Data. *J. Comput. Netw. Commun.* **2021**, *2021*, 1–16. [CrossRef]
100. Dutta, N.; Tanwar, S.; Patel, S.K.; Ghinea, G. SVM-based Analysis for Predicting Success Rate of Interest Packets in Information Centric Networks. *Appl. Artif. Intell.* **2021**, 1–22. [CrossRef]
101. Zhang, Y.; Xu, K.; Bai, B.; Lei, K. IFS-RL: An intelligent forwarding strategy based on reinforcement learning in named-data networking. In Proceedings of the 2018 Workshop on Network Meets AI and ML, Part of SIGCOMM 2018, Budapest, Hungary, 24 August 2018; pp. 54–59. [CrossRef]
102. Fu, B.; Qian, L.; Zhu, Y.; Wang, L. Reinforcement learning-based algorithm for efficient and adaptive forwarding in named data networking. In Proceedings of the 2017 IEEE/CIC International Conference on Communications in China, ICCC 2017, Qingdao, China, 22–24 October 2017; pp. 1–6. [CrossRef]
103. Akinwande, O. Interest forwarding in named data networking using reinforcement learning. *Sensors* **2018**, *18*, 3354. [CrossRef]
104. Duarte, J.M.; Braun, T.; Villas, L.A. Receiver mobility in vehicular named data networking. In Proceedings of the 2017 Workshop on Mobility in the Evolving Internet Architecture, Part of SIGCOMM 2017, Los Angeles, CA, USA, 25 August 2017; pp. 43–48. [CrossRef]
105. Mun, J.H.; Lim, H. On Sharing an FIB Table in Named Data Networking. *Appl. Sci.* **2019**, *9*, 3178. [CrossRef]
106. Mastorakis, S.; Chan, K.; Ko, B.; Afanasyev, A.; Zhang, L. Experimentation With Fuzzy Interest Forwarding in Named Data Networking. *arXiv* **2018**, arXiv:2010.07832.
107. Sofia, R.C. Guidelines towards information-driven mobility management. *Future Internet* **2019**, *11*, 111. [CrossRef]

108. Chen, M.; Ong Mau, D.; Zhang, Y.; Taleb, T.; Leung, V.C. VENDNET: VEhicular Named Data NETwork. *Veh. Commun.* **2014**, *1*, 208–213. [CrossRef]
109. Yang, N.; Chen, K.; Liu, Y. Towards Efficient NDN Framework for Connected Vehicle Applications. *IEEE Access* **2020**, *8*, 60850–60866. [CrossRef]
110. Benedetti, P.; Piro, G.; Grieco, L.A. A softwarized and MEC-enabled protocol architecture supporting consumer mobility in Information-Centric Networks. *Comput. Netw.* **2021**, *188*, 107867. [CrossRef]
111. Hussaini, M.; Naeem, M.A.; Kim, B.S. Opmss: Optimal producer mobility support solution for named data networking. *Appl. Sci.* **2021**, *11*, 4064. [CrossRef]
112. Choi, J.H.; Cha, J.H.; Han, Y.H.; Min, S.G. A dual-connectivity mobility link service for producer mobility in the named data networking. *Sensors* **2020**, *20*, 4859. [CrossRef]
113. Meddeb, M.; Dhraief, A.; Belghith, A.; Monteil, T.; Drira, K. Producer Mobility support in Named Data Internet of Things Network. *Procedia Comput. Sci.* **2017**, *109*, 1067–1073. [CrossRef]
114. Augé, J.; Carofiglio, G.; Grassi, G.; Muscariello, L.; Pau, G.; Zeng, X. MAP-Me: Managing Anchor-Less Producer Mobility in Content-Centric Networks. *IEEE Trans. Netw. Serv. Manag.* **2018**, *15*, 596–610. [CrossRef]
115. Li, Y.; Piao, X.; Zhang, H.; Lei, K. NDN Producer Mobility Management Based on Echo State Network: A Lightweight Machine Learning Approach. In Proceedings of the International Conference on Parallel and Distributed Systems-ICPADS, Singapore, 11–13 December 2018; pp. 275–282. [CrossRef]
116. Rao, Y.; Gao, D.; Zhang, H.; Foh, C.H. Mobility support for the user in NDN-based cloud storage service. In Proceedings of the 2015 IEEE Globecom Workshops, GC Wkshps, San Diego, CA, USA, 6–10 December 2015; pp. 1–6. [CrossRef]
117. Azgin, A.; Ravindran, R.; Chakraborti, A.; Wang, G.Q. Seamless producer mobility as a service in information centric networks. In Proceedings of the 2016 3rd ACM Conference on Information-Centric Networking, Kyoto, Japan, 26–28 September 2016; pp. 243–248. [CrossRef]
118. Ali, I.; Lim, H. Anchor-Less Producer Mobility Management in Named Data Networking for Real-Time Multimedia. *Mob. Inf. Syst.* **2019**, *2019*, 3531567. [CrossRef]
119. Liu, D.; Huang, C.; Chen, X.; Jia, X. Supporting producer mobility via named data networking in space-terrestrial integrated networks. In Proceedings of the 12th International Conference, WASA 2017, Guilin, China, 19–21 June 2017; pp. 829–841. [CrossRef]
120. Lei, K.; Fang, J.; Zhang, Q.; Lou, J.; Du, M.; Huang, J.; Wang, J.; Xu, K. Blockchain-Based Cache Poisoning Security Protection and Privacy-Aware Access Control in NDN Vehicular Edge Computing Networks. *J. Grid Comput.* **2020**, *18*, 593–613. [CrossRef]
121. Amadeo, M.; Campolo, C.; Molinaro, A.; Rottondi, C.; Verticale, G. Securing the mobile edge through named data networking. In Proceedings of the IEEE World Forum on Internet of Things, WF-IoT 2018, Singapore, 5–8 February 2018; pp. 80–85. [CrossRef]
122. Chen, N.; Zhu, H.; Yin, J.; Fei, Y.; Xiao, L.; Zhu, M. Modeling and verifying NDN-based IoV using CSP. *J. Softw. Evol. Process* **2021**, e2371. [CrossRef]
123. Ullah, S.; Khan, M.A.; Ahmad, J.; Jamal, S.S.; e Huma, Z.; Hassan, M.T.; Pitropakis, N.; Arshad; Buchanan, W.J. HDL-IDS: A Hybrid Deep Learning Architecture for Intrusion Detection in the Internet of Vehicles. *Sensors* **2022**, *22*, 1340. [CrossRef] [PubMed]
124. Hao, B.; Wang, G.; Zhang, M.; Zhu, J.; Xing, L.; Wu, Q. Stochastic Adaptive Forwarding Strategy Based on Deep Reinforcement Learning for Secure Mobile Video Communications in NDN. *Secur. Commun. Netw.* **2021**, *2021*, 6630717. [CrossRef]
125. Sena, Y.A.B.D.; Dias, K.L.; Zanchettin, C. DQN-AF: Deep Q-Network based Adaptive Forwarding Strategy for Named Data Networking. In Proceedings of the 2020 IEEE Latin-American Conference on Communications, LATINCOM 2020, Santo Domingo, Dominican Republic, 18–20 November 2020. [CrossRef]
126. Ansari, M.R.; Monteuuis, J.P.; Petit, J.; Chen, C. V2X Misbehavior and Collective Perception Service: Considerations for Standardization. In Proceedings of the 2021 IEEE Conference on Standards for Communications and Networking (CSCN), Thessaloniki, Greece, 15–17 December 2021; pp. 1–6. [CrossRef]

future internet

MDPI

Article

An In-Network Cooperative Storage Schema Based on Neighbor Offloading in a Programmable Data Plane

Shoujiang Dang [1,2,*] and Rui Han [1,2]

1 National Network New Media Engineering Research Center, Institute of Acoustics, Chinese Academy of Sciences, No. 21, North Fourth Ring Road, Haidian District, Beijing 100190, China; hanr@dsp.ac.cn
2 School of Electronic, Electrical and Communication Engineering, University of Chinese Academy of Sciences, No. 19(A), Yuquan Road, Shijingshan District, Beijing 100049, China
* Correspondence: dangsj@dsp.ac.cn; Tel.: +86-19801254996

Abstract: In scientific domains such as high-energy particle physics and genomics, the quantity of high-speed data traffic generated may far exceed the storage throughput and be unable to be in time stored in the current node. Cooperating and utilizing multiple storage nodes on the forwarding path provides an opportunity for high-speed data storage. This paper proposes the use of flow entries to dynamically split traffic among selected neighbor nodes to sequentially amortize excess traffic. We propose a neighbor selection mechanism based on the Local Name Mapping and Resolution System, in which the node weights are computed by combing the link bandwidth and node storage capability, and determining whether to split traffic by comparing normalized weight values with thresholds. To dynamically offload traffic among multiple targets, the cooperative storage strategy implemented in a programmable data plane is presented using the relative weights and ID suffix matching. Evaluation shows that our proposed schema is more efficient compared with end-to-end transmission and ECMP in terms of bandwidth usage and transfer time, and is beneficial in big science.

Keywords: in-network storage; load balancing; locally cooperative storage; an IP-compatible ID based protocol

Citation: Dang, S.; Han, R. An In-Network Cooperative Storage Schema Based on Neighbor Offloading in a Programmable Data Plane. *Future Internet* **2022**, *14*, 18. https://doi.org/10.3390/fi14010018

Academic Editors: José Carlos Lopez-Ardao, Miguel Rodríguez Pérez and Sergio Herrería Alonso

Received: 13 December 2021
Accepted: 28 December 2021
Published: 30 December 2021

1. Introduction

In many scientific disciplines, a huge amount of data is being generated. In the X-ray Free Electron Laser device (XFEL), the average data rate has reached more than 2 GB/s, the peak rate has reached 100 GB/s, and the data storage capacity has exceeded 100 PB [1]. Significant challenges exist in the collection and storage of data by large scientific facilities. The ever-increasing data generation speed and the sheer quantity of data require effective tools to collect and filter the data, save the data through high-speed networks, and prevent the loss of valuable data. Due to the storage capability limitations of existing computer systems, it is too late to completely record and save rapidly generated data [2]. However, there may be an opportunity for more nodes to participate in storage via high-speed networks. In addition, a large amount of data not only needs to be stored properly, but also needs to be able to be efficiently accessed. Due to the increase in the quantity of data, it is becoming increasingly difficult to read and move data. To deal with these challenges, a common approach is to build a data center on site, connect the experimental station with an optical fiber network/InfiniBand, and equip the data center with high-speed storage and computing equipment (CPU, GPU) for scientific computing. However, a bottleneck remains in network throughput and storage capacity in these solutions, because they still use end-to-end transmission, and provide limited improvements using the shortest routing path or multiple paths in the transport layer. Furthermore, distributing the large quantity of raw data and making this data available to scientists around the world in a scalable manner are additional typical challenges.

Recently, Named Data Networking (NDN) [3,4] and big science communities have combined cache and forwarding strategies, and used the cache capability to accelerate the distribution and management for big science experimental data [5–9]. In [5], the author proposes the federating of distributed catalog systems that store and manage NDN names to increase the speed of discovery of desired data. The work of [6] presented a preliminary study addressing opportunities and challenges to enable NDN-based intelligent data retrieval in networks for high-energy physics (HEP). However, these studies focused on the management of a huge quantity of scientific data and data retrieval, and rarely involved the persistent storage of collected data in networks.

As a result of the development of networking and storage, the forwarding rate of the interface commonly exceeds the storage bandwidth in a single network node. However, other switches in the network can have high speeds and large persistent storage capabilities. To better handle high-speed data injection, multiple switches can cooperate to store heavy data storage traffic. Using the network as a means of storage, the data can be stored along the path, which facilitates edge processing and avoids sending data to the cloud for reading, processing, and analysis, thus saving network bandwidth and increasing the responsiveness of the applications. However, no research has been conducted on directly using the network as a form of permanent storage, which would allow producers to simply offload their data to the network and let the network manage the storage and access to data.

This paper proposes the realization of data storage in the forwarding process, and to cooperatively offload the data beyond the local storage capacity to neighbor storage nodes. The main contributions of our work are as follows:

(1) To address high-speed data storage, we proposed an in-network storage service node structure to support cooperative storage in neighbor nodes, and designed an identifier (ID)-based cooperative storage protocol.
(2) To support locally cooperative storage and the discovery of more in-network storage service nodes, we proposed a neighbor selection mechanism based on the Local Name Mapping and Resolution System, in which the node weights are computed by combining the link bandwidth and node storage capability, and determining whether to split traffic by comparing normalized weight values with thresholds.
(3) To dynamically split the traffic among multiple targets, the cooperative storage strategy implemented in a programmable data plane is presented using the relative weights and ID suffix matching. Evaluation shows that our proposed schema is more efficient compared with end-to-end transmission and ECMP [10] in term of bandwidth usage and transfer time.

The remainder of the paper is organized as follows. In Section 2, we review the related work in distributed storage, edge storage, current scientific data management systems, and NDN-based big science. The overall system design and supported protocol are introduced in Section 3. Section 4 presents our core cooperative storage mechanism. Then, the experimental results and analysis are shown in Section 5. Finally, conclusions of this paper are presented in Section 6.

2. Background

Numerous studies have been conducted on distributed storage in academia and industry, but most of storage systems are constructed at the application layer. As a result of the recent development of the programmable network technology, solutions have been developed that use in-network computing to offload part of the storage function into the network [11]. In this section, we present a comprehensive review of the current relevant literature.

2.1. Distributed Storage

Existing data storage services, such as HDFS [12], GFS [13], and CEPH [14], are mostly clustering systems in data centers. Due to monitoring of the load status of each storage node, data requests are routed to storage nodes by centralized scheduling to support load balancing. When load balancing is performed, generally only the storage space of the node is considered, and the network link usage between the requester and the node is ignored. In a CDN network, contents are spread based on requests and the popularity of contents. It dispatches requests to the nearest content replica via central scheduling, and resolves download bottlenecks through caching, and is not suitable for real-time data injecting. The cloud storage system is a core cloud-based service offered by most cloud providers, such as Google, Amazon, and Microsoft. However, cloud storage system is located far from users, resulting in a high response delay for applications. Moreover, a massive amount of data generated by IoT devices places significant pressure on the core network for its transmission to remote clouds. For this purpose, providing storage as close to users as possible or at the edge of the network is beneficial.

2.2. *Storage at the Edge*

The existing literature tackles the problem of edge data storage with strategies for the optimal placement of data storage, aiming to maximize the QoS offered by the edge [15,16]. In [17], the authors explored the advantages and challenges of edge data storage to reduce the pressure on the core network from the massive data generated by IoT devices when transferring these data to cloud storage. In [18], the authors developed and assessed a preliminary design for the management of popular data at the edge. The Reverse-CDN proposed an architectural design vision to combine both Fog Computing and Information Centric Networking (ICN) to process IoT data locally at the edge [19]. Decentralized content-addressed storage systems, such as IPFS [20], are essentially the product of an end-to-end communication mechanism where retrieval of data needs to first establish a connection.

2.3. *Current Scientific Data Management Systems*

Scientific communities have designed and developed various customized data management software packages to satisfy their needs. The climate community has developed ESGF [21] to manage CMIP5 [22] data. Similarly, the HEP community uses Xrootd [23] for its data access and storage [8]. These applications are all based on IP networking protocols, inherit the defects caused by the end-to-end communication principle of the existing IP network, and cannot provide an appropriate network service model to efficiently facilitate data discovery and retrieval.

2.4. *NDN Based Big Science*

In [7], the authors studied federated distributed catalog systems that store and manage NDN names to support climate data discovery in multiple domains using Named Data Networking. The authors in [8] use a VIP forwarding and caching mechanism [24] and designed an NDN-based Large Hadron Collider (LHC) network. This work shows that NDN with VIP forwarding and caching can achieve a large reduction in the average delay per chunk compared to a no-caching case in the simulated LHC network [8]. The work of [9] uses NDN-based primitives to present a design of the deadline-based data transfer protocol for reserved bandwidth data transfers. However, these studies focused on the metadata management of huge scientific data or complex scientific workflow, rarely involved the collected data persistence storage.

Combined with the existing work above, we propose the use of in-network storage to support high-speed scientific data collection, retrieval, and sharing via ICN architecture. We designed the structure of an in-network storage service node, and cooperatively store the data based on an ID protocol.

3. Proposed System Architecture

3.1. *System Architecture*

To better implement a new network architecture or network protocol on the existing network infrastructure, and reduce deployment or upgrade costs, the system architecture is usually required to be compatible with the existing network infrastructure [25]. To be compatible with the existing network infrastructure, the proposed in-network storage solution is based on name resolution ICN where ID is resolved into an address for routing. Under the architecture of routing and forwarding in ICN based on control and user plane separation, the ID/IP integrated ICN reconstruction module has the capability of asking the Global Name Mapping and Resolution System (GNMRS) or Local Name Mapping and Resolution System (LNMRS) for resolving an ID to address(es) [26]. The ID/IP integrated ICN reconstruction module can also process the address(es) in the packet according to the rules and regulations generated by the local computing module. The rules and regulations may include deleting some addresses from the address field, or changing the destination address [27]. Then, a new ID/IP integrated ICN packet is reconstructed and finally forwarded by the IP forwarding module [27]. The focus of the current study is extending the local computing module to cooperatively support storing data chunk traffic in neighbors.

The designed in-network storage process is as follows:

- An intermediate node with storage service capability, which has been deployed in the network, registers a mapping of a specific identifier indicating a storage service and its Network Address (NA) (selecting any one of the interface IP addresses as the node's NA) with the LNMRS (step1). As shown in Figure 1, there are two local resolution areas marked with a dashed circle.
- Pull-based communications offer several advantages over push-based communications, such as built-in multicast delivery, receiver-oriented congestion control, and native support for client mobility [28]. To support proactive in-network storage, the producer initiates the sending of a "request for pull" message to a storage service node instance, which, in turn, triggers a pull request to be sent back to the producer by the instance. When the producer requests the in-network storage services, the producer first resolves the specific storage service identifier from the LNMRS to obtain the IP address list of storage service nodes (step2), which contains nearby nodes, such as R1 and R2. Then, the closest node (R1) is selected as the target according to latency, and the target (R1) is messaged to ask for the storage service (step3). The message includes the IDs of written data.
- After receiving the "request for pull" messages, the selected node parses the IDs from the messages and requests data chunks based on the IDs from the producer (step4).
- Then, the producer starts to send chunk data to the storage service node based on an ID-based protocol described in detail below (step5).
- Then, the extended cooperative storage schema begins to play a role. Based on the obtained information, the decision state of the current storage service node and the corresponding neighbor information are calculated. After the storage service node receives the data chunk, the storage service node first closes the context of the corresponding ID request, and decides to forward the data chunk to local storage or its neighbor storage node. If the storage service node decides to store in a neighbor node (R2), it selects the neighbor, modifies the destination address field in the packet, and then forwards the chunk data based on the destination address (step6). Otherwise, if the storage service node decides to store locally, it directly forwards the chunk data packets to the local storage module for chunk data storage.
- The selected neighbor node receives the chunk data and stores it locally. After chunk data is stored in the storage service node, the mapping of the ID of the data chunk and the NA of the storage service node is registered to LNMRS for further retrieval (step7).
- If there are no available neighbor nodes and the current node has no capability to deal with the data chunk packets, it will forward the packets to the least loaded neighbor and subtract 1 from the TTL field, which indicates the longest storage node path it can forward along. It is currently considered that the chunk data can only be offloaded once; thereafter, there is a need to wait for the data to be stored locally or forwarded to a unified flood discharge area in the cloud. The storage service nodes are assumed to have been deployed ahead of time, which is outside the scope of this paper.

- Similar processes can be executed in another local resolution area. It is assumed that traffic offloading is not possible between different resolution domains. Every storage service node can execute the offload process mentioned above in its own local resolution area, so that the entire network iteratively achieves load balancing.
- Then, the consumer queries the GNMRS or LNMRS for the data chunk ID, obtains the NA of R2, and obtains chunk data from R2.

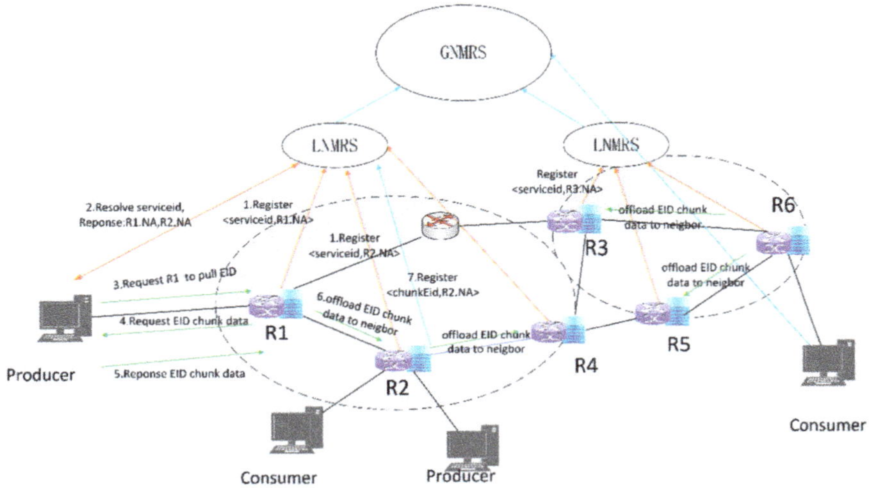

Figure 1. The overview of the chunk storage process.

3.2. *An ID-Based Cooperative Storage Protocol*

An identifier is a series of digits, characters, and symbols or any other form of data used to identify subscribers, users, network elements, functions, network entities providing services/applications, or other entities [29]. Entities are named using a persistent name, which does not change with mobility. This name may also be used by the network to locate, replicate, cache, and access the data. Due to the use of non-semantic identifiers as data names, other attributes of data, including signatures, life cycles, and preferences, can be maintained by the application, and the relationship between the attributes and the ID can also be maintained at the application layer.

In ICN, the data is a first-class citizen. To facilitate data transmission, caching, and storage, the data must be named by an identifier. Many methods can be used to assign the name to a data chunk [29], such as the hierarchically human-readable naming method, the self-certifying flat naming method, the attribute-based naming method, and the hybrid method of the multiple naming mechanism. In this paper, the name is composed by the hash value of the URI and the hash value of the data chunk. To process IDs more efficiently in the network, identifier-related fields are added into the packet header when the packet formats are designed. By also referring to the ID length in MobilityFirst [30], we choose the same length as that of the MobilityFirst GUID, which has a length of 20 bytes. The concept of an end-to-end connection does not apply in ICN, which has multisource routing, so our proposed transport protocol is connectionless [26]. In addition, due to some defects of IPv4, our proposal is designed to expand the next header field based on the IPv6 protocol. We name the proposed schema the identifier protocol (IDP), where a set of rules and regulations that specifies how the locators of a data packet are manipulated based on the ID in the network layer under the ID/locator separation of ICN.

Figure 2 shows the IDP protocol layer layout, which is based on the current TCP/IP protocol (IPv6), including mainly the network layer and transport layer. In the network layer, an ID header acts as the last header using the extension header defined in IPv6 [31]. The ID header contains the Next Header field, the Source ID type field, the Destination

ID type, the Source ID field, and the Destination ID field. Because the intermediate router, who cannot recognize the ID header, will ignore the unsupported extension headers, the use of an extension header in IPv6 to host the ID is compatible with existing TCP/IP architectures [25].

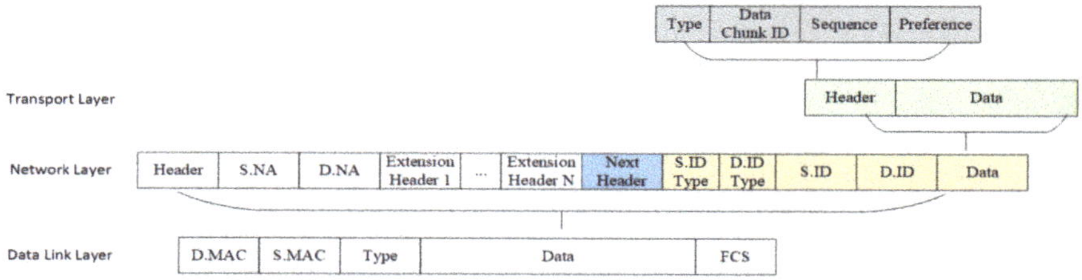

Figure 2. IDP protocol layer layout.

The "Next Header" field of the ID header is used to specify which transport protocol the packet will be passed to for processing. It can include our proposed transport protocol SEADP (Data Protocol in SEANet [32]), and can also point to an existing transport layer, such as TCP or UDP. We select 0×99 as our transport protocol value from the unsigned range 144–252 in the IANA registry ipv6-parameters [25].

Because various entities in the network can be labelled an identifier (including data and services), it may be necessary to classify IDs. We designed two fields to respectively identify the source ID type and the destination ID type.

For in-network data storage service, there are at least two types of packets, namely storage service request packet and data chunk request/response packet. We mainly combine source device ID and storage service ID, and source device ID and data chunk ID, to deal with in-network data storage service. The main ID combination relationship is shown in Figure 3. First, the producer requests the storage service through the producer's device ID as the source ID, the storage service ID as the destination ID, and the data chunk ID contained in the transport layer header. The storage service node requests the data chunk through the storage service node's device ID as the source ID and the data chunk ID as the destination ID. Then, the producer sends the data packet with the producer's device ID as the source ID, the storage service ID as the destination ID, and the data chunk ID and data contained in the payload. If the storage service node offloads the packets, the packets' destination IP is modified and the ID fields are unchanged, and are then forwarded.

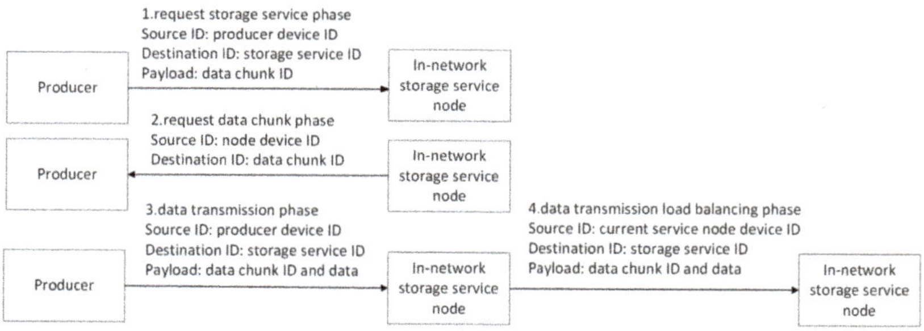

Figure 3. ID combination relationship in an in-network storage service scenario.

All IPv6-compliant hosts should be able to dynamically determine the packet length to be transmitted, for example, by using the path MTU discovery process. Briefly, when a host sends a very large IPv6 packet, the router does not segment the packet if it cannot forward such a large packet, but returns an error message to the sending host. This message tells the host that all future packets sent to the destination address will be segmented. It is far more efficient for hosts to send packets of the right size in the first place than for routers along the way to dynamically fragment each packet. Because IPv6 is an end-to-end transmission protocol and the concept of an end-to-end connection does not apply in ICN, our proposed transport protocol is connectionless. There is no handshake throughout the transmission, thus reducing the latency associated with establishing the connection [25]. Considering the optimal chunk size for efficient transmission in ICN and the large overhead that a small data chunk imposes on transmission and caching [33], the data chunk size is set to several MBs in our proposal. The designated data chunk size is larger than that of the path MTU, so the data chunk needs to be segmented. The segmentation information, including data chunk ID and the segment number, remains in the data chunk transport layer header. In addition, a new field, "preference", was designed, and is mapped from application-related requirements, including storage QoS, specific storing positions, caching strategies, security strategies, multipath transmission, and one-to-many transmission based on multicasting. The intermediate node may execute the corresponding action according to the "preference" field. Our proposed transport protocol SEADP header mainly contains "type", "data chunk ID", "sequence", and "preference" fields. The "type" field currently contains the data request packet and the data chunk response packet. The "data chunk ID" field represents the identifier of the data to be transmitted and stored. The "sequence" field of the packet indicates the segment number of the data chunk segmented by data source.

3.3. In-Network Storage Service Node Structure

According to the ID-based protocol described in Section 3.2, the intermediate router's in-network storage function is enhanced to take advantage of the power of the ID to support in-network storage. We designed its structure based on a programmable data plane. Figure 4 shows the inner structure of the enhanced network node. This can be an instance of a local computing model in ICN based on control and user plane separation. Currently the node structure is assumed to be deployed on every network node. The deployment is outside the scope of our paper and will be studied in other articles.

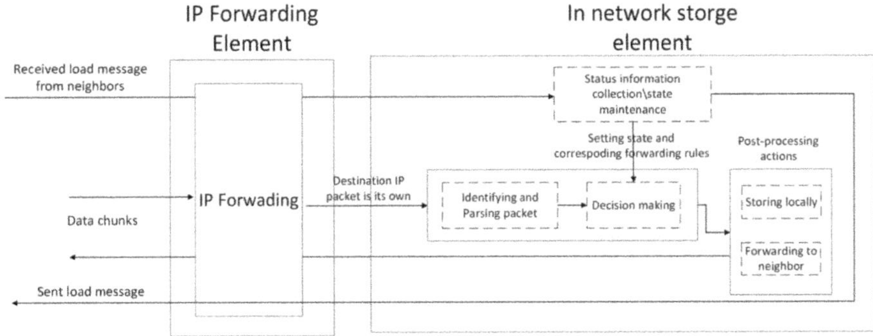

Figure 4. The extended network intermediate node structure.

The network intermediate nodes first need to ensure line-speed forwarding. In order not to affect the forwarding performance of the network intermediate nodes, the relevant status information collection, calculation, and maintenance are placed in the network storage processing element using local computing capacity. In addition, the calculated state value and decision result are inserted into the forwarding processing pipeline for ID-based protocol processing through the programming interface. The structure includes: the Identifying and Parsing packet, Decision making, Post-processing action, and Status information collection/state maintenance.

- The "Identifying and Parsing packet" module needs to correctly identify our ID-based transport layer packet, parse packet fields, and obtain the storage service ID/data chunk ID, sequence, and preference;

- The "Decision making" module executes one of the "post-processing actions" according to the state sets. If the state is set to "forwarding to neighbor node", the "Decision making" module executes the "Forwarding to neighbor" action; otherwise, the "Storing locally" action is invoked;
- The "Post-processing actions" module currently contains "Storing locally" and "Forwarding to neighbor". The "Forwarding to neighbor" action will change the destination IP of the data chunk packet to a new destination according to the rules from the computation result of the "Status information collection/state maintenance" module and then forward it. The "Storing locally" action will ACK the received packet, and then request the next segment and reassemble the data chunk from segments for storage in the local file system;
- The core module is "Status information collection/state maintenance". This module periodically collects neighbor status information and local status information, and parses packet information, then computes the state threshold and maps forwarding rules based on the ID. These results are set to shared memory that the forwarding pipeline can access according to the programmable interface. The time interval should be set at least as high as the maximum average RTT in the network [34].

4. In-Network Cooperative Storage Mechanism

4.1. Neighbor Selection Mechanism

LNMRS is usually deployed at the edge of the network and close to the forwarding devices, and maintains local ID–locator pairs. As a result, LNMRS may be more efficient or provide on-site service for users. LNMRS is also a hierarchical service system with a distributed architecture, which can synchronize part of the mapping of the ID and NA to GNMRS to be accessed globally [35].

The in-network storage service node registers the mapping storage service ID and its NA to LNMRS on startup. Considering the locality of the storage and a fast response, the neighbor nodes should be selected from nearby nodes. In addition, LNMRS is a suitable choice to select candidate neighbors by resolving the service ID.

In the case of the storage, congestion occurs where the transmission bandwidth is limited or storage IO is insufficient. Only considering link bandwidth or node storage capability is not enough, especially in high throughput traffic scenarios. Thus, we combine the two pieces of information as the criteria for node selection. The load information exchange packet contains the current input interface's total available input and output bandwidth (interface_total_bandwidth), and the current node's available writing throughput (w_{IO}) and message identification. When the neighbor node receives a load information exchange packet, it will respond with an ACK. Based on the received ACK time, we estimate the latency (RTT). RTT is computed as an exponentially weighted moving average (EWMA) of RTT in Equation (1). rtt_i is the time from the transmission of the i-th packet until receipt of the ACK of the i-th packet. RTT_i is estimated as the average round trip time after the i-th packet. We assume $RTT_0 = 0$ and $\alpha = 0.125$ (which is a typical value in TCP [36]).

$$RTT(k)_i = (1-\alpha) * RTT(k)_{i-1} + \alpha * rtt(k)_i \tag{1}$$

The neighbor node's weight is computed based on the neighbor's interface_total_band width and available writing throughput in Equation (2).

$$\begin{gathered}\text{neighbor}_{\text{weight(k)}} = \\ (\beta * bandwidth(k) * RTT(k) + (1-\beta) * w(k)_{IO} * RTT(k))/2\end{gathered} \tag{2}$$

The weight of the node itself is computed in Equation (3) based on the node's available writing throughput and the average RTT of all neighbor nodes.

$$\text{self\_weight} = (w_{IO} * averag\_RTT)/2 \tag{3}$$

Because the chunk data needs to be shifted from nodes with heavy load to those with less load, determining when to shift the load is important. We define a normalized value γ in Equation (4).

$$\gamma = \frac{self\_weight(i) - \min(weight(k), k \in neighbors\_list|i)}{max(weight(k), k \in neighbor|i) - \min(weight(k), k \in neighbors\_list|i)} \quad (4)$$

If γ is smaller than the defined threshold, the chunk data should be offloaded to its neighbors; otherwise, the traffic is not offloaded to its neighbors and it is better to store the chunk itself. We defined the threshold as 0.5 in simulation. When the threshold is set to 0, it means it has no cooperative storage mechanism and is just like a common node.

As shown in Algorithm 1, when the node is in the cooperative state, it sorts the neighbors' weight and selects a certain number of nodes (N) that exceed the current node's weight value as its final offloaded targets. Then, the traffic is split by the data chunk ID to these selected targets according to their weights.

Algorithm 1 Neighbor selection algorithm.

```
Input:
rtt = {rtt(1)_i, rtt(2)_i, ... rtt(k)_i}, current rtt of each neighbor interface;
bandwidth = {bandwidth(1)_i, bandwidth(2)_i, ... bandwidth(k)_i}, current bandwidth of each neighbor interface;
w_IO = {w_IO(1)_i, w_IO(2)_i, ... w_IO(k)_i,}, current writing throughput of each neighbor;
Output: state, target_list
Initialize RTT0 = 0, α = 0.875, β = 0.1
While k < K do
   calculate RTT(k) ← Equation (1)
   calculate neighbor_weight(k) ← Equation (2)
EndWhile
calculate averag_RTT = 1/K * Σ_{1}^{K} RTT(k)
calculate self_weight ← Equation (3)
calculate normalized value γ ← Equation (4)
if γ < threshold, then
     state ← "cooperative"
     sortedlist ← sort (neighbor_list | i) by weight desc
     sublist ← subsist (0, number_threshold-1) from sortedlist
     target_list ← sublist
else
     state ← "local_storage"
EndIf
Return state, target_list
```

4.2. Cooperative Storage Strategy

Many load-balancing strategies have been proposed in academia and industry, such as the round robin strategy, random strategy, hash strategy, or weight-based strategy [37], and traffic splitting in programmable switches [38–41]. The round robin strategy distributes traffic equally among the instances by forwarding traffic to each instance in turn. The random strategy will select a random instance to forward traffic. The hash strategy will compute the hash value x of a field, such as the source IP or destination IP, compute index y as × mod K (the count of instances), then obtain the corresponding target based on y in K targets. Traffic splitting in programmable switches mainly focuses on the weight approximation using flow table entries, but is still subject to flow-based traffic and does not consider how to obtain the weight. The strategies above do not consider the weight of targets and will cause load unbalancing if flow-based splitting is used. We consider the weight-based strategy and compute the relative weight in Equation (5).

$$\text{relativeWeight(n)} = weight(n) \Big/ \sum_{1}^{N} weight(n) \tag{5}$$

According to the relative weight ratio of K nodes, we calculate the number of nodes corresponding to the constraint of storage space having a length under M (each index occupies 2 bytes) in Equation (6).

$$\text{Count(n)} = floor(\ relativeWeight(n)\ *\ M\ /2) \tag{6}$$

Then, two linear storage spaces are constructed separately for storing the index of targets and targets' IPs. Linear Storage Space 1 is used to store the index of the neighbors' IP, which is stored in Linear Storage Space 2. The flow entries will match the L bits of the ID's suffix; then, the actions are executed to determine whether it is in the cooperative state. If this is the case, the corresponding index x is searched for in Linear Storage Space 1 based on the value of the L bits of the ID's suffix; the target IP in Linear Storage Space 2 is obtained with the value of index x as the index in Linear Storage Space 2; and the "Forwarding to neighbor" action is executed. Figure 5 shows a simple example.

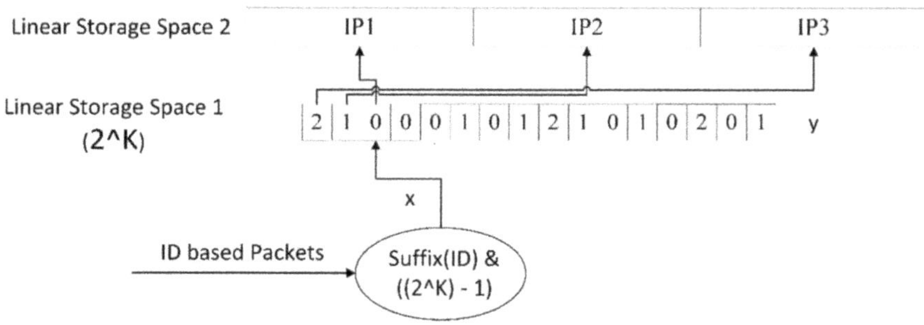

Figure 5. The load balancing strategy procedure.

However, the ping-pong loops problem may exist. When a producer P1 begins to store chunks, the corresponding in-network storage service node A enters the load balancing state, and may offload a portion of the chunks to its neighbor B according the load balancing rules. Immediately afterwards, the in-network storage service B receives the data storage from another producer P2. Assuming that A is also a neighbor of B, B also enters the cooperative state after calculating the normalized value of its weight. A portion of the chunks may be offloaded to A according the offloading rules in B. After A receives these chunks, they may be offloaded to B according to A's offloading rules, and a loop occurs. To avoid ping-pong loop problems, the input port avoidance mechanism is adopted. The input port avoidance mechanism is used to exclude the input port of the packet and to prevent forwarding to the input port where the packet comes. It is currently considered that the chunk data can only be offloaded once, before waiting to be stored locally or forwarded to a unified flood discharge area in the cloud. The detailed algorithm is described in Algorithm 2.

Algorithm 2 Cooperative storage algorithm.

Input:
weight = $\{weight(1), weight(2), \ldots, weight(n)\}$, node weight of selected targets;
Output:
Initialize
While n < N **do**
    calculate relativeWeight(n) ← Equation (5)
    calculate Count(n) ← Equation (6)
  EndWhile
  fill up linear storage space 1
  fill up linear storage space 2
  calculate index1=hash(ID) % N
  get index2 = linear_storage_space_1(index1)
get destination = linear_storage_space2(index2)
get Out_port from destination
if packet.In_port == Out_port, then
    Local Storage
else
    Forward to neighbor
Return

5. Evaluation

The rationality and performance of the algorithm were verified and evaluated using an actual deployed prototype system in a simulation environment for a large-scale scientific facility.

5.1. Network Environment Setup

The network topology is a simple FatTree topology, as shown in Figure 6. All the switches, except a common layer 3 Huawei router between R1 and R3, are our own programmable switches implemented based on a Protocol Oblivious Forwarding (POF) switch [42]. The controller is implemented based on an ONOS controller. We implemented the whole routing process based on the POF protocol and corresponding instruction sets. These applications can support hybrid routing with the current IP infrastructure through OSPF.

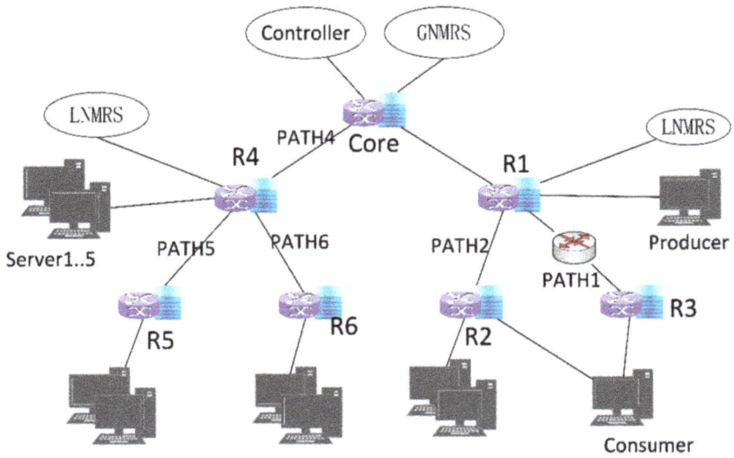

Figure 6. Network topology in the evaluation.

The key components are GNMRS and, in particular, LNMRS. We implemented LNMRS, which can maintain name and NA mapping for resolving the ID to addresses locally, and synchronize a portion of the mappings of the ID and NA to GNMRS to be accessed globally.

In the network topology, our in-network storage modules are deployed in every POF switch, which are connected with 10 Gbps optical fiber. In addition, there is a common layer-3 IP router in the path from R1 to R3. This can simulate the hybrid network environment, and is also used for traffic control, because our new protocol is implemented based on the raw socket in a Linux environment and the network throughput in the Linux server can only achieve 4 Gbps on average. Two servers, namely the producer and the consumer, are connected to the network using 10 Gbps optical fiber. All POF switches are configured for in-network storage with a size of 10 TB.

5.2. Performance Results

We mainly focus on the throughput and transfer time influenced by the load balancing schema in the ID-based data chunk transmission.

5.2.1. Evaluation of the Selection Mechanism

To evaluate our neighbor selection mechanism, we use the left part of the topology to compare our proposal with the ECMP mechanism under a constructed flow. We use three servers to simulate two 4 Gbps data flows and one 2 Gbps data flow. We limit the bandwidth between R4 and R5, and the bandwidth between R4 and R6, to 4 Gbps. In addition, the bandwidth between R4 and the core is limited to 2 Gbps. According to the ECMP mechanism, the paths R4 $\rightarrow$ R5, and R4 $\rightarrow$ R6 are equal cost paths.

First, server one writes 4 Gbps data flows to R4 for 30 s. After 10 s, server two writes 4 Gbps data flows to R4 for 30 s. Then, server three writes 2 Gbps data flows to R4 for 30 s. We compared our proposed neighbor selection mechanism with ECMP and compared statistics for the throughput of the two mechanisms. Figure 7 shows the throughput of our selection mechanism and ECMP. The data flow generated by our mechanism is basically consistent with the expected data flow. However, the ECMP mechanism cannot achieve the expected data flow and has a long flow completion time. Our peak data traffic reached 10.43 Gbps and ECMP's peak data traffic was only 7.51 Gbps. In addition, during the period of 20 to 30 s of the whole experiment, the ECMP mechanism could not use more paths to offload the excess traffic. The network started to become congested, resulting in a drop in throughput. According to the ECMP mechanism, two paths can only be used for storing data and the peak throughput absorbed by R4 is about 8 Gbps. Our neighbor selection mechanism can discover three paths and split the data traffic, thereby improving the flow completion time.

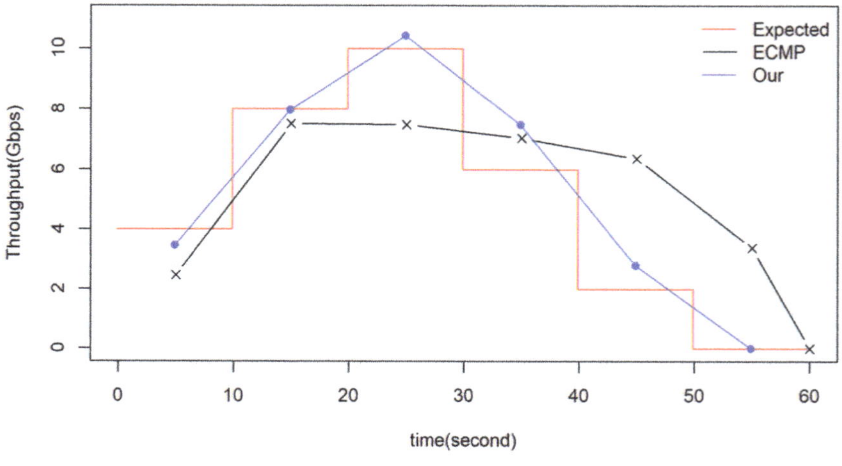

Figure 7. Results of the selection mechanism.

5.2.2. Comparison with End-to-End Transmission

Bbcp [43] is a point-to-point network file copy application, which is capable of transferring files approaching line speeds in a wide area network (WAN). We tested this under our simulated network environment and found that it can almost occupy the whole bandwidth when the link is idle. However, it can use the single shortest path and has no awareness of multipath. Our in-network storage can discover multiple neighbors to offload its traffic based on the LNRMS. If there is congestion in the shortest path, the efficiency of Bbcp is degraded. However, our in-network storage is barely influenced because of its selection of multiple neighbors and the traffic splitting mechanism.

To evaluate the end-to-end transmission and in-network storage, we carried out four experiments, including an end-to-end remote file transfer test and an in-network storage test under no constraint on the bandwidth, and an end-to-end remote file transfer test and an in-network storage test under a network bandwidth constraint. We used the right-hand part of the topology in Figure 6. The bandwidth from the producer to the consumer was 10 Gbps in the physical link and the network throughput of our current transport layer protocol stack implemented in the raw socket only reached 4 Gbps on average. Therefore, to ensure a fair evaluation between the end-to-end transmission and in-network storage, we limited the bandwidth of the link between R1 and R3 to 4 Gbps.

First, we constrained the bandwidth between R1 and R3 to 4 Gbps. In Figure 8a, the transmission bandwidth of Bbcp is shown between the producer and the consumer along the path of producer → R1 → R3 → consumer under the constraint of 4 Gbps controlled by the common router in the path of R1 to R3. It shows that the average bandwidth of Bbcp is 4.53 Gbps and Bbcp almost uses all the available bandwidth. From Figure 8b, the average transmission bandwidth of our proposed algorithm is 4.6 Gbps when our proposed algorithm is not in a cooperative state, and it only stores the data in R1. The bandwidth utilization of our algorithm is almost the same as that of Bbcp. The file transfer time is compared in Table 1.

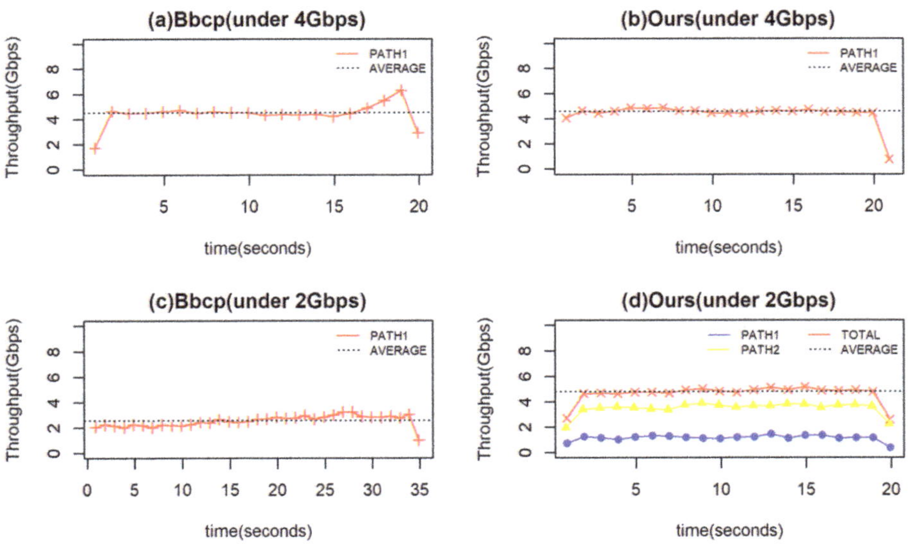

Figure 8. Results of comparison with end-to-end transmission.

Table 1. Comparison of transfer time between Bbcp and our mechanism.

Experiments	Time(s)
Bbcp (under 4 Gbps)	19
Ours (under 4 Gbps)	18.5
Bbcp (under 2 Gbps)	33.5
Ours (under 2 Gbps)	17.5

Then, the bandwidth between R1 and R3 was limited to 2 Gbps. In Figure 8c, the average transmission bandwidth of Bbcp is only 2.56 Gbps because it can only use the single path of producer → R1 → R3 → consumer. It can also be seen in Figure 8d that the average transmission bandwidth of our proposed algorithm is 4.82 Gbps, which is barely influenced by the constraint, because the proposed algorithm can use R2 to cooperatively store its data and the bandwidth from R1 to R2 is 10 Gbps [44]. In all our tests, we transferred the same file having a size of 10 GB. In addition, the transfer time of Bbcp was also lengthened due to bandwidth rate degradation. The transfer time of our proposed algorithm showed almost no increase although there was a slight truncation of the traffic.

From the comparison, we can see that our proposed algorithm is useful in improving the bandwidth usage, especially in multiple path environments.

5.2.3. Influence of Cooperative Storage Schema

Traffic is spilt based on the flow or flowlet in common load-balancing strategies. The flow-based load balancing strategies must distinguish between elephant and mice flows, because these two flows have an obvious influence on the load-balancing strategies. If the traffic is split based on an elephant flow, it may overload the target while the other servers are underloaded. Conversely, if the traffic is a mice flow, it may be not beneficial to split it over multiple servers due to the cost of packet reordering. Our proposed cooperative storage schema is based on the splitting of the data chunk by the application, which can be stored in a disordered state between data chunks and in different locations. The only requirement is that the packets in a chunk are delivered in order. This can be realized using an in-network cache to temporarily hold complete chunks in hop-by-hop transmission [45], or by utilizing the suffix of the ID to match flow entries to maintain the consistency of the packet forwarding path. We used the second method to implement our schema considering the latency of the hop-by-hop transmission.

To verify our cooperative storage strategy, we evaluated our algorithm and the ECMP algorithm under equal weights and under changed weights. We varied the weights by changing the network bandwidth between R1 and R3.

We undertook the experiments using two split ratios under our network environment with no bandwidth constraint. First, we used R1 to split data chunks equally to R3 and R2, and gathered statistics of bandwidth usage and transfer time. Figure 9a shows 4.77 Gbps on average was achieved where there is no cooperative storage strategy. Second, we limited the bandwidth of the path from R1 to R3 to 1 Gbps, and the split ratio of PATH2 to PATH1 was changed to 3:1. From Figure 9c, we can see that we obtained total bandwidth of 4.69 Gbps, which is comparable to the bandwidth of 4.77 Gbps in Figure 9a. Furthermore, the transfer time in the two cases is nearly the same from Table 2. Our cooperative storage strategy has no influence on transmission bandwidth because of the chunk ID-based transmission.

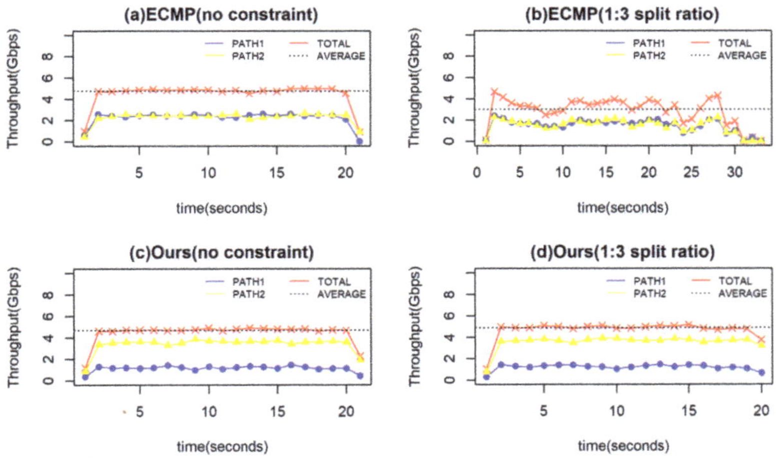

Figure 9. Results of the influence of the load balancing schema.

Table 2. Comparison of the transfer time between ECMP and our mechanism.

Experiments	Time(s)
ECMP (no constraint)	18
ECMP (1:3 split ratio)	28.5
Ours (no constraint)	18.3
Ours (1:3 split ratio)	17.6

Then, we limited the bandwidth of PATH1 to 1 Gbps and undertook the experiments to load balance the traffic between PATH1 and PATH2. The actual transmission bandwidth was reduced to 3.0 Gbps, compared with 4.77 Gbps under no constraint in Figure 9b. The incorrect split ratio causes congestion on PATH1, degrades the transmission speed of PATH1, can resulting in a decrease in the total bandwidth. The transfer time of ECMP under the split ratio of 1:3 was increased to 28.5 s, compared to the transfer time of 17.6 s of our schema in a split ratio of 1:3. The congestion and fluctuation of traffic in PATH1 result in heavy partial packet transmission, and the transfer time is even more significantly affected from Table 2. Thus, our weight-based schema is more efficient than the schema without an awareness of the multipath state.

5.2.4. The Accuracy of Traffic Split Ratio

Given a limited rule capacity at the switch, we need to make a tradeoff between the accuracy of the traffic split ratio and the number of flow entries. We measured the throughput under different lengths of the ID's suffix (M) under an expected split ratio of 4:1. We adjusted the bandwidth ratio between PATH1 and PATH2 to 1:4 through the three-layer router.

When the length of the matched ID's suffix is 2, we built four flow entries for every interface. The mask of the corresponding field is 0×11, and the IDs are split in the ratio of 3:1. If the length of the matched ID's suffix is 3, the IDs are split in the ratio of 3:1, and the IDs are split in the ratio of 4.3:1 when the length of the matched ID's suffix is 4. As shown in Figure 10, the measured traffic ratio (4.4:1) is the most accurate when the length of the ID's suffix is 4. However, the number of flow entries is also the greatest. If the split ratio needs to be updated frequently, the cost of maintaining the accuracy of the traffic split ratio and the resulting traffic fluctuation cannot be neglected.

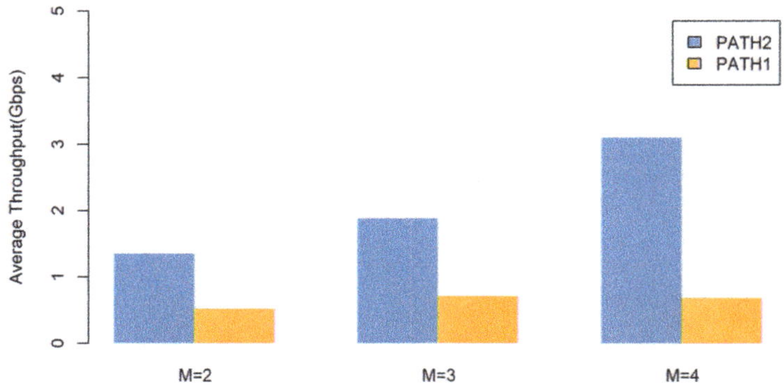

Figure 10. Results of the influence of the number of flow entries.

5.2.5. Comparison with Common File Systems

We compared our in-network cooperative storage system with IPFS and CEPH, which are common distributed storage systems. We wrote different files having sizes of 1, 5 and 10 GB to these file systems under different cluster sizes (including three and five nodes). Our experiments show that the average throughput of the three-node CEPH cluster is about 8 Gbps. When the number of CEPH cluster nodes increases to five, the average throughput is about 9.15 Gbps for a single client. The average throughput of the five-node CEPH cluster drops to about 8.05 Gbps when two clients write simultaneously. It is shown that CEPH achieved a relatively high performance through load-balancing strategies. However, this is still implemented in the application layer based on the current IP infrastructure. CEPH has inherent flaws, such as application layer protocol processing and scheduling overhead. IPFS is another popular distributed storage system based on DHT. It takes about 2 min to upload a 10 GB file for the first time, but the time for subsequent uploads is reduced by half. The average upload rate is about 1.5 Gbps in IPFS.

We used five servers to write simultaneously to R4, and the bandwidth of PATH4 was limited to 2 Gbps. The traffic of the R4 output interface is shown in Figure 11. Four servers can write at an average of 4 Gbps, whereas one server can write at only 1.6 Gbps. The peak throughput reaches 16.3 Gbps when five servers write simultaneously, and the average throughput is about 9.3 Gbps. Through experiments, it is shown that our storage system co-designed with the network is more suitable for real-time data writing. By comparison, IPFS is more suitable for file sharing, and CEPH is a cluster system, which is limited by its scale.

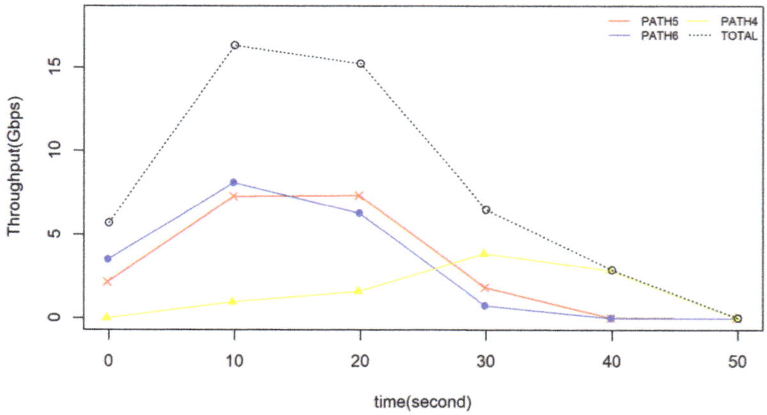

Figure 11. Throughput when five servers write simultaneously.

6. Conclusions

This paper proposes an in-network cooperative storage schema based on neighbor offloading, in which the ID-based traffic is sequentially dynamically offloaded to neighbors and multiple neighbor nodes are utilized to detour the congestion path. First, we proposed an in-network storage service node structure to support cooperative storage in neighbor nodes, and designed an ID-based cooperative storage protocol. Then, a neighbor selection mechanism based on LNRMS was introduced in which the node weights are computed by combining the link bandwidth and node storage capability, and determining whether to split the traffic by comparing normalized weight values with a threshold. In addition, a cooperative storage strategy in a programmable data plane is presented using the relative weights and ID suffix matching to approximate the traffic split ratio. Finally, the experimental results show that our proposed schema is more efficient compared with end-to-end transmission and ECMP in terms of transfer bandwidth and transfer time. In the future, we will study the influence of fluctuation caused by dynamic updating due to load changes in traffic splitting under different traffic characteristics, and consider recording historical neighbor forwarding information in the versions.

Author Contributions: Conceptualization, S.D., and R.H.; methodology, S.D., and R.H.; software, S.D.; writing—original draft preparation, S.D.; writing—review and editing, R.H.; supervision, R.H.; project administration, R.H.; funding acquisition, R.H. All authors have read and agreed to the published version of the manuscript.

Funding: This work was funded by Strategic Leadership Project of Chinese Academy of Sciences: SEANET Technology Standardization Research System Development (Project No. XDC02070100).

Data Availability Statement: Not Applicable, the study does not report any data.

Acknowledgments: We would like to express our gratitude to J.W. and R.H. for their meaningful support for this work.

Conflicts of Interest: The authors declare no conflict of interest.

References

1. Ping, H. An Overview of SHINE Data System. Available online: https://indico.ihep.ac.cn/event/13035/contribution/4/material/slides/0.pdf (accessed on 10 December 2021).
2. Chen, G. Challenges of big data in science researches. *Chin. Sci. Bull.* **2015**, *60*, 439–444. (In Chinese) [CrossRef]
3. Jacobson, V.; Smetters, D.K.; Thornton, J.D.; Plass, M.F.; Briggs, N.H.; Braynard, R.L. Networking named content. In Proceedings of the 5th International Conference on Emerging Networking Experiments and Technologies, Rome, Italy, 1–4 December 2009; pp. 1–12.
4. Zhang, L.; Alexander, A.; Jeffrey, B.; Jacobson, V.; Claffy, K.; Crowley, P.J.; Papadopoulos, C.; Wang, L.; Zhang, B. Named data networking. *ACM SIGCOMM Comput. Commun. Rev.* **2014**, *44*, 66–73. [CrossRef]
5. Fan, C.; Shannigrahi, S.; DiBenedetto, S.; Olschanowsky, C.M.; Papadopoulos, C.; Newman, H.B. Managing scientific data with named data networking. In Proceedings of the Fifth International Workshop on Network-Aware Data Management, Austin, TX, USA, 15 November 2015; pp. 1–7.
6. Dabin, K.; Inchan, H.; Vartika, S.; Young-Bae, K.; Huhnkuk, L. Implementation of a front-end and back-end NDN system for climate modeling application. In Proceedings of the 2015 International Conference on Information and Communication Technology Convergence (ICTC), Jeju, Korea, 28–30 October 2015; pp. 554–559.
7. Catherine, O.; Susmit, S.; Christos, P. Supporting climate research using named data networking. In Proceedings of the IEEE 20th International Workshop on Local & Metropolitan Area Networks (LANMAN), Reno, NV, USA, 21–23 May 2014; pp. 1–6.
8. Huhnkuk, L.; Alexander, N.; Dabin, K.; Young-Bae, K.; Susmit, S.; Christos, P. NDN Construction for Big Science: Lessons Learned from Establishing a Testbed. *IEEE Netw.* **2018**, *32*, 124–136.
9. Susmit, S.; Chengyu, F.; Christos, P. Named Data Networking Strategies for Improving Large Scientific Data Transfers. In Proceedings of the IEEE International Conference on Communications Workshops, Kansas City, MO, USA, 20–24 May 2018; pp. 1–6.
10. Christian, E. Analysis of an Equal-Cost Multi-Path Algorithm. Available online: https://datatracker.ietf.org/doc/html/rfc2992 (accessed on 10 December 2021).
11. Liu, M.; Luo, L.; Nelson, J.; Ceze, L.; Krishnamurthy, A.; Atreya., K. IncBricks: Toward In-Network Computation with an In-Network Cache. In Proceedings of the Twenty-Second International Conference on Architectural Support for Programming Languages and Operating Systems, Xi'an, China, 8–12 April 2017; Volume 52, pp. 795–809.

12. Stathis, M.; Bianca, S. The Evolution of the Hadoop Distributed File System. In Proceedings of the 2018 32nd International Conference on Advanced Information Networking and Applications Workshops (WAINA), Krakow, Poland, 16–18 May 2018; pp. 67–74.
13. Ghemawat, S.; Gobioff, H.; Leung, S.-T. The Google File System. *SIGOPS Oper. Syst. Rev.* **2003**, *37*, 29–43. [CrossRef]
14. Weil, S.A.; Brandt, S.A.; Miller, E.L.; Long, D.D.E.; Maltzahn, C. Ceph: A Scalable, High-Performance Distributed File System. In Proceedings of the 7th Symposium on Operating Systems Design and Implementation, Seattle, DC, USA, 6–8 November 2006; pp. 307–320.
15. Onur, A.; Truong, K.P. On uncoordinated service placement in edge-clouds. In Proceedings of the 2017 IEEE International Conference on Cloud Computing Technology and Science (CloudCom), Hong Kong, 11–14 December 2017; pp. 41–48.
16. Tao, O.; Zhi, Z.; Xu, C. Follow me at the edge: Mobilityaware dynamic service placement for mobile edge computing. *IEEE J. Sel. Areas Commun.* **2018**, *36*, 1–10.
17. Carofiglio, G.; Morabito, G.; Muscariello, L.; Solis, I.; Varvello, M. From content delivery today to information centric networking. *Comput. Netw.* **2013**, *57*, 3116–3127. [CrossRef]
18. Adrian-Cristian, N.; Spyridon, M.; Ioannis, P. Store Edge Networked Data (SEND): A Data and Performance Driven Edge Storage Framework. In Proceedings of the IEEE INFOCOM 2021-IEEE Conference on Computer Communications, Virtual Conference, 10–13 May 2021; pp. 1–10.
19. Eve, M.; David, Z.; Jeff, S.; Moustafa, H.; Brown, A.; Ambrosin, M. An Architectural Vision for a Data-Centric IoT: Rethinking Things, Trust and Clouds. In Proceedings of the 2017 IEEE 37th International Conference on Distributed Computing Systems (ICDCS), Atlanta, GA, USA, 5–8 June 2017; pp. 1717–1728.
20. Benet, J. IPFS—Content Addressed, Versioned, P2P File System. *arXiv* **2014**, arXiv:1407.3561.
21. Cinquini, L.; Crichton, D.; Mattmann, C.; Harney, J.; Shipman, G.; Wang, F.; Ananthakrishnan, R.; Miller, N.; Denvil, S.; Morgan, M.; et al. The earth system grid federation: An open infrastructure for access to distributed geospatial data. *Future Gener. Comput. Syst.* **2014**, *36*, 400–417. [CrossRef]
22. Karl, E.; Ronald, J.; Gerald, A. An overview of cmip5 and the experiment design. *Bull. Am. Meteorol. Soc.* **2012**, *93*, 485–498.
23. Alvise, D.; Peter, E.; Fabrizio, F.; Andrew, H. Xrootd-a highly scalable architecture for data access. *WSEAS Trans. Comput.* **2005**, *4*, 348–353.
24. Ying, C.; Fan, L.; Edmund, Y.; Ran, L. Enhanced VIP Algorithms for Forwarding, Caching, and Congestion Control in Named Data Networks. In Proceedings of the 2016 IEEE Global Communications Conference (GLOBECOM), Washington, DC USA, 4–8 December 2016; pp. 1–7.
25. Zeng, L.; Ni, H.; Han, R. An Incrementally Deployable IP-Compatible-Information-Centric Networking Hierarchical Cache System. *Appl. Sci.* **2020**, *10*, 6228. [CrossRef]
26. Xu, Y.; Ni, H.; Zhu, X. An Effective Transmission Scheme Based on Early Congestion Detection for Information-Centric Network. *Electronics* **2021**, *10*, 2205. [CrossRef]
27. You, J.; Ji, G.; Xiao, Z.; Jin, L. ITU-T Y.3075 Requirements and Capabilities of ICN Routing and Forwarding based on Control and User Plane Separation in IMT-2020. Available online: https://www.itu.int/rec/dologin_pub.asp?lang=e&id=T-REC-Y.3075-202009-I!!PDF-E&type=items (accessed on 12 December 2021).
28. Psaras, I.; Ascigil, O.; Rene, S.; Pavlou, G.; Afanasyev, A.; Zhang, L. Mobile Data Repositories at the Edge. In Proceedings of the USENIX Workshop on Hot Topics in Edge Computing HotEdge '18, Boston, MA, USA, 10 July 2018.
29. Dang, S.; You, J.; Li, Y.Y. ICN-DOS, Requirements and Capabilities of Data Object Segmentation in Information Centric NETWORKING for IMT-2020. Available online: https://www.itu.int/md/T17-SG13-C-1318 (accessed on 12 December 2021).
30. Raychaudhuri, D.; Nagaraja, K.; Venkataramani, A. MobilityFirst: A robust and trustworthy mobility-centric architecture for the future internet. *ACM SIGMOBILE Mob. Comput. Commun. Rev.* **2012**, *16*, 2–13. [CrossRef]
31. Ye, X.; Cao, J.; Zhu, X.Y. ICN-TL Requirements and Mechanisms of Transport Layer for Information Centric Networking in IMT-2020. Available online: https://www.itu.int/md/T17-SG13-200720-TD-WP1-0589 (accessed on 12 December 2021).
32. Wang, J.; Chen, G.; You, J.; Sun, P. SEANet: Architecture and Technologies of an On-site, Elastic, Autonomous Network. *New Media* **2020**, *9*, 1–8.
33. Song, Y.; Ni, H.; Zhu, X. Analytical modelling of optimal chunk size for efficient transmission in information-centric network. *Int. J. Innov. Comput. Inf. Control* **2020**, *16*, 1511–1525.
34. Schneider, K.; Zhang, B.; Mai, V.S.; Benmohamed, L. The Case for Hop-by-Hop Traffic Engineering. *arXiv* **2020**, arXiv:2010.13198.
35. You, J.; Zhang, J.; Li, Y.Y. ICN-NMR Framework of Locally Enhanced Name Mapping and Resolution for Information Centric Networking in IMT-2020. Available online: https://www.itu.int/md/T17-SG13-C-1319/ (accessed on 12 December 2021).
36. Jacobson, V. Congestion avoidance and control. *SIGCOMM Comput. Commun. Rev.* **1988**, *18*, 314–329. [CrossRef]
37. Zhou, J.; Tewari, M.; Zhu, M.; Kabbani, A.; Poutievski, L.; Singh, A.; Vahdat, A. WCMP: Weighted cost multipathing for improved fairness in data centers. *EuroSys* **2014**, *5*, 1–14.
38. Qadir, J.; Ali, A.; Yau, K.L.; Sathiaseelan, A.; Crowcroft, J. Exploiting the Power of Multiplicity: A Holistic Survey of Network-Layer Multipath. *IEEE Commun. Surv. Tutor.* **2015**, *17*, 2176–2213. [CrossRef]
39. Rottenstreich, O.; Kanizo, Y.; Kaplan, H.; Rexford, J. Accurate Traffic Splitting on SDN Switches. *IEEE J. Sel. Areas Commun.* **2018**, *36*, 2190–2201. [CrossRef]

40. Tuncer, D.; Charalambides, M.; Clayman, S.; Pavlou, G. Flexible Traffic Splitting in OpenFlow Networks. *IEEE Trans. Netw. Serv. Manag.* **2016**, 3, 407–420. [CrossRef]
41. Kang, N.; Ghobadi, M.; Reumann, J.; Shraer, A.; Rexford, J. Efficient traffic splitting on commodity switches. In Proceedings of the 11th ACM Conference on Emerging Networking Experiments and Technologies, CoNEXT 2015, Heidelberg, Germany, 1–4 December 2015.
42. Li, S.; Hu, D.; Fang, W.; Ma, S.; Chen, C.; Huang, H.; Zhu, Z. Protocol Oblivious Forwarding (POF): Software-Defined Networking with Enhanced Programmability. *IEEE Netw.* **2017**, *31*, 58–66. [CrossRef]
43. Andy, H. Bbcp. Available online: https://www.slac.stanford.edu/~abh/bbcp/ (accessed on 10 December 2021).
44. Liu, Y.; Qin, X.; Zhu, T.; Chen, X.; Wei, G. Improve MPTCP with SDN: From the perspective of resource pooling. *J. Netw. Comput. Appl.* **2019**, *141*, 73–85. [CrossRef]
45. Klaus, S.; Bei, C.; Lotfi, B. Hop-by-Hop Multipath Routing: Choosing the Right Nexthop Set. In Proceedings of the IEEE INFOCOM 2020—IEEE Conference on Computer Communications, Virtual Conference, 6–9 July 2020; pp. 2273–2282.

future internet

Article

Controlling the Trade-Off between Resource Efficiency and User Satisfaction in NDNs Based on Naïve Bayes Data Classification and Lagrange Method

Abdelkader Tayeb Herouala [1], Chaker Abdelaziz Kerrache [1], Benameur Ziani [1], Carlos T. Calafate [2,*], Nasreddine Lagraa [1] and Abdou el Karim Tahari [1]

[1] Laboratoire d'Informatique et de Mathématiques, Université Amar Telidji de Laghouat, Laghouat 03000, Algeria; aek.herouala@lagh-univ.dz (A.T.H.); kr.abdelaziz@gmail.com (C.A.K.); bziani@lagh-univ.dz (B.Z.); n.lagraa@lagh-univ.dz (N.L.); k_tahari@yahoo.fr (A.e.K.T.)
[2] Computer Engineering Department (DISCA), Universitat Politècnica de València, 46022 Valencia, Spain
* Correspondence: calafate@disca.upv.es

Abstract: This paper addresses the fundamental problem of the trade-off between resource efficiency and user satisfaction in the limited environments of Named Data Networks (NDNs). The proposed strategy is named RADC (Resource Allocation based Data Classification), which aims at managing such trade-off by controlling the system's fairness index. To this end, a machine learning technique based on Multinomial Naïve Bayes is used to classify the received contents. Then, an adaptive resource allocation strategy based on the Lagrange utility function is proposed. To cache the received content, an adequate content placement and a replacement mechanism are enforced. Simulation at the system level shows that this strategy could be a powerful tool for administrators to manage the trade-off between efficiency and user satisfaction.

Keywords: named data network; cache strategy; machine learning; placement strategy; name classification; resource allocation

Citation: Herouala, A.T.; Kerrache, C.A.; Ziani, B.; Calafate, C.T.; Lagraa, N.; Tahari, A.e.K. Controlling the Trade-Off between Resource Efficiency and User Satisfaction in NDNs Based on Naïve Bayes Data Classification and Lagrange Method. *Future Internet* **2022**, *14*, 48. https://doi.org/10.3390/fi14020048

Academic Editor: Vijayakumar Varadarajan

Received: 23 December 2021
Accepted: 29 January 2022
Published: 31 January 2022

1. Introduction

In recent years, new network architectures under the name of Information-Centric Networking (ICN) have attracted the attention of many academics and enterprises. ICN proposes new structures and concepts that can solve the traffic congestion problems caused by massive content distribution in the network [1]. The content is considered a very important factor in ICN, which is the basis for content demand and reception. Among the ICN architectures, NDN is a promising active research project that aims to develop a candidate architecture for the future Internet [2]. NDN follows the same design as the IP system, the difference being that the names of the data replace the IP address. NDN data names are similar to the URL names structure, which better matches the current shift in intranet usage from site-specific to content-specific searches. Routing and data transmission also depend on the content names by implementing the longest prefix corresponding to the requested name prefix [3]. In-network caching is also an important key in NDN, where each router has a cache, and the received data are cached according to a specific caching strategy. The caching mechanism enables future requests to be satisfied by the nearest available routers instead of reaching the end producer of the data [4].

It is obvious that the in-network caching saves limited bandwidth resources and improves network performance; however, in practice, the size of the caches remains limited due to the huge amount of data circulating in the network; therefore, the main challenge is how to use these limited caching resources for massive content while ensuring Quality of Service (QoS). In addition, the administration and allocation of caching resources in the network is also an important issue. These problems have been the subject of several research papers in terms of caching strategies and algorithms; however, many of the proposed strategies focus on the large network environment and simply consider user satisfaction as a basic factor for caching important data. In practice, these methods are not always effective when dealing with a small limited network, such as a university network, a company network, etc. This type of network usually has a limited set of nodes, and its own data representing its direction, which is often stored in its own servers. In fact, the huge demand for content from social networks and video-on-demand (VoD) has become a large part of the data consumed in everyday life [5,6]. This poses a problem for this type of network, since the owners of these networks will not be able to take advantage of the cached data using existing caching strategies. As an example, the analysis of data traffic at the University Amar Telidji Laghouat shows that the consumed data belonging to the university represents only 4% of the total resource consumption. In this case, with the existing strategies in the NDN, the data related to this network has no advantage in terms of delay, satisfaction, and load on the servers. This looks to be contrary to the primary goals of NDN; therefore, controlling the trade-off between maximum user satisfaction and resource efficiency is one of the most fundamental issues in a limited NDN environment. From the network administrator's perspective, it is very essential to use the network caches efficiently, and the benefits must be maximized. Similarly, from the users' perspective, it is more important that their Quality of Service (QoS) requirements are met adequately. Hence, the question is how to manage this trade-off?

In this paper, we use a proven Bayesian classification-based machine learning method to classify the received data in such a network. Then, using the Lagrange method, we propose a utility function based on a preference parameter that allocates different partitions to the content classes in the routers. A new placement and replacement strategy is also defined for processing the data received by the routers. Finally an exhaustive simulation of the results is presented to demonstrate the performance of the proposed model.

This paper is organized as follows. In Section 2, we review some NDN background. In Section 3, the related work on the NDN caching solution is presented. In Section 4, the proposed solution for resource allocation is described. Simulation results are then presented and discussed in Section 5. Finally, we conclude this study and discuss future research in Section 6.

2. NDN Background

In an NDN communication process, as illustrated in Figure 1, there are two types of packets called interest and data packet. On the one hand, the interest packet mainly contains the name of the content; on the other hand, the data packet mainly contains the name of the content, the encrypted information, and the data. As Figure 1 indicates, NDN is based on three essential modules for processing the requested and received packets. These modules are called Content Store (CS), Pending Interest Table (PIT), and Forwarding Information Base (FIB). The CS has a role of caching the data packets with some caching strategies in order to satisfy the next requests for the same data. PIT is used to record the interfaces of the incoming interest packets until the interest packet is satisfied, or the recording time of the same packet is expired. The PIT also has a content name aggregation function, in which the interest packets for the same content received from different interfaces will be registered as a single request to avoid the retransmission of similar requests for the same data. The role of the FIB module is to transmit the packets of interest to the producer(s) of the data using packet transmission mechanisms, which are mainly based on the names of the contents.

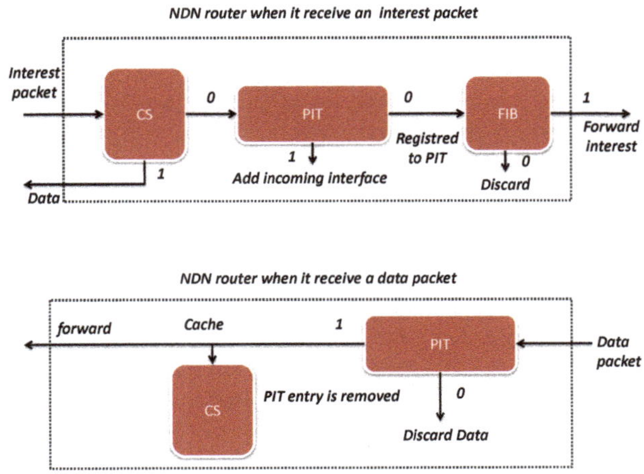

Figure 1. NDN background.

3. Related Work

Based on previous studies on resource allocation in ICNs, we can see a strong similarity between these strategies and content placement strategies. Arguably, the first and best known resource allocation strategy is Leave Copy Every Were (LCE) [3], where resources are shared and cached in all nodes in the return path. There are also other strategies proposed in this area, which can be divided into two categories. Strategies that allocate resources over the entire network or path, and strategies that allocate resources over the node itself.

The first category refers to caching strategies that use coordination between nodes to allocate resources fairly, or to allocate popular resources to nodes that are considered important in the network. In [7], the authors propose a strategy called ProbCache, which approximates the capacity of paths to cache content using the hop count. In this strategy, nodes connected near the content source will be given priority to cache content over other nodes along the path. In [8], the authors propose a strategy that combines routing and resource allocation in the NDN network. This strategy uses a Q-learning algorithm based on the basic components of a semi-Markov decision process called SMDP to prove the best routing and resource allocation path. In [9], the authors propose a joint routing strategy with a resource allocation algorithm for wireless environments, which they called dynamic routing and resource allocation. In [10], the authors propose a resource allocation strategy for a limited network environment. They formulate an optimization problem that attempts to maximize the caching of popular content to reduce latency. At the same time, they minimize the hit variance to achieve a balanced resource allocation among the network nodes. In [11], the authors attempt to divide the network topology into different tiers based on the demands from different interfaces and video applications. The goal is to minimize redundancy and cache popular content closer to the consumers. In [12], the authors studied the object allocation problem using game theory as a distributed many-to-one in ISP networks to improve cache utilization in NDN. In [13], the authors measure the node utilization ratio (NUR), which considers the consumers distribution, the server distribution, and the content routing paths. Then they take the proportion between the NUR and the cache size of the node, along with the transmission ratio of the node, to allocate more contents in the relevant nodes.

Despite the importance of these strategies, they do not offer much to the limited network community, as most of these strategies only rely on the popularity of the data in the network distribution. Further, they do not provide an effective solution for controlling the data that should be cached.

Regarding the second category of proposed strategies, we find that researchers have been more interested in classifying the data to give each class a chance to obtain space to cache its data. The main reason is that sharing data in caches by prioritizing caching for the most popular data prevents the rest of the data from having a chance to be cached. In [14], the authors propose a method for dynamic allocation of cache space. They divide the content into five categories based on the packet characteristics, where each category has a different priority to decide how many items should be cached in each category. In [15], the authors propose a caching strategy called Cache Based on Popularity and Class (CAPIC). This strategy caches a copy of the data one hop closer to the consumer for each request. In addition, when the CS of a node is full, the router checks the class of the received data and replaces it in the portion of the same class. Each class has a portion that changes periodically depending on the calculated popularity of the classes. In [16], the authors divide the received data into two categories of popular and non-popular data based on a calculated threshold where the non-popular data are placed near the consumers, and the popular data are placed at the shortest path that contains a maximum of users that request the same data.

We believe that this type of strategy is better than the first type for limited network environments. The main reason is that, in these strategies, many types of data can be cached, especially the less requested data. Yet, these strategies remain limited because they do not provide a robust classification model for the data, as they classify the data only using their frequencies, and do not focus on the content. Furthermore, the proposed strategies do not provide a control factor that matches the demands of consumers and the needs of the managers of these networks.

4. RADC: Resource Allocation Based Data Classification

4.1. System Model

The system model is much more focused on limited network environments, such as academic and industrial networks, etc. In these types of networks, each area consists of a network administrator, and a set of nodes and server(s). For instance, Figure 2 illustrates a network with four areas denoted $Area = \{Area_i | i \in [1.4]\}$. Routers in the same network area sent periodically the sum of each consumed class to the administrator side. Similarly, the network administrator will calculate the allocated partitions for each class according to the utility function given in Section 4.3. The communication between administrator/routers of the same area can be performed using the IP layer in the NDN hourglass [3]. To classify the received contents, each router in the network will use the proposed classification model (Section 4.2) to assign the contents to the corresponding classes. The nodes outside those networks use the default caching mechanism used in NDN.

4.2. Content Classification

As mentioned above, most of the proposed methods in the NDN do not provide a robust model for content classification, as they typically use simple parameters to classify their content (e.g., request frequency). One of the most important keys introduced by NDN that can be exploited to classify contents efficiently is the data naming, which adopts the same structure as URL naming. The names in NDN are significant sources of information and knowledge; however, processing names is a difficult task regarding their unstructured format. We use a Naïve Bayes (NB) classifier [17], which has been widely used to analyze and solve many scientific problems due to its simplicity, efficiency, and effectiveness [18,19]. Naive Bayes is computationally inexpensive and also needs a very low amount of memory [20]. The role of the NB in this study is to classify the received content into two classes, where the first class represents the consumed data that belong to the server or domain of the addressed network, and the second class is the consumed data that do not bring any benefit to the addressed network. Before starting to use the BN model, a step concerning the preparation of the dataset for the training and testing of the model should be performed. The dataset used should contain a sufficient number of names of the requested contents in each selected class. Once the preparation of the dataset is complete, the conversion of the dataset into a word count matrix is performed, with each row representing a data name, and the columns representing the features that represent the entire vocabulary existing in the dataset. The model uses the n-gram model [21] to extract more features, and also uses the Term-Frequency Inverse Document Frequency (TF-IDF) model [22] to weigh the selected features.

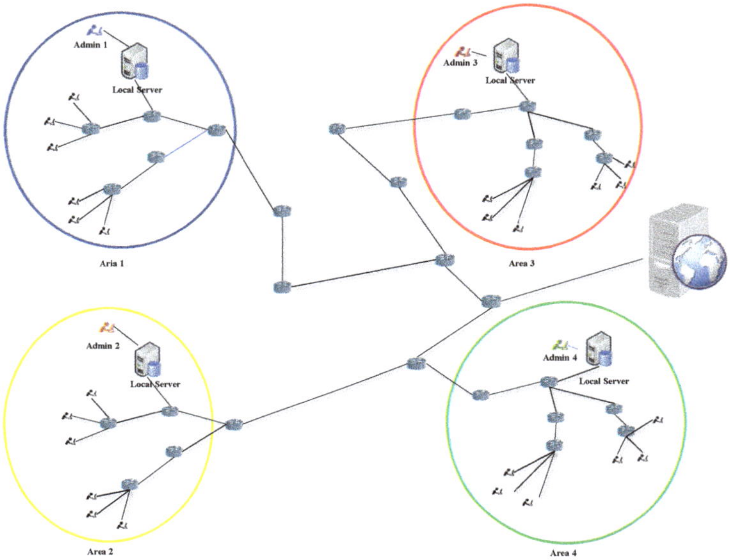

Figure 2. Example of targeted networks.

4.2.1. The High-Level Description of the Naïve Bayes Classifier

If the number of content names (L) fit into s classes where $s \in \{c1, c2, \ldots, cs\}$, the predicted output class is $c \in C$. The Naïve Bayes algorithm can be described as follows:

$$P(c \mid l) = \frac{P(l \mid c)P(c)}{P(l)} \tag{1}$$

where l is the content name, and c indicates classes.

$$C_{\text{Predicted}} = \arg\max_{c \in C} P(l \mid c)P(c) \tag{2}$$

4.2.2. Multinomial Naïve Bayes Classifier

Naive Bayes classifiers differ mainly in the assumptions they make about the probability distribution of $P(l \mid c)$. in text classification such as content names of NDN the multinomial distribution performs well compared to other distributions such as the Gaussian distribution or the Bernoulli distribution [17]. The Multinomial Naïve Bayes algorithm can be written as:

$$P(c \mid l) = \frac{P(c) \prod_{w \in l} P(l \mid c)^{n_{\text{wl}}}}{P(d)} \tag{3}$$

where n_{wl} denotes the number of times that word w occurs in the name, and $P(w \mid c)$ is the probability of observing word w given class c.

$P(w \mid c)$ is calculated as:

$$P(c \mid w_i) = \frac{\text{count}(w_i, c) + 1}{\text{count}(w, c) + V} \tag{4}$$

where count(w_i, c) is the number of times the word appears in class c in the training dataset, count(w, c) is the total number of all words appearing for class c, and V is a constant value that represents the number of features (words) that represent the model.

4.3. Resource Allocation

One of these solutions, which is very simple, efficient, and suitable for solving problems of allocation of limited resources between a set of independent activities, is that of Lagrange multipliers. The use of this technique does not guarantee that a solution will necessarily be found for all problems, but it is safe in the sense that any solution found by using them is a real solution [18]. To this end, the aim of this work is to optimize the resource allocation of the two selected classes of data by giving each of them an associated space taking into account two factors. The first one is the priority level of the space associated with data that is not important to the targeted network compared to the second space. The second factor is the frequency of requests received in each class. This is performed within the constraint of the maximum cache space in each router as formulated by the equation below:

$$P \mid \begin{array}{ll} \max\limits_{\tilde{x}} & f(\tilde{x}) \\ \text{u.c} & g(\tilde{x}) = c \end{array} \tag{5}$$

Hence, our problem can be formulated as:

$$P \mid \begin{array}{ll} \max\limits_{C_1, C_2} & f(C_1, C_2) \\ \text{u.c} & g(C_1, C_2) = \text{MaxC} \end{array} \tag{6}$$

where C_1 and C_2 represent, respectively, the space associated to the data that belong to the domain of the targeted network, and the second to the other data; $MaxC$ represents the maximum cache space in the router.

The chosen objective function $f(C_1, C_2)$ and the constrained $g(C_1, C_2)$ that can give the best allocation to the two spaces are formulated according to:

$$P \mid \begin{array}{l} \max\limits_{C_1, C_2} -\alpha C_1 - \beta C_2^k \\ \text{u.c } C_1 + C_2 = \text{MaxC} \end{array} \tag{7}$$

where α and β represent, respectively, the frequency of requests that belong to the organization, and the requests that do not belong to the organization, These requests are classified using the Naive Bayes classifier. k represents the priority level of the space C_2 with respect to the space C_1, where the value of factor k takes the scale from 0.1 to 0.9 inspired from the work presented in [14]. The meaning of the scale is presented in Table 1 (Note: The values 0.2, 0.4, and 0.8 of the factor k mean that the importance of the space C_1 in respect to C_2 is situated between the two adjacent levels shown in the Table 1).

Table 1. Meaning of the scale from 0.1 to 0.9.

Value	Meaning
0.1	The data in the C_1 space have an obvious importance in the network compared to the C_2 space.
0.3	The data in the C_1 space have a higher importance in the network compared to the C_2 space.
0.6	The data in the C_1 space have a low importance in the network compared to the C_2 space.
0.9	The data in the C_1 space have the same importance in the network compared to the C_2 space.

The Lagrangian function associated with this program is written as shown below:

$$L(C_1, C_2, \lambda) = -\alpha C_1 - \beta C_2^k - \lambda(C_1 + C_2 - \text{Max}\,C) \tag{8}$$

where λ is called the Lagrange multiplier associated with the constraint.

We obtain the desired spaces by calculating the derivatives of our variable as follows:

$$\begin{cases} \frac{\partial L}{\partial C_1} = 0 \Leftrightarrow -\alpha - \lambda = 0 \\ \frac{\partial L}{\partial C_2} = 0 \Leftrightarrow -\beta k C_2^{k-1} - \lambda = 0 \\ \frac{\partial L}{\partial \lambda} = 0 \Leftrightarrow -C_1 - C_2 + \text{Max}\,C = 0 \end{cases} \quad (9)$$

Finally we obtain the two spaces C_1 and C_2, where C_1 is equal to $\left\lceil \text{Max}\,C - \left(\frac{\alpha}{k\beta}\right)^{\frac{1}{k-1}} \right\rceil$, and C_2 is equal to $\left\lfloor \left(\frac{\alpha}{k\beta}\right)^{\frac{1}{k-1}} \right\rfloor$. We can add the Laplace smoothing to the spaces to avoid zero in the denominator, so C_1 and C_2 will equal, respectively:

$$\begin{cases} C_1 = \left\lceil \text{Max}\,C - \left(\frac{\alpha+1}{k\beta+1}\right)^{\frac{1}{k-1}} \right\rceil \\ C_2 = \left\lfloor \left(\frac{\alpha+1}{k\beta+1}\right)^{\frac{1}{k-1}} \right\rfloor \end{cases} \quad (10)$$

In certain cases when the margin between the frequencies of requests belonging to the two classes is very large, we can find that C_1 is equal to a negative number to satisfy our constraint. Since the Lagrange is limited by the constraints of equality, we can fix this problem by affecting $C_1 = 0$, and $C_2 = MaxC$ when $C_1 \leqslant 0$.

4.4. Placement and Replacement of Data

Unlike strategies that share the same space for all content, our strategy allocates the desired space for each content class. It is known that content placement and replacement in NDN is also an important feature that can improve the overall network performance. As shown in Figure 3, when the routers in the network receive the content, they classify it using the classification model. Then, the received content is placed in the corresponding space that was allocated to the same class. If the corresponding space is full, a replacement strategy will be applied to insert the received content by evicting another from the same class. We note that the design of our strategy allows us to adapt any replacement strategy to replace data of the same class, but for simplicity we use the Least Recently Used (LRU) method to replace the received data. In this strategy, the received data will replace the least recently used data [19].

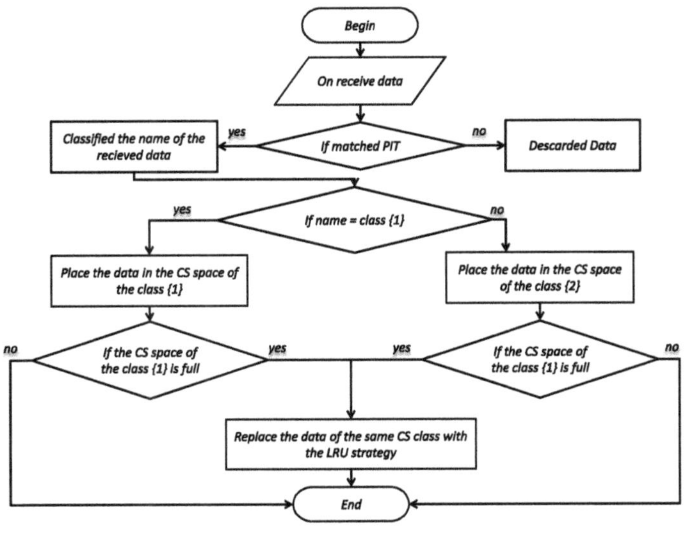

Figure 3. Data placement and replacement process.

5. Performance Results of RACD Strategy

In this section we present and discuss the different performance results obtained. To this end we start by detailing the simulation environment used, and next we proceed to show the actual results achieved. Further, for the sake of reproducibility, the detailed working code of RACD is available at: https://github.com/Herouala/RADC.git (accessed on 12 December 2021).

5.1. Simulation Environment

The performance evaluation of the proposed strategy was performed in two steps. First, we implemented the classification model to classify the dataset used in the simulation. Then, the proposed strategy was evaluated through simulation. Due to the lack of standard datasets in NDN, we constructed our dataset by combining the DMOZ (DMOZ: https://web.archive.org/web/20170312160530/http://rdf.dmoz.org/rdf/content.rdf.u8.gz (accessed on 12 December 2021)) dataset with links extracted from the website of the University Amar Telidji Laghouat (UATL). DMOZ is a well-known dataset for classification, and it is used in many works as in [20,21]. The names of the DMOZ dataset are labeled as data belonging to the class $\{0\}$ (unimportant), and the links extracted from the UATL are labeled as data belonging to class $\{1\}$ (important). The dataset used is divided into two parts, one for the training and the other for testing the model. The implemented Multinomial Naïve Bayes classifier model has achieved 98% accuracy in classifying the content names. After that, the RADC strategy is evaluated through ndnSIM [22], which represents the official ns-3-based simulator deployed by the NDN community. ndnSIM enables reliable simulation experiments that can be replicated in real environments without having to modify the source code [23]. To make our simulation as realistic as possible, we chose a real dataset from the University Amar Telidji Laghouat (UATL) as a realistic interest consumption. The UATL dataset contains more than one million links consumed during a week at this university.

The strategies are simulated using a non-complete K-ary tree as in [24–26]. This topology is based on two parameters, D and k, where they represent consecutively the depth of the tree and the number of children of each node in the tree. We chose $D \in [3, 7]$ and k in the interval $[0, 5]$, where k is randomly generated for each node as in [26]. For the simulation scenario, we assume that there are two root nodes, where the first is a server that belongs to the university, and it contains all its data. The second represents a server containing all other data requests. Furthermore, we assume that all requests are sent by leaf nodes.

For each test run, users are assumed to express interests from the used UATL dataset. For simplicity, we set a similar cache size for all nodes in the network. Each link in our topology has a bandwidth of 10 Gbps, which is greater than the traffic demand, and a propagation delay of 1 ms. Table 2 shows the choice of the main parameters for our simulation.

Table 2. Parameter settings for the simulation.

Parameter	Default Value	Range
Tree-ary k	-	[0;5]
Tree depth D	5	[3;7]
node cache capacity	100	[10;100]
Priority level	0.4	[0.1;0.9]
Delay	1 ms	-
Bandwidth	10 Gbps	-

For the evaluation method, we studied the impact of parameters priority level, cache size, and tree depth on the caching performance in the network. We compare the results of our resource allocation mechanism with the mechanism that promotes content sharing in the same cache. The sharing mechanism represents the majority of the proposed strategies, with a difference in the choice of the placement and replacement strategy. Since RADC does not apply any particular technique to manage the cache content, any strategy could easily be adopted. We compare the two strategies in the same environment, applying the same placement and replacement technique. The two strategies represent, consecutively, LCE [3] and LRU [19] for placement and replacement.

We quantify the performance of the in network caching from two metric aspects:

Cache hit ratio, a metric that measures the ratio between the served content requests and the number of received requests. It also reflects the load saving of the servers, and the delay of the data delivery, since the higher the cache hit ratio is, the faster the data delivery becomes before the user's request reaches the data source. In our case, we study the cache hit of the served data that belong to the university, and also the hit of the other data for both strategies; we can define it as follows:

$$Cachehitratio = \frac{r}{|R|} \tag{11}$$

where r represent the number of requests satisfied by in-network caches, and $|R|$ is the total number of requests.

The hop reduction ratio indicates the variation of the data delivery distance in the network. In our study, we investigate the hop reduction ratio of the data belonging to the university against the other data in both tested strategies. The hop reduction ratio is defined as follows:

$$Hop\ reduction\ ratio = 1 - \frac{\sum_{r \in R} h_r}{\sum_{r \in R} H_r} \tag{12}$$

where h_r and H_r are the hop counts from the requester of $r \in R$ to the node, which serves the request, and to the original content server, respectively.

5.2. *Performance Results*

In the comparison of results, we named the results obtained for the two classes with our strategy as UD-RA and OD-RA, where the UD-RA acronym refers to the results obtained for the data belonging to the university using our strategy, and the OD-RA acronym refers to the obtained results for the data outside the university using our strategy. Further, for the sharing strategy, we name the results obtained for the two classes as UD-sharing and OD-sharing.

5.2.1. The Impact of the Priority Level on Performance

Figure 4 shows the effect of the priority level (k) on the performance of the resource allocation strategy in terms of hit ratio and hop reduction ratio. We can see that, when the priority level k increases, the results of the university data (UD-RA) decrease, and the results of the other data (OD-RA) increase. Here we observe that, from 0.1 to 0.4, there is an increase in the hit ratio and in the hop reduction ratio for university data compared to non-university affiliated data. The reason is that, as the priority coefficient increases, the space given to non-university data increase until it reaches nearly the same size for both types of data when $k = 0.4$. Starting from $k = 0.5$, We can see a clear superiority for OD-RA reaches almost 90% as a difference in the hit ratio, and also for the hop reduction ratio. Notice that, as the k coefficient increases, the space allocated to OD-RA also increases. The reason why OD-RA takes more storage space compared to UD-RA even with the priority coefficient k in the ranges from 0.5 to 0.9 is because there is a large difference between the amounts of data consumed in the two classes, as the amount of data that belong to the university represent only 4% of the total resource consumption. This, in turn, has an impact on space allocation.

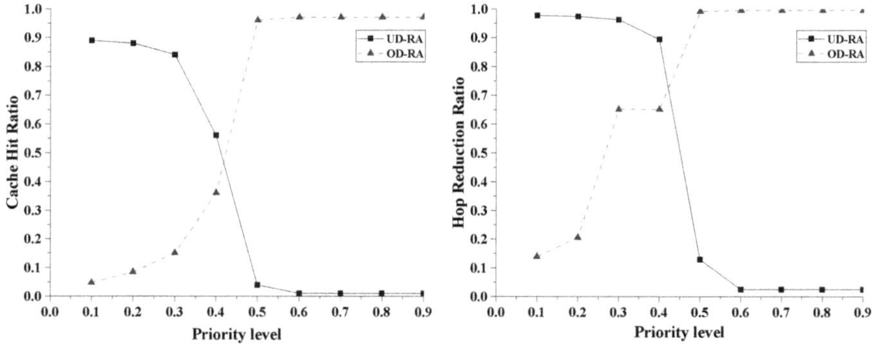

(a) Priority level (k) vs Cache Hit Ratio **(b) Priority level (k) vs Hop Reduction Ratio**

Figure 4. Caching performance vs. priority level in terms of (**a**) cache hit ratio, and (**b**) hop reduction ratio.

5.2.2. Impact of the Cache Size on the Performance of Our Strategy

Figures 5 and 6 show the effect of cache size on the performance of the data sharing and data resource allocation strategy in terms of hit ratio and hop reduction ratio. From the two figures, we can see that, for the sharing strategy, the hop reduction and hit ratio values increase with the cache size, with a large margin for the OD-Sharing results compared to the UD-Sharing. The improvement in results when cache size increases makes sense because now more data are cached, which increases the chances of having the data searched to the closest possible node. The difference in the results between the two classes is due to the fact that the requests in the OD-Sharing class are more numerous than those of UD-Sharing.

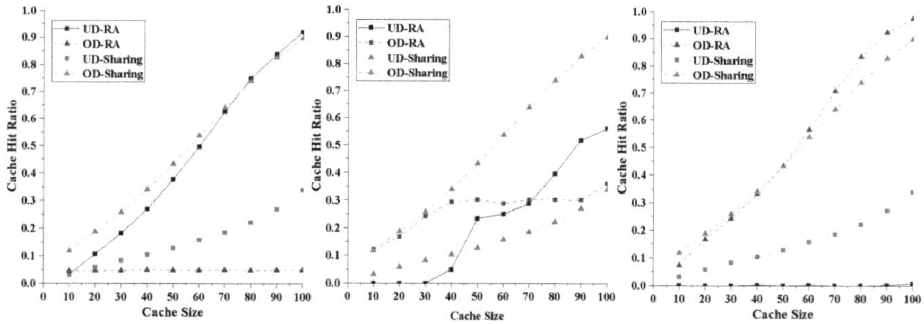

(a) Cache Size vs Cache Hit Ratio (□□ □□) **(b) Cache Size vs Cache Hit Ratio (□□ □□)** **(c) Cache Size vs Cache Hit Ratio (□□ □□)**

Figure 5. Caching performance vs. cache size in terms of: cache hit ratio.

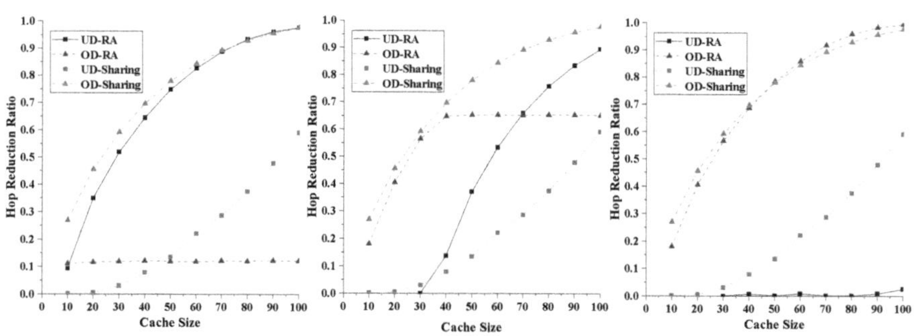

(a) Cache Size vs Hop Reduction Ratio (□□ □□) **(b) Cache Size vs Hop Reduction Ratio (□□ □□)** **(c) Cache Size vs Hop Reduction Ratio (□□ □□**

Figure 6. Caching performance vs. cache size in terms of hop reduction ratio when varying k.

For the resource allocation strategy, with $k = 0.1$, the university data (UD-RA) achieves a higher hit ratio and hop reduction ratio as the cache size increases. The OD-RA achieves low and constant values compared to UD-RA. This is because, when using a very small value for coefficient k, the cache partition size given to UD-RA is much larger than the size given to OD-RA, which was small and fixed in all sizes tested. When $k = 0.9$, we find that our strategy performs inversely to $k = 0.1$ where, as the cache size increases, OD-RA performs well compared to UD-RA in terms of hit ratio and hop reduction ratio. The reason is that, when the coefficient is $k = 0.9$, the coefficients of the two data classes become almost equivalent; therefore, the allocations of each partition become related to the amount of data received from both classes. Since the consumed university data represents only 4% of the total consumption, all the cache is allocated to OD-RA data. We also notice that, when $k = 0.4$, from 10 to 30, the UD-RA obtains a null hit and hop reduction ratio, in contrast to the OD-RA, which increases with the increase in cache size. From 40 to 60 we notice an increase in the results of both classes of data, with their convergence until they are equal in the capacity 70. From 80 to 100, the results show a continuous increase in both classes, with dominance of UD-RA over OD-RA. The reason for the continuous increase in results for both classes is that both classes obtain more space when the total cache space increases. The reason why these results are obtained when $k = 0.4$ is that, when the cache size is between 10 and 30, the Lagrange utility function gives the total memory space for the OD-RA. Then, beyond size 40, the function starts to give more memory space for UD-RA, and OD-RA space remains fixed until they reach equal results for size 70. Then, the increase in space continues for UD-RA, which takes more memory space, and thus achieves better results than OD-RA.

5.2.3. Impact of Tree Depth on Performance

Figures 7 and 8 show the effect of tree depth on the performance of the sharing and resource allocation strategies by calculating the hit ratio and hop reduction ratio for each strategy. For the sharing strategy, we observe a slight increase in the results for both classes of data in terms of hit and hop reduction ratio as depth increases. We also observe a large difference in the results for the two classes, with a 50% difference in hits, and a difference ranging from 60% to 35% in hop reduction results between OD-Sharing and UD-Sharing. The increase in the results obtained is due to the number of nodes that increase with length, allowing more cache space to be allocated. This allows more data to be cached, including low-consumption data, such as university-owned data (UD-Sharing), which results in a higher hit ratio and less hop ratio.

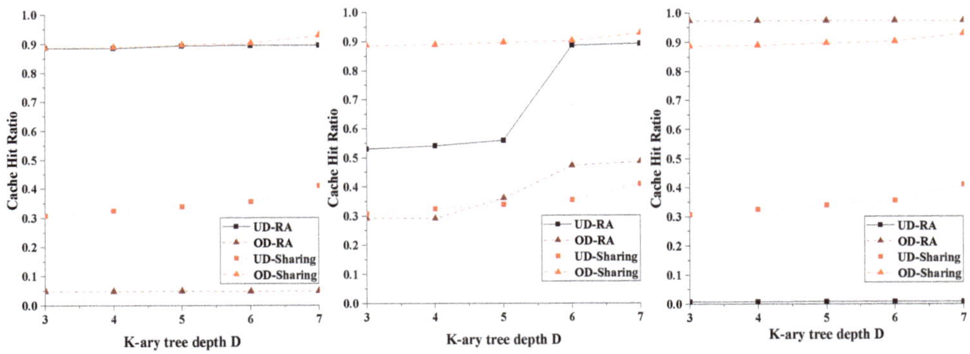

Figure 7. Caching performance vs. tree depth in terms of cache hit ratio when varying k.

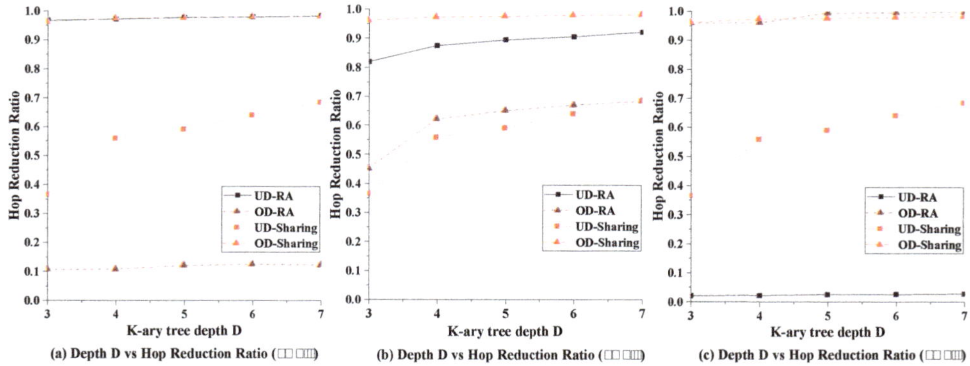

Figure 8. Caching performance vs. tree depth in terms of hop reduction ratio when varying k.

Regarding the resource allocation strategy, the results differ depending on the k factor. When $k = 0.1$, we notice a very slight increase in the hit and hop reduction results with the increase in depth D. We also notice a big difference between the results obtained for the two classes of data, being that this difference grows up to 85% between the UD-RA and the OD-RA, where OD-RA achieves the lowest percentage. Indeed, the use of a coefficient $k = 0.1$ leads to a very high priority for the data that belong to the university compared to the other data. With the use of $k = 0.9$, we see the opposite of the results given with the use of $k = 0.1$. This is because the coefficients between the two classes of data become closer, and thus the priority becomes dependent on the amount of consumed data for both classes. Since the amount of consumed data that are not affiliated with the university is much larger than the others, OD-RA gained a significant advantage within the given space, leading to the obtained results. At $k = 0.4$, there is an improvement in the results of both classes as the depth D increases. The difference between the two classes of data is between 20% and 60% for the hit ratio, and between 40% and 25% for the hop reduction ratio, with a preference for the UD-RA. The convergence of the results obtained for the two classes is due to the fact that the space given to them is similar, with a bias towards the space reserved for the university data.

Through this study and the results obtained, we can say that the proposed strategy offers the possibility to control the data in an excellent way, even at the levels where there is a big difference between the consumed data classes. Indeed, by using this strategy, we are able to determine the desired priority level, which in turn can allocate space for the desired data resources.

6. Conclusions and Future Work

The proposed strategies for resource allocation in NDNs are known to be effective in optimizing the overall network performance; however, these strategies may suffer in a limited network environment. That is because a fundamental trade-off problem between resource efficiency and user satisfaction might arise. To this end, a resource allocation scheme based on data classification has been proposed with a weighting factor that is used to decide at which level we want to control the aforementioned trade-off. Simulation results show that our strategy outperforms the sharing strategies, since it proves its ability to control any desired trade-off point of their respective efficiency and satisfaction levels.

As future work, further experiments should be performed to evaluate the robustness of our approach against other state-of-the-art approaches. We also want to add other factors to the objective function of the resource allocation which may help at enhancing and improving network performance.

Author Contributions: Data curation, C.A.K.; Investigation, A.T.H. and N.L.; Methodology, C.A.K., B.Z., C.T.C. and N.L.; Software, A.T.H.; Supervision, C.A.K., B.Z. and A.e.K.T.; Validation, C.T.C.; Writing—original draft, A.T.H.; Writing—review & editing, C.A.K., B.Z., C.T.C., N.L. and A.e.K.T. All authors have read and agreed to the published version of the manuscript.

Funding: This work is derived from R&D project RTI2018-096384-B-I00, funded by MCIN/AEI/ 10.13039/501100011033 and "ERDF A way of making Europe".

Institutional Review Board Statement: Not applicable.

Informed Consent Statement: Not applicable.

Data Availability Statement: Not applicable.

Conflicts of Interest: The authors declare no conflict of interest.

References

1. Vasilakos, A.V.; Li, Z.; Simon, G.; You, W. Information centric network: Research challenges and opportunities. *J. Netw. Comput. Appl.* **2015**, *52*, 1–10. [CrossRef]
2. Abdullahi, I.; Arif, S.; Hassan, S. Survey on caching approaches in information centric networking. *J. Netw. Comput. Appl.* **2015**, *56*, 48–59. [CrossRef]
3. Zhang, L.; Afanasyev, A.; Burke, J.; Jacobson, V.; Claffy, K.; Crowley, P.; Papadopoulos, C.; Wang, L.; Zhang, B. Named data networking. *ACM SIGCOMM Comput. Commun. Rev.* **2014**, *44*, 66–73. [CrossRef]
4. Ahmad, F.; Kerrache, C.A.; Kurugollu, F.; Hussain, R. Realization of blockchain in named data networking-based internet-of-vehicles. *IT Prof.* **2019**, *21*, 41–47. [CrossRef]
5. Cisco Annual Internet Report (2018–2023) White Paper. Available online: https://www.cisco.com/c/en/us/solutions/collateral/executive-perspectives/annual-internet-report/white-paper-c11-741490.html (accessed on 12 December 2021).
6. Kerrche, C.A.; Ahmad, F.; Elhoseny, M.; Adnane, A.; Ahmad, Z.; Nour, B. Internet of vehicles over named data networking: current status and future challenges. In *Emerging Technologies for Connected Internet of Vehicles and Intelligent Transportation System Networks*; Springer: Cham, Switzerland, 2020; pp. 83–99.
7. Psaras, I.; Chai, W.K.; Pavlou, G. In-network cache management and resource allocation for information-centric networks. *IEEE Trans. Parallel Distrib. Syst.* **2013**, *25*, 2920–2931. [CrossRef]
8. Yao, J.; Yin, B.; Lu, X. A novel joint adaptive forwarding and resource allocation strategy for named data networking based on SMDP. In Proceedings of the 2016 12th IEEE International Conference on Control and Automation (ICCA), Kathmandu, Nepal, 1–3 June 2016; pp. 956–961.
9. Li, C.; Xie, R.; Huang, T.; Huo, R.; Liu, J.; Liu, Y. Joint Forwarding Strategy and Resource Allocation in Information-Centric HWNs. In Proceedings of the GLOBECOM 2017-2017 IEEE Global Communications Conference, Singapore, 4–8 December 2017; pp. 1–6.
10. Yuan, D.; Xu, Y.; Ran, J.; Hu, H.; Liu, Y.; Li, X. An optimal fair resource allocation strategy for a lightweight content-centric networking architecture. In Proceedings of the 2017 3rd IEEE International Conference on Computer and Communications (ICCC), Chengdu, China, 13–16 December 2017; pp. 573–577.
11. Zhang, Y.; Tan, X.; Li, W. In-network cache size allocation for video streaming on named data networking. In Proceedings of the 2017 VI International Conference on Network, Communication and Computing, Kunming, China, 8–10 December 2017; pp. 18–23.
12. Ehsanpour, M.; Bayat, S.; Hemmatyar, A.M.A. On Efficient and Social-Aware Object Allocation in Named Data Networks Using Matching Theory. In Proceedings of the 2018 IEEE Symposium on Computers and Communications (ISCC), Natal, Brazil, 25–28 June 2018; pp. 298–303.
13. Zhang, M.; Xie, P.; Zhu, J.; Wu, Q.; Zheng, R.; Zhang, H. NCPP-based caching and NUR-based resource allocation for information-centric networking. *J. Ambient. Intell. Humaniz. Comput.* **2019**, *10*, 1739–1745. [CrossRef]
14. Huo, R.; Xie, R.; Zhang, H.; Huang, T.; Liu, Y. What to cache: differentiated caching resource allocation and management in information-centric networking. *China Commun.* **2016**, *13*, 261–276. [CrossRef]
15. Yovita, L.V.; Syambas, N.R.; Edward, I.Y.M. CAPIC: Cache based on popularity and class in named data network. In Proceedings of the 2018 International Conference on Control, Electronics, Renewable Energy and Communications (ICCEREC), Bandung, Indonesia, 5–7 December 2018; pp. 24–29.
16. Naeem, M.A.; Nor, S.A.; Hassan, S.; Kim, B.S. Compound popular content caching strategy in named data networking. *Electronics* **2019**, *8*, 771. [CrossRef]
17. McCallum, A.; Nigam, K. A comparison of event models for naive bayes text classification. In Proceedings of the AAAI-98 Workshop, Madison, WI, USA, 26–27 July 1998; Volume 752, pp. 41–48.
18. Dehghan, M.; Massoulie, L.; Towsley, D.; Menasche, D.S.; Tay, Y.C. A utility optimization approach to network cache design. *IEEE/ACM Trans. Netw.* **2019**, *27*, 1013–1027. [CrossRef]
19. Situmorang, H.; Syambas, N.R.; Juhana, T.; Edward, I.Y.M. A Simulation of Cache Replacement Strategy on Named Data Network. In Proceedings of the 2018 12th International Conference on Telecommunication Systems, Services, and Applications (TSSA), Yogyakarta, Indonesia, 4–5 October 2018; pp. 1–4.

20. Kim, J.; Ko, M.C.; Kim, J.; Shin, M.S. Route Prefix Caching Using Bloom Filters in Named Data Networking. *Appl. Sci.* **2020**, *10*, 2226. [CrossRef]
21. Zaeem, R.N.; Barber, K.S. A Large Publicly Available Corpus of Website Privacy Policies Based on DMOZ. In Proceedings of the Eleventh ACM Conference on Data and Application Security and Privacy, Virtual, 26–28 April 2021.
22. Mastorakis, S.; Afanasyev, A.; Zhang, L. On the evolution of ndnSIM: An open-source simulator for NDN experimentation. *ACM SIGCOMM Comput. Commun. Rev.* **2017**, *47*, 19–33. [CrossRef]
23. Amadeo, M.; Ruggeri, G.; Campolo, C.; Molinaro, A. IoT services allocation at the edge via named data networking: From optimal bounds to practical design. *IEEE Trans. Netw. Serv. Manag.* **2019**, *16*, 661–674. [CrossRef]
24. Chai, W.K.; He, D.; Psaras, I.; Pavlou, G. Cache "less for more" in information-centric networks (extended version). *Comput. Commun.* **2013**, *36*, 758–770. [CrossRef]
25. Ren, J.; Qi, W.; Westphal, C.; Wang, J.; Lu, K.; Liu, S.; Wang, S. Magic: A distributed max-gain in-network caching strategy in information-centric networks. In Proceedings of the 2014 IEEE conference on computer communications workshops (INFOCOM WKSHPS), Toronto, ON, Canada, 27 April–2 May 2014; pp. 470–475.
26. Hu, X.; Gong, J.; Cheng, G.; Fan, C. Enhancing in-network caching by coupling cache placement, replacement and location. In Proceedings of the 2015 IEEE International Conference on Communications (ICC), London, UK, 8–12 June 2015; pp. 5672–5678.

future internet

Article

Weighted-CAPIC Caching Algorithm for Priority Traffic in Named Data Network

Leanna Vidya Yovita [1,*], Nana Rachmana Syambas [2] and Ian Joseph Matheus Edward [2]

1 School of Electrical Engineering, Telkom University, Bandung 40257, Indonesia
2 School of Electrical Engineering and Informatics, Bandung Institute of Technology, Bandung 40116, Indonesia; nana@stei.itb.ac.id (N.R.S.); ian@stei.itb.ac.id (I.J.M.E.)
* Correspondence: leanna@telkomuniversity.ac.id; Tel.: +62-22-7564108

Abstract: Today, the internet requires many additional mechanisms or protocols to support various ever-growing applications. As a future internet architecture candidate, the Named Data Network (NDN) offers a solution that naturally fulfills this need. One of the critical components in NDN is cache. Caching in NDN solves bandwidth usage, server load, and service time. Some research about caching has been conducted, but improvements can be made. In this research, we derived the utility function of multiclass content to obtain the relationship between the class's weight and cache hit ratio. Then, we formulated it into the Weighted-CAPIC caching algorithm. Our research shows that Weighted-CAPIC provides a higher cache hit ratio for the priority class and the whole system. This performance is supported while the algorithm still provides the same path-stretch value as Dynamic-CAPIC. The Weighted-CAPIC is suitable to used in mobile nodes due to its ability to work individually without having to coordinate with other nodes.

Keywords: Named Data Network; cache; weight; class; priority

Citation: Yovita, L.V.; Syambas, N.R.; Edward, I.J.M. Weighted-CAPIC Caching Algorithm for Priority Traffic in Named Data Network. *Future Internet* **2022**, *14*, 84. https://doi.org/10.3390/fi14030084

Academic Editors: José Carlos Lopez-Ardao, Miguel Rodríguez Pérez and Sergio Herrería Alonso

Received: 20 February 2022
Accepted: 11 March 2022
Published: 12 March 2022

Publisher's Note: MDPI stays neutral with regard to jurisdictional claims in published maps and institutional affiliations.

1. Introduction

Global internet traffic continues to increase every year. Based on Cisco's internet report in 2020, it is estimated that 67% of the world's population will be connected to the internet by 2023, meaning there will be 5.3 billion internet consumers. This number is an increase from a total of 3.9 billion consumers in 2018 [1]. Based on the Global Internet Phenomena Report 2019 by Sandvine, traffic related to streaming media is the highest, followed by user-generated content (UGC) and web access [2].

The current internet uses TCP/IP, a host-based network. In this architecture, consumer requests are addressed to a specific server address. This address is included in the request packet that the consumer sends. Only the destination server can respond to the consumer's request. The traffic growth cannot be supported efficiently by TCP/IP network. The network becomes burdened by the communication process.

Named Data Network (NDN) is a content-based architecture that focuses on the name of the content. Content names are used at the application level, and the other layer is used for caching, forwarding, routing, and other data processing [3]. The NDN is suitable for communication needs that are basically focused on content. In contrast to IP that cannot support this naturally, NDN makes communication run more efficiently. NDN is ready for a consumer character with dynamic demand. Another advantage is that we can treat traffic classes differently to give higher performance to the higher-priority classes based on the caching rules.

Caching as one of the primary and crucial components in NDN is a solution of bandwidth usage, server load, and service time [4]. Previous research has showed that algorithms with a different treatment for contents provide better performance. Kim et al. proposed treatment techniques for different service classes, both with forwarding techniques and caching techniques [5]. The service is divided into four classes, and there are different policies for each class. Sourlas et al. proposed partition-based caching that gives separated space for contents and simulated some cases of partition with a particular forwarding algorithm [6]. Dehghan et al. proposed utility-driven caching that provides different allocation for content providers and content publishers based on the maximum utility [7]. Split-caching presented by Majd et al. used the technique to divide the content store into two parts, for priority and nonpriority content. The algorithm decides whether the content needs to be cached or not [8]. Yovita et al. proposed Dynamic-CAPIC, a caching algorithm that distinguishes the cache portion based on the popularity and class of the content in real-time [9]. This algorithm increases the network cache hit ratio (CHR) and provides performance that matches a particular class on the system. However, we need to develop a technique to give the best performance for the main priority class while increasing total system performance by efficiently using all available resources on the network, to provide a better experience for the consumer. In this research, we propose Weighted-Cache based on Popularity and Class (Weighted-CAPIC) as a solution to provide a proper performance for priority class. We derive the utility function formula for multiclass content to obtain the relationship between weight and cache hit ratio (CHR). This algorithm is suitable for the NDN mobile node because every router node can run this algorithm individually without coordinating with other nodes.

This research is written in the following order. Section 2 explains the basics of Named Data Networking. Section 3 describes the research related to caching algorithms on Named Data Network. Section 4 discusses our proposed caching algorithm, Weighted Cache based on Popularity and Class (Weighted-CAPIC) algorithm. Section 5 explains the system model, while Section 6 concludes the simulation results and analysis. Section 7 is about algorithm complexity. Section 8 is the conclusion, and Section 9 is the future research.

2. Named Data Network (NDN)

Named Data Network (NDN) is one of the candidates for future internet architecture because it offers a more efficient communication mechanism than the current internet (TCP/IP). The network must adaptively support various technological development and various consumer characteristics. Unlike NDN, which has these features naturally, the internet today (TCP/IP) requires many additional mechanisms or protocols to support multiple evolved applications [10]. IP networks add some mechanisms to maintain end-to-end connections and map the data with a specific address. On an IP network, the consumer sends a request for content to a particular server. This request message contains the address of the server. Therefore, the response to this consumer request is only made by a server with a specific address.

NDN has two types of packet: an interest packet containing a consumer's request and a data packet containing the information as a response to the interest packet. Every node on NDN has three components, namely Content Store (CS), Pending Interest Table (PIT), and Forwarding Information Based (FIB) [11]. When an NDN router node receives an interest message containing a content request from customers, the node checks its content store (CS) for whether the node has the requested data in its cache. If the content exists in the router, it will be sent directly to the consumer. If not, then the router node will continue to check its PIT and update this request's information. The router will add information about the face from which the request came. The router also writes the requested content's name if this data does not exist before in the PIT. If there is no previous information about the requested content on the PIT, then the interest message will be forwarded to another node based on the information in the FIB. The information on FIB contains the next hop to reach the other node with content.

Content storage is a limited resource in NDN. Even if we can provide high memory due to the affordable price, using too large memory on the router node will cause high processing time and poor system performance. Therefore, it is essential to regulate the mechanism in the content store to obtain optimal performance.

3. Related Works: Caching Algorithms on Named Data Network

Caching is a fundamental feature available on NDN. In the IP (TCP/IP) networks, the buffer on the router also can store data, but after forwarding the data, it can no longer be used by the router [12]. This character causes a big difference between IP networks and NDN.

Until now, there have been many studies on NDN caching algorithms. The most common caching algorithm stored content on all nodes in the content delivery path. Regarding the content replacement in the content store, some basic algorithms commonly used in NDN are FIFO (first in, first out), Least Recently Used (LRU), and Least Frequently Used (LFU) [13]. FIFO stores content in the content store in the order it was entered and deletes the last order when the content store is full. The LRU removes content that has been unsolicited for a long time. LFU removes content that consumers rarely request. When the content store is full, it is necessary to select which content to delete so that the place can be used to store new content. According to research conducted by Jing et al., LFU is the most effective cache replacement strategy for networks with demand patterns following the Zipf–Mandelbrot distribution [14]. Hence we use LFU in this research.

Research on the content store's rule has been carried out previously, such as in utility-driven caching by Dehghan et al. [7]. In this scheme, content is modeled by generating it from different parties. Each party has a certain proportion in the NDN router's content store to keep the data based on their utility. The content is given a particular time to live. When the content has expired, it will be removed from the content store.

The other method was proposed by Kim et al. [5]. They presented the diff-caching model in NDN. There are four classes in this model, i.e., Dedicated Caching (DC), Assured Caching (AC), Best-Effort (BE), and Bypass Caching (BC). The caching process works by marking the data packet based on the service class. The data is always stored for the DC class content (up to a certain period). According to the producers' budget, the AC class content is stored within certain resource limits. The BE content class is stored if there are resources, and it should be ready to be deleted if the content store is full [5].

The caching algorithm in the previous studies mostly differentiates content treatment based on its popularity. The content store is partitioned to store popular and less popular content separately, as in the split-cache algorithm [8]. The split-caching proposed by Majd et al. selects the content to be cached based on the number of requests, and the number of hops traveled from the consumer to the router that stores the content [8].

The partition of content stores can also be done based on the content provider [15]. Sourlas et al. explained the partition-based caching that cached content based on its delivery rate [6]. The priority content is stored separately to avoid deleting priority content in the cache to be replaced with nonpriority content.

The partitions of the content store can be static, or they can be changed every time. One of the caching algorithms with static divisions in the content store is Static-CAPIC [16]. The authors classify traffic into three classes with different characters. In this algorithm, a different treatment has been carried out on each content class. Each class has a different portion of the storage. The Static-CAPIC algorithm has improved the cache hit ratio of the network. Still, it is deficient in providing performance for the priority class when the consumer's demand changes dynamically [9].

Yovita et al. developed this into Dynamic-CAPIC, where the cache portion for each content class can change according to demand conditions [9]. This increases the network cache hit ratio and provides a performance that matches the content class character in the system. However, we need to improve the performance of priority classes while increasing total system performance to give the best experience for priority consumers by using the system's resources efficiently. If we exploit all resources efficiently, this will provide enormous value to the development of caching techniques. Because of this issue, in this research, we propose a Weighted Cache based on Popularity and Class (Weighted-CAPIC) to improve the performance of the main priority class and the whole system in the mobile environment.

4. Weighted Cache Based on Popularity and Class (Weighted-CAPIC) Algorithm

The utility function for content without content class distinction was carried out by Dehghan et al. [7]. In this section, we develop this equation to accommodate multiclass content. Internet traffic has different requirements; therefore, we should provide different treatment for each class of traffic. We derived the utility function formula for multiclass content to obtain the relationship between weight and cache hit ratio. This was necessary to determine the proper treatment for the priority content class.

The utility function for multiclass content uses the utility relationship to the hit probability for each content class, d, which can be written as in Equation (1)

$$U_d(p_d) = \begin{cases} \omega_d \frac{p_d}{1-\gamma}^{1-\gamma}, \gamma \geq 0 \text{ and } \gamma \neq 1 \\ \omega_d . \log_{p_d} \end{cases} \quad (1)$$

where:
p_d = hit probability for content class d
$U_d(p_d)$ = utility for class d with a hit probability p_d
ω_d = weight of class d

To maximize the utility of the content store, we can write the Equation (2)

$$max_{p_d} \sum_{d=1}^{D} U_d(p_d) \quad (2)$$

Since there are D number of content classes and we used three content classes with different requirement as in Table 1, it can be written that the relationship between the portion of each class, c_d, and the total size of content storage, C, is as Equations (3) and (4).

$$c_1 + c_2 + c_3 = C \quad (3)$$

and

$$\sum_{i=1}^{c_1} r_{i1} + \sum_{i=1}^{c_2} r_{i2} + \sum_{i=1}^{c_3} r_{i3} \leq D \quad (4)$$

Table 1. Class of content.

Class	Example	Request Duration	Request Rate
1	Real-time apps	Long (years)	Low–mid
2	User-Generated Content (UGC)	Mid (weeks)	Mid
3	Web access	Mid (weeks)	High
4	email, VoNDN	Short(days)	Not cached

Hit probability of class d with cache proportion c_d and using LFU cache replacement can be defined as Equation (5).

$$p_d = \sum_{i=1}^{c_d} r_{di} \quad (5)$$

where:
r_{di} = hit probability of content i on class d;
p_d = hit probability of class d;
c_d = cache proportion of class d.

The constraint for the utility function in the content store that stores multiclass content is as in Equation (6).

$$\sum_{d=1}^{D} (p_d . N_d) = C \quad (6)$$

Due to the neutrality for making the utility linear in using p_d, we use the value $\gamma = 0$. Therefore, the maximization of utility can be written as Equations (7) and (8).

$$max_{p_d} \sum_{d=1}^{D} \omega_d \frac{p_d}{1-\gamma}^{1-\gamma} \tag{7}$$

$$max_{p_d} \sum_{d=1}^{D} \omega_d . p_d \tag{8}$$

The number of utilities will be maximized if we set a more significant weight for the class d with a greater hit probability. When applied to the system, this will cause the maximum overall hit rate in the system. Based on the formula that has been derived from the utility function of multiclass content, the total utility will be maximized if we give the most significant weight to the priority class with the highest hit probability. Weighted-CAPIC adds this concept to Dynamic-CAPIC from previous research. We add the weight parameter for determining the cache portion of each class, where ω_d is the weight for class d. The Weighted-CAPIC formula determines the cache portion for class d, as in Equation (9).

The Weighted-CAPIC formula ensures that the weight is assigned to the class that requires it according to real-time conditions. The cache portion of one class can change over time, and the Weighted-CAPIC formula assigns a certain weight to the class according to what it needs in a certain real-time period. These algorithms can be implemented in the NDN mobile node because every router node can run this algorithm individually without coordinating with other nodes.

As in Figure 1, when the interest message enters, the router will check its content store. If the content requested exists in the content store, it will be sent to the consumer by the router, and the copy will be kept in the one-hop downstream router. For every request, content gets closer to the consumer. The popularity of content determines on which nodes it will be stored in the network.

The next step is router checks how much cache portion for keep the content in the router. This portion is determined based on the content class, using the Weighted-CAPIC formula. The Weighted-CAPIC algorithm calculates the total content requested over a specific frequency limit factor for each content class. The value of this parameter is described as X_d. After that, the router will sort the values and assign a weight, ω_d, corresponding to them. The router further calculates the cache portion for the content class according to Equation (9). If the cache capacity for a class is full, the algorithm will run LRU as a cache replacement method. If the cache capacity for a class is full, the algorithm will run LRU as a cache replacement method. The router performs this two-step mechanism every time there is an incoming packet of interest and data (content).

$$c_d = \frac{X_d.\omega_d}{\sum_{y=1}^{D} X_y . \sum_{m=1}^{D} \omega_m} C \tag{9}$$

where:
X_d = the number of variations in content class k that is requested more than a certain Frequency value;
c_d = cache proportion for class d;
C = total content store capacity;
D = number of content classes.

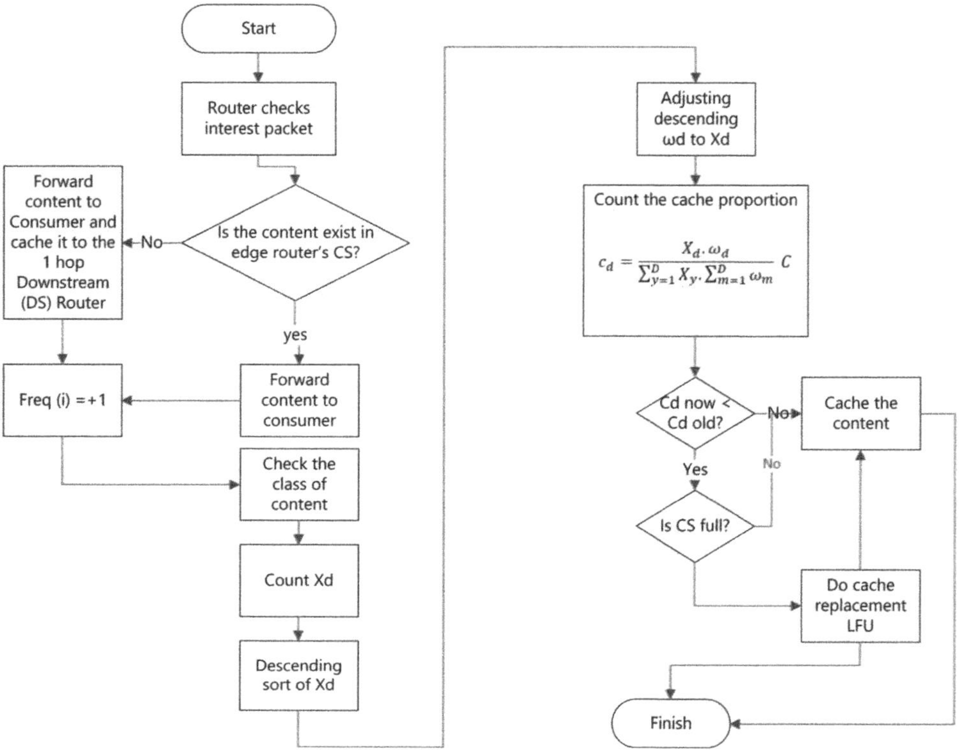

Figure 1. Flowchart of Weighted CAPIC.

5. System Model

Based on the Global Internet Phenomena Report 2019, internet traffic accessed by consumers can be divided into several groups with different characteristics [2]. Abane et al. conducted research using the Zipf–Mandelbrot distribution to model content popularity [10]. Following the RFC 7945, the Zipf–Mandelbrot distribution is suitable for the network where nodes can store content locally [17]. The average of consumer request rate is modeled using Poisson distribution.

In this research, the probability of connection between nodes is 50%. This modeling causes topology changes in a network for every simulation round. This effect the cache conditions and change dynamically. Content popularity is modeled by Zipf–Mandelbrot with exponential factor 0.8 and flattened factor 3. The average number of requests for each class follows the Poisson distribution and traffic class, according to Table 1. Simulation parameters are in detail according to Table 2.

Two experimental scenarios were carried out, with variations in the number of request levels and variations in the request rate. In the scenario of variation in number request level, classes 1, 2, and 3 each have a demand average of 10, 20, and 30, respectively. In the scenario of variations in the request rate, the number of request levels is set to 6. This means that the user request pattern changes six times during the simulation time.

In the first scenario that simulated the various number of request levels, there were three types of request levels, i.e., number of requests levels 2, 4, and 6. The number of requests level 2 means the pattern of consumer demand changed two times. For this scenario, there are two requests patterns: (10, 20, 30) and (20 30 10). This means that at half the simulation time, we used the (10, 20, 30) pattern. The (10, 20, 30) pattern implies that the number of requests for the first class is 10 requests per simulation round, the second class is 20 requests per second, and the third class is 30 requests per simulation round. As for the other half time of the simulation, we used (20, 30, 10) pattern. The number of requests levels 4 and 6 also worked in the same way as the previous mechanism. In the scenario of 'number of requests level' 4, there were four changes in the request pattern during the simulation. Similarly, for the 'number of request level' 6, there were six changes in the pattern of requests during the simulation.

Table 2. Simulation parameters.

Parameter	**Value**
Number of routers	50
Content store proportion (class 1, 2, and 3)	1. Weighted-CAPIC and Dynamic-CAPIC: dynamic according to the formula 2. Static-CAPIC: 5%, 10% and 15% of total content number, normalized to total capacity of 150
Number of content	500 (1st class) 700 (2nd class) 1000 (3rd class)
Number of rounds	10.000
Frequency Limit Factor	1600
Number of request level	2 times: (10, 20, 30) and (20 30 10) 4 times: (10, 20, 30), (20 10 30), (30 20 10), and (10 30 20) 6 times (10, 20, 30), (20 10 30), (30 20 10), (20 30 10), (30 10 20) and (10 30 20)
Difference in request rate	10, 20, 30 (difference: 10) 10, 30, 50 (difference: 20) 10, 40, 70 (difference: 30) 10, 50, 90 (difference: 40)

The second scenario simulates the different request rate gaps between content classes. The term "Difference in request rate" means a difference (gap) between the request rates for classes 1, 2, and 3. In this scenario, the values are tested starting from 10, 20, 30, and 40. A 'differences in request rate' of 10 means that class 1 has a request rate of 10 requests per simulation round, then the second class increases by 10 compared to the first class, which is 20 requests per simulation round, the third class increased by 10 compared to the second class, which is 30 requests per simulation round. Likewise, the 'difference in request rate' values of 20, 30, and 40 are modeled as more significant gaps between the request rates for each content class.

System performance is seen from the cache hit ratio and path stretch parameters. The cache hit ratio parameter compares the number of requests that can be responded directly by the router node to the total number of requests. The path stretch parameter indicates the number of hops taken until a request receives data (content) as a response. Weighted-CAPIC performance is compared to Dynamic-CAPIC, Static-CAPIC, and LCD+Sharing schemes. LCD+Sharing scheme combines two typical cache schemes in NDN, namely LCD (leave copy down) and sharing scheme. We test the Weighted-CAPIC, Dynamic-CAPIC, Static-CAPIC, and LCD+sharing method in the same environment to obtain the appropriate comparison.

6. Simulation Results and Analysis

The number of request levels shows how often the pattern of consumer demand changes during the simulation time of request pattern change. Request level 6 means that the pattern of consumer demand changes six times. The higher the request level, the more often the demand pattern changes. Weighted-CAPIC provides the highest cache hit ratio for class 1, which is on average 46% larger than Dynamic-CAPIC, four times larger than Static-CAPIC, and two times larger than LCD+Sharing, as in Figure 2. This result is in accordance with the goal of the Weighted-CAPIC algorithm, which provides the highest performance for the class with the highest priority. The result also proves that Weighted-CAPIC can accommodate dynamically changing traffic patterns in the system. Class 1 is a priority class. It has a longest request duration as in Table 1. The information has a long time to go before it becomes obsolete. The content is usually still requested over the years. Therefore, storing content in the network will provide an advantage for these repeated content requests.

The Weighted-CAPIC path stretch has the same value as Dynamic-CAPIC, and its value is the smallest compared to Static-CAPIC and LCD+Sharing schemes. With the same path stretch, Weighted-CAPIC provides the larger cache hit ratio compared to Dynamic-CAPIC.

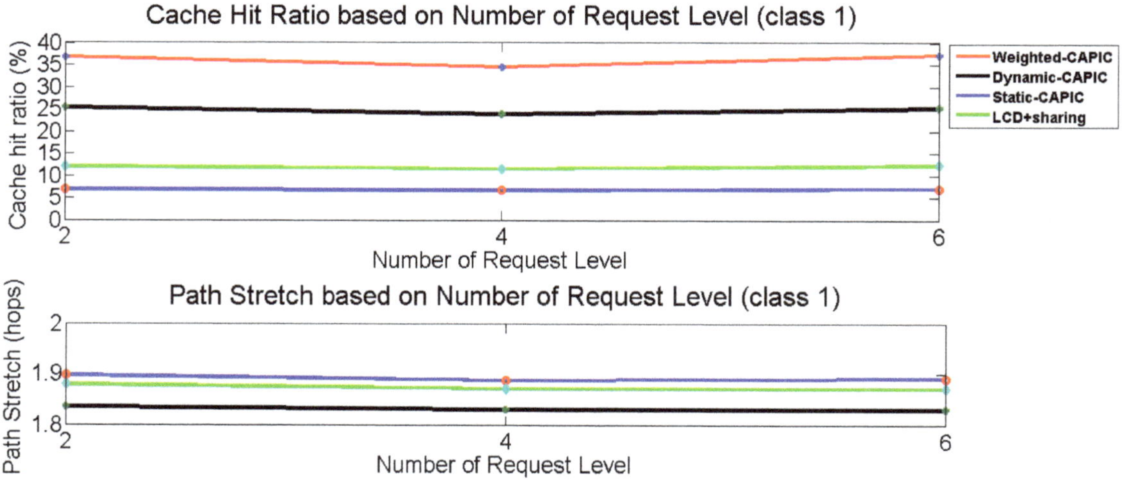

Figure 2. Cache hit ratio and path stretch of class 1 for various level of request.

The second class cache hit ratio for Weighted-CAPIC is 14.6% smaller than LCD+Sharing, 7% smaller than Dynamic-CAPIC, and 2% larger than Static-CAPIC, as in Figure 3. The second class with UGC traffic type is the second priority, after real-time traffic. So, based on the formula, Weighted-CAPIC provides a smaller portion of storage for this second class compared to the first class. This can also be regarded as compensation, because Weighted-CAPIC has given a more significant cache portion for the first class, which is a priority class. The content store capacity is fixed in every NDN router, and the cache portion must be adjusted based on the content class's needs. In Weighted-CAPIC, this adjustment is made using a formula that is ran on each NDN router.

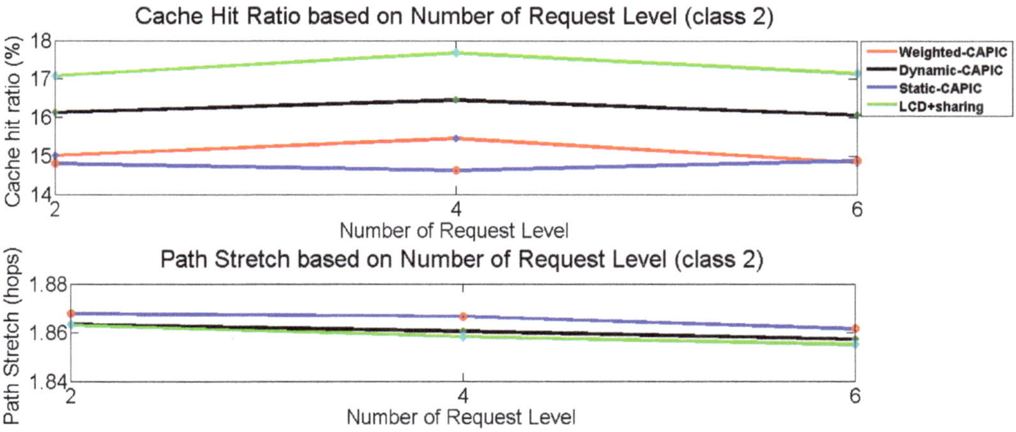

Figure 3. Cache hit ratio and path stretch of class 2 for various level of request.

The Weighted-CAPIC path stretch has the same value as Dynamic-CAPIC, having a smaller value of about 0.2% than Static-CAPIC. In this second class, the Weighted-CAPIC path-stretch value is higher than the first class, which means that it is necessary to pass longer distances to obtain the desired content for this second class.

Weighted-CAPIC gives the smallest cache hit ratio value for class 3, as in Figure 4, because the priority of the third class is the lowest. In addition, by looking at Figures 2–4, it can be seen that Weighted-CAPIC provides the lowest cache hit ratio for the third class compared to first and second. Again, this corresponds to the priorities of classes 1, 2, and 3. Path stretch decreases when the request level increases for all caching schemes.

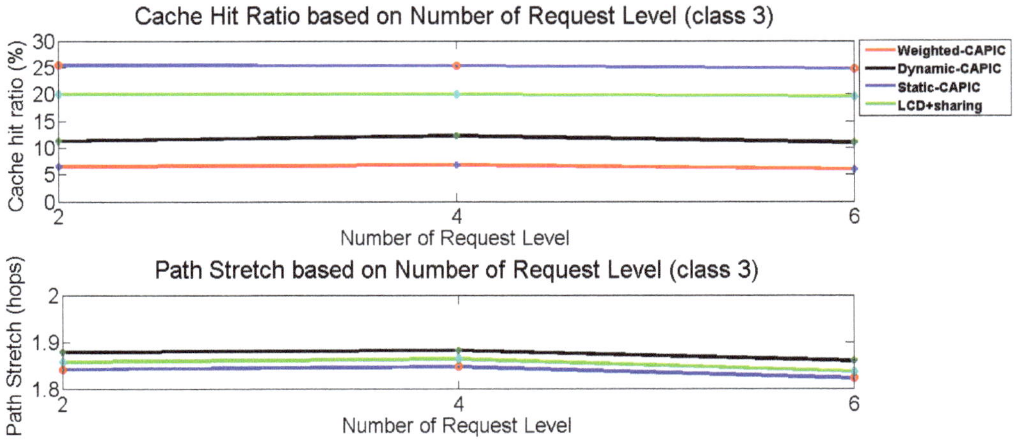

Figure 4. Cache hit ratio and path stretch of class 3 for various level of request.

Weighted-CAPIC provides the highest network cache hit ratio parameter. This parameter involves the overall cache hit ratio for all traffic classes in the network/system. Figure 5 shows that the higher the number of request levels, the higher the cache hit ratio of the Weighted-CAPIC and Dynamic-CAPIC schemes. Weighted-CAPIC cache hit ratio is higher than Dynamic-CAPIC, and at request level 6, Weighted-CAPIC gives the most significant cache hit ratio, which is 11% higher than Dynamic-CAPIC. Static-CAPIC provides the smallest cache hit ratio for the higher request levels. This condition happens because Static-CAPIC cannot adjust its rules when requests come in arbitrary patterns.

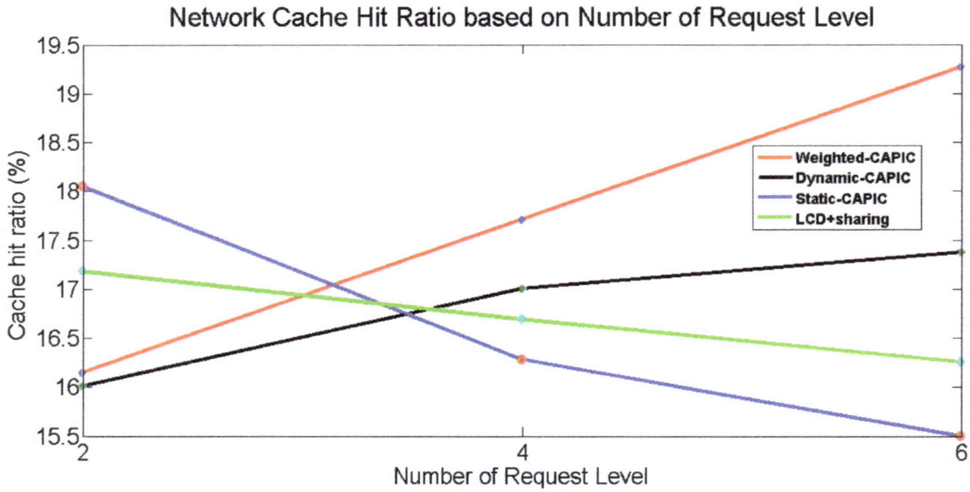

Figure 5. Network cache hit ratio for various level of request.

Algorithm 1: FunctionWeighted-CAPIC

Initialization;
D: number of content class ;
i: content ;
C: capacity of content store ;
Function WeightedCAPIC(input: D, C, i) ;
if *request=i* **then**
    Check last-location i;
    if hop_i!=1 **then**
        hop_i=hop_i-1 ;
    else
        hop_i=1
else
    end process
% calculate number of content variation for each class;
for $m = 1 : D$ **do**
    $Total_X$ =+ X_m;
Sorting(Xm) **for** $d = 1 : D$ **do**
    give the ω_d based on X_m
% calculate cache portion of content class d;
$c_d = \frac{X_d.\omega_d}{\sum_{i=1}^{D} X_i . \sum_{m=1}^{d} \omega_m} C$;
Check capacity of c_d;
if *capacity not full* **then**
    save i in part-of-Content-store;
else
    do replacement algorithm;
    save i in part-of-Content-store;
send i to consumer ;

In the 'difference in request rate' scenario, Weighted-CAPIC provides the most significant value compared to other schemes for class 1. As in Figure 6, its cache hit ratio is 45% higher than Dynamic-CAPIC. The cache hit ratio of Weighted-CAPIC is fluctuates 1–4% around its average value and decreases further for more considerable request rate differences. This means that Weighted-CAPIC can accommodate the various consumer request rates, with the low or high gap between each traffic class. Dynamic-CAPIC gives the cache hit ratio variation 1–3% around the average value and falls for the more significant request rate difference value. It drops dramatically compared to the average, with the difference in request rate increasing up to 50%. At the same time, Static-CAPIC is more stable with a difference of only 1% compared to the average value.

Figure 6. Cache hit ratio and path stretch of class 1 for various request rate difference.

For the path stretch parameter, Weighted-CAPIC has the same value as Dynamic-CAPIC. Weighted-CAPIC provides the lowest path stretch for this first class. It means that content can be obtained with the lowest hop distance, shorter than the other strategies. Weighted-CAPIC gives the smallest path stretch compared to the all comparison scheme.

LCD+Sharing provides the largest cache hit ratio for the second class, followed by Dynamic-CAPIC and Weighted-CAPIC. This happens because the second class has a lower priority than the first class, so Weighted-CAPIC provides a smaller cache portion and results in a smaller cache hit ratio. The Weighted-CAPIC path stretch is the same as Dynamic-CAPIC and is between the Static-CAPIC and LCD+Sharing stretch path values, as shown in Figure 7. In the second class, Weighted-CAPIC provides a relatively more extensive path stretch for increasing the demand gap between classes. This means that the greater the request rate, the higher the number of hops that must be passed to obtain the desired data. This happens because the greater the number of requests, the greater the variety of data requested.

The largest cache hit ratio is given by the Static-CAPIC scheme for class 3, as shown in Figure 8, because it has the highest rate compared to other classes. Static-CAPIC has accommodated this condition by providing the largest cache portion in class 3. The cache hit ratio of Weighted-CAPIC fluctuates around 3–8% of the average value. The cache hit ratio of Dynamic-CAPIC fluctuates about 1–4% of the average value. The LCD+Sharing cache hit ratio increases along with the rise in request rate difference, up to 12% of the average for the largest request gaps. Weighted-CAPIC provides the lowest cache hit ratio compared to other comparison schemes for the third class. This happens because, in Weighted-CAPIC, the third class is the lowest priority class; therefore, it is given the minor cache portion compared to the first and second class. This is different from the Static-CAPIC and LCD+Sharing schemes, which provide greater performance for this third class. Weighted-CAPIC provides a lower cache hit ratio than Dynamic-CAPIC in this third class as compensation for delivering greater performance to the classes with higher priority. The

Weighted-CAPIC path stretch is the same as Dynamic-CAPIC, which is the largest stretch path, followed by LCD+sharing and Static-CAPIC scheme.

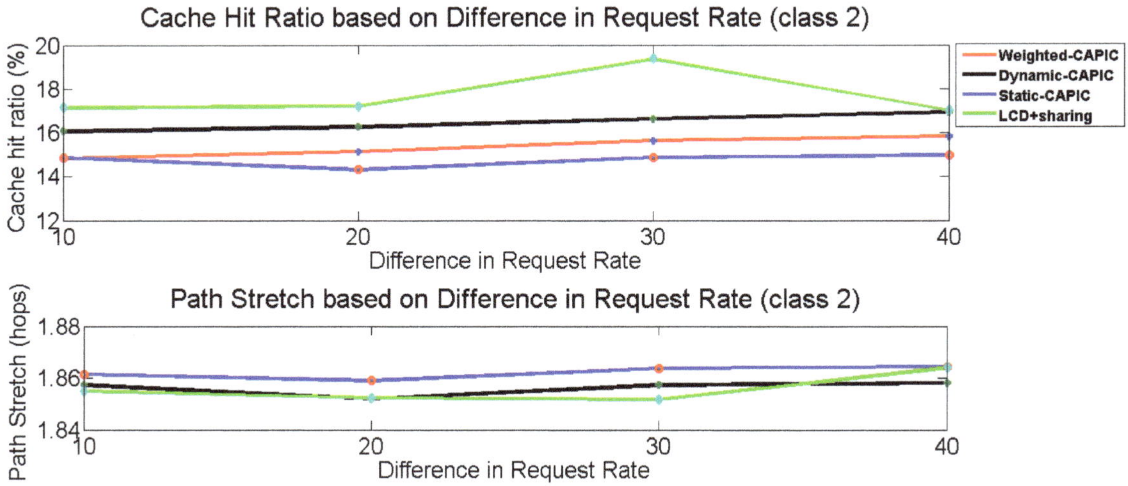

Figure 7. Cache hit ratio and path stretch of class 2 for various request rate difference.

Figure 8. Cache hit ratio and path stretch of class 3 for various request rate difference.

The simulation scenario with variations in the request rate difference gives the largest network cache hit ratio value for Weighted-CAPIC, followed by Dynamic-CAPIC, LCD+Sharing, and Static-CAPIC. As in Figure 9, Weighted-CAPIC provides an average network cache hit ratio of 9.7% higher than Dynamic-CAPIC, 22% higher than LCD+Sharing, and 24% higher than Static-CAPIC scheme. The network cache hit ratio is the total cache hit ratio for the system. Figure 9 shows that Weighted-CAPIC not only provides the higher performance to the priority class but also increases the total system cache hit ratio, compared to the Dynamic-CAPIC, Static-CAPIC, and LCD+sharing schemes.

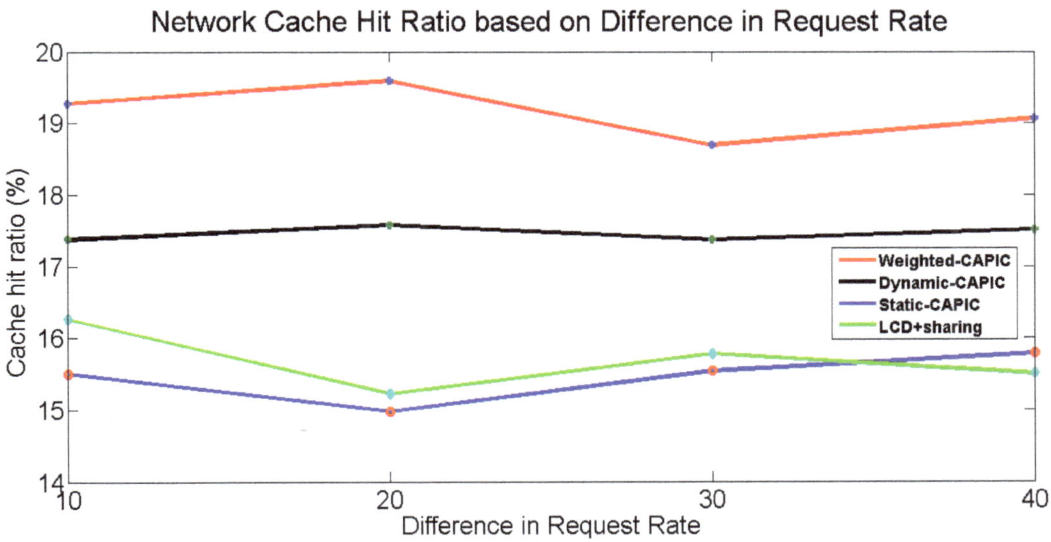

Figure 9. Network cache hit ratio for various request rate difference.

From the simulation results, Weighted-CAPIC gives the higher cache hit ratio to the priority class. Weighted-CAPIC improved the method in the Dynamic-CAPIC algorithm. Dynamic-CAPIC can dynamically allocate cache portions for each traffic class based on consumer demand. This is done using the Dynamic-CAPIC formula, which provides the cache portion based on the variations in content for certain classes. However, based on our derived multiclass-content utility function, we must treat priority classes differently besides just considering the traffic variation. The Weighted-CAPIC formula is also as dynamic as the Dynamic-CAPIC formula, which can provide different portions at any time, according to consumer demand patterns. However, Dynamic-CAPIC has not provided the most extensive resource for the highest priority class. The highest priority is given to the class which has the highest opportunity to be requested by consumers at one time.

For this reason, Dynamic-CAPIC was improved into Weighted-CAPIC, which focused more on providing larger resources to the appropriate class. The simulation results for Weighted-CAPIC show that this algorithm successfully provides a higher cache hit ratio for the priority class and provides a higher cache hit ratio for the network system. This performance is supported while the algorithm still provides the same path stretch value as Dynamic-CAPIC.

7. Algorithm Complexity

To analyze the Weighted-CAPIC performance, we conducted several tests and also explored it in terms of algorithm complexity. We used the Time Complexity parameter to measure the complexity of the algorithm. Measurement of complexity based on time is carried out for a particular input [18]. However, the processing time is highly dependent on various factors such as the type of machine used, the parallel processing performed, and so on. Big-O notation is used to represent it to avoid biased algorithm complexity values.

Weighted-CAPIC, according to the pseudocode in the algorithm 1, can be written as the input correlation of total class content D, equal to 5D+2, so that the Big-O notation is O(D). From the previous research, Dynamic-CAPIC caching algorithm as input correlation can be written mathematically as 4D+2, and its Big-O notation is O(D) [9]. This means that the complexity of the Weighted-CAPIC algorithm is the same as Dynamic-CAPIC but provides a better cache hit ratio (CHR) than Dynamic-CAPIC for the priority class and the overall network.

8. Conclusions

This research proposed the Weighted-CAPIC caching algorithm to improve the priority class performance. Weighted-CAPIC works by adjusting different weights for every content class besides considering the variation of the content in providing a cache portion in the NDN router. Weighted-CAPIC provides the highest cache hit ratio for the priority class and the network, outperforming the Dynamic-CAPIC, Static-CAPIC, and LCD+Sharing scheme. The performance evaluation shows that the greater the number of request levels (the more dynamic consumer demand), the greater the total cache hit ratio provided by Weighted-CAPIC. Weighted-CAPIC can accommodate dynamic consumer demands more than all comparison algorithms in this study, i.e., Dynamic-CAPIC, Static-CAPIC, and LCD+Sharing. It is also very good at accommodating the various consumer request rates, with the low or high gap between each traffic class, and provides better performance with the same complexity as Dynamic-CAPIC. Weighted-CAPIC is also suitable for systems with dynamic consumer demand patterns and has a high priority gap for each traffic class. In this condition, Weighted-CAPIC can provide higher performance on certain priority classes. Weighted-CAPIC also provides a higher network/system cache hit ratio than Dynamic-CAPIC and all comparison algorithms, which means that Weighted-CAPIC allocates the cache portion for every class efficiently. But as compensation, Weighted-CAPIC reduces the cache portion of other classes that are of a lower priority.

9. Future Research

We will combine the Weighted-CAPIC with machine learning to determine the weight of each content class as the future research.

Author Contributions: Conceptualization, L.V.Y. and N.R.S.; methodology, L.V.Y. and N.R.S.; software, L.V.Y. and I.J.M.E.; formal analysis, L.V.Y. and N.R.S.; writing original draft, L.V.Y.; validation, N.R.S. and I.J.M.E.; writing—review editing, N.R.S.; supervision, N.R.S. and I.J.M.E.; resources, L.V.Y. and I.J.M.E. All authors have read and agreed to the published version of the manuscript.

Funding: This research received no external funding.

Data Availability Statement: Not applicable, the study does not report any data.

Acknowledgments: This research was supported by Telkom University.

Conflicts of Interest: The funders had no role in the design of the study; in the collection, analyses, or interpretation of data; in the writing of the manuscript, or in the decision to publish the results.

Abbreviations

The following abbreviations are used in this manuscript:

NDN	Named Data Network
UGC	User-Generated Content
CHR	Cache Hit Ratio
CS	Content Store
PIT	Pending Interest Table
FIB	Forwarding Information Based
FIFO	First In, First Out
LRU	Least Recently Used
LFU	Least Frequently Used
LCD	Leave Copy Down

References

1. Cisco. *Cisco Annual Internet Report (2018–2023)*. Technical Report. 2020. Available online: https://www.cisco.com/c/en/us/solutions/collateral/executive-perspectives/annual-internet-report/white-paper-c11-741490.html (accessed on 19 February 2022).
2. Sandvine. *The Global Internet Phenomena Report*; Technical Report. September 2019. Available online: https://www.sandvine.com/hubfs/Sandvine_Redesign_2019/Downloads/Internet%20Phenomena/Internet%20Phenomena%20Report%20Q32019%2020190910.pdf (accessed on 19 February 2022).

3. Shannigrahi, S.; Fan, C.; Partridge, C. What's in a Name?: Naming Big Science Data in Named Data Networking. In Proceedings of the ICN 2020—Proceedings of the 7th ACM Conference on Information-Centric Networking, Virtual Event, Canada, 29 September–1 October 2020; pp. 12–23. [CrossRef]
4. Jin, H.; Xu, D.; Zhao, C.; Liang, D. Information-centric mobile caching network frameworks and caching optimization: A survey. *Eurasip J. Wirel. Commun. Netw.* **2017**, 33, 1–32. [CrossRef]
5. Kim, Y.; Kim, Y.; Bi, J.; Yeom, I. Differentiated forwarding and caching in named-data networking. *J. Netw. Comput. Appl.* **2016**, *60*, 155–169. [CrossRef]
6. Sourlas, V. Partition-based Caching in Information-Centric Networks. In Proceedings of the Seventh IEEE International Workshop on Network Science for Communication Networks (NetSciCom 2015), Hong Kong, China, 27 April 2015; pp. 396–401.
7. Dehghan, M.; Massoulie, L.; Towsley, D.; Menasche, D.S.; Tay, Y.C. A Utility Optimization Approach to Network Cache Design. *IEEE/ACM Trans. Netw.* **2016**, *27*, 1013–1027. [CrossRef]
8. Majd, N.E.; Misra, S.; Tourani, R. Split-Cache: A holistic caching framework for improved network performance in wireless ad hoc networks. In Proceedings of the 2014 IEEE Global Communications Conference, GLOBECOM 2014, Austin, TX, USA, 8–12 December 2014; pp. 137–142. [CrossRef]
9. Yovita, L.V.; Syambas, N.R.; Joseph, I.; Edward, M.; Kamiyama, N. Performance Analysis of Cache Based on Popularity and Class in Named Data Network. *Future Internet* **2020**, *12*, 227. [CrossRef]
10. Abane, A.; Daoui, M.; Bouzefrane, S.; Banerjee, S.; Abane, A.; Daoui, M.; Bouzefrane, S.; Banerjee, S.; Realistic, P.M.A.; Abane, A.; et al. A Realistic Deployment of Named Data Networking in the Internet of Things to cite this version: HAL Id: Hal-02920555 a Realistic Deployment of Named Data Networking in the Internet of Things. *J. Cyber Secur. Mobil.* **2020**, *9*, 1–27.
11. Afanasyev, A.; Burke, J.; Wang, L.; Zhang, B. A Brief Introduction to Named Data Networking. In Proceedings of the MILCOM 2018—2018 IEEE Military Communications Conference (MILCOM), Los Angeles, CA, USA, 29–31 October 2018; IEEE: Piscataway, NJ, USA, 2018.
12. Conti, M.; Member, S.; Gangwal, A.; Hassan, M.; Lal, C.; Losiouk, E. The Road Ahead for Networking: A Survey on ICN-IP Coexistence Solutions. *IEEE Commun. Surv. Tutor.* **2020**, *22*, 2104–2129. [CrossRef]
13. Shailendra, S.; Sengottuvelan, S.; Rath, H.K.; Panigrahi, B.; Simha, A. Performance evaluation of caching policies in NDN-an ICN architecture. In Proceedings of the IEEE Region 10 Annual International Conference, Proceedings/TENCON, Penang, Malaysia, 5–8 November 2017; pp. 1117–1121. [CrossRef]
14. Jing, Y. *Evaluating Caching Mechanisms in Future Internet Architectures*; Technical report; Massachusetts Institute of Technology: Cambridge, CA, USA, 2016.
15. Chu, W.; Dehghan, M.; Towsley, D.; Zhang, Z.L. On allocating cache resources to content providers. In Proceedings of the ACM-ICN 2016—Proceedings of the 2016 3rd ACM Conference on Information-Centric Networking, Kyoto, Japan, 26–28 September 2016; pp. 154–159. [CrossRef]
16. Yovita, L.V.; Syambas, N.R.; Matheus Edward, I.Y. CAPIC: Cache based on Popularity and Class in Named Data Network. In Proceedings of the 2018 International Conference on Control, Electronics, Renewable Energy and Communications, ICCEREC 2018, Bandung, Indonesia, 5–7 December 2018; pp. 24–29. [CrossRef]
17. Pentikousis, K.; Bari, P. *RFC7945: Information-Centric Networking: Evaluation and Security Considerations.* Technical Report. 2016. Available online: https://www.hjp.at/doc/rfc/rfc7945.html (accessed on 19 February 2022).
18. Mala, F.A.; Ali, R. The Big-O of Mathematics and Computer Science. *J. Appl. Math. Comput.* **2022**, *6*, 1–3. [CrossRef]

future internet

MDPI

Article

Insights from the Experimentation of Named Data Networks in Mobile Wireless Environments

Luís Gameiro [1], Carlos Senna [1] and Miguel Luís [1,2,*]

[1] Instituto de Telecomunicações, 3810-193 Aveiro, Portugal; luismiguelgameiro@ua.pt (L.G.); cr.senna@av.it.pt (C.S.)
[2] ISEL—Instituto Superior de Engenharia de Lisboa, Instituto Politécnico de Lisboa, 1959-007 Lisboa, Portugal
* Correspondence: nmal@av.it.pt

Abstract: The Information-Centric Network (ICN) paradigm has been touted as one of the candidates for the Internet of the future, where the Named Data Network (NDN) architecture is the one leading the way. Despite the large amount of works published in the literature targeting new implementations of such architecture, covering different network topologies and use cases, there are few NDN implementations in real networks. Moreover, most of these real-world NDN implementations, especially those addressing wireless and wired communication channels, are at a small scale, in laboratory environments. In this work, we evaluate the performance of an NDN-based implementation in a mobile wireless network, as part of a smart city infrastructure, making use of multiple wireless interfaces. We start by showing how we have implemented the NDN stack in current network nodes of the smart city infrastructure, following a hybrid solution where both TCP/IP and NDN paradigms can coexist. The implementation is evaluated in three scenarios, targeting different situations: mobility, the simultaneous use of different wireless interfaces and the network characteristics. The results show that our implementation works properly and insights about the correct NDN parameterization are derived.

Keywords: mobile networks; Named Data Networks; performance evaluation; smart city

Citation: Gameiro, L.; Senna, C.; Luís, M. Insights from the Experimentation of Named Data Networks in Mobile Wireless Environments. *Future Internet* **2022**, *14*, 196. https://doi.org/10.3390/fi14070196

Academic Editors: José Carlos Lopez-Ardao, Miguel Rodríguez Pérez and Sergio Herrería Alonso

Received: 8 June 2022
Accepted: 24 June 2022
Published: 27 June 2022

1. Introduction

Information Centric Networking (ICN) has been considered an alternative to the current Internet architecture, based on the TCP/IP stack, to overcome the drawbacks of host-centric architectures when applied to mobile Internet-of-Things (IoT) networks [1]. Such drawbacks are the frequent topology changes and intermittent connectivity caused by the mobility of the network nodes. Conceptually, in ICN each piece of data has a unique, persistent and location-independent name that is directly used by the applications for content search and retrieval. Therefore, ICN enables the deployment of in-network caching and content replication thus facilitating the efficient and timely delivery of information. Additionally, the publish-subscribe concept adopted by the ICN architecture, allied to a location-independent data identification, provides an efficient support for users' mobility.

Amongst the several ICN architectures, the one receiving more attention by the research community is the Named Data Network (NDN) [2], a pull-based network system where the Consumer requests the Content by sending an Interest packet. This packet is relayed by the network nodes until it reaches the Content, either at the Producer or cached at some node on the network. It is up to the node that owns the Content to send it along the reverse path, taken by the Interest, until it reaches the interested Consumer. Such architecture has been designed for infrastructured network topologies, with point-to-point links, and its application in mobile wireless environments requires the modification of some of the heuristics used in the NDN, especially those related with the Content forwarding [3]. Similarly, the use of the NDN architecture in a publish-subscribe communication paradigm is not straightforward because each each data packet must precede an Interest packet [4].

Despite the touted virtues of the NDN paradigm, it is rare to find implementations or testbeds where such architecture can be tested under the same conditions as TCP/IP-based networks. The majority of the evaluation and comparison works based on non-simulation scenarios consider solely wired network infrastructures, or in a small scale. To the best of our knowledge there are no ICN/NDN implementations at city-scale with both wired and wireless communication links.

To address this gap, we present a city-scale implementation of an NDN architecture, optimized to support both single-data-request and publish-subscribe communication paradigms, and the support for multiple radio access technologies in a mobile environment. Such implementation is currently installed in the communications infrastructure of the Aveiro Tech City Living Lab (ATCLL) (https://aveiro-living-lab.it.pt/ accessed on 7 June 2022), and is open for experimentation, development, test and/or demonstration of concepts, products or services targeting the Information and Communications Technology (ICT) topics. In this work, we explain the steps needed to implement an NDN architecture in such type of environments, where network nodes share multiple radio access technologies, such as ITS-G5, IEEE 802.11, LoRa and cellular 4G/5G, and we discuss its performance in real-world conditions, targeting the dissemination of a given Content to mobile users.

In summary, the main contributions of this work are:

- The deployment of an NDN architecture over a real city-scale wireless communication infrastructure;
- The performance assessment of the NDN architecture in the presence of multiple radio access communication technologies;
- The analysis of how several network characteristics impact the performance of the NDN architecture in wireless mobile environments.

The remainder of this paper is organized as follows. Section 2 overviews the related work about deployment of ICN-based solutions in real hardware devices, and consequently, in non-simulated environments. Section 3 presents the main features of our NDN-based solution and describes how we have deployed it on the ATCLL network. Section 4 presents and discusses the performance results of our solution targeting the dissemination of contents in a mobile network of the ATCLL. Finally, Section 5 concludes the paper and discusses future work.

2. Related Work

In the literature one can find several works on the deployment of solutions based on the ICN architecture, but a small number addresses large-scale infrastructures [5]. In this section, we overview the most relevant related works focused on the experimentation of ICN and NDN architectures in testbeds.

Jacobson et al. [2,6,7] introduced in 2009 the Content-Centered Network (CCN), a communication architecture built on named data, and compared it with an IP-based networks. In their comparison, several metrics were considered such as data transfer efficiency, content delivery efficiency, and Voice-over-CCN. Still, the scenario in use was a small controlled testbed in a laboratory environment.

In [8], the authors discuss the comparison between the content Cache-supported NDN architecture cloudlets with location-oriented TCP/IP architecture. The work considers the placement of cloudlets at the edge of an NDN network, close to end users. They used a virtual tour guide that provides audio and visual content to tourists in popular locations as an application scenario. However, the testbed, consisting of a standard Dell Ubuntu 16.04 laptop to host the NDN cloudlet server and five Android phones as clients, seems very simple being far from representing a city's communications infrastructure, without considering any multi-radio access technology.

In [9], the authors have repeated the experiments in [6] on a more complex communications infrastructure. They interconnect through the Internet some Consumers and a router to a Producer that is placed geographically far. They deploy the CCN over UDP to establish the connection between the two environments (CCN and Internet). The results are similar to those obtained by [6].

The solution proposed in [10], the Named-data Link State Routing protocol (NLSR), is a protocol for intra-domain routing in NDN. The NLSR, an application level protocol similar to many IP routing protocols, uses NDN's Interest/Data packets to disseminate routing updates over network. The authors made an evaluation in terms of CPU processing time, routing convergence time, and forwarding plane performance using Mini-NDN, an emulation tool based on Mininet.

In [11] the authors propose and evaluate a scalable framework for lightweight authentication and hierarchical routing in NDN IoT. They demonstrate the scalability of their solution in a simulated dense environment with up to 40,000 nodes/km^2 whose results show an average onboarding convergence time of around 250 s and overhead of less than 20 kibibytes per node. However, they do not test their solution in a real environment.

In [12], the authors present NDN as a solution for in-vehicle communication. The solution was deployed on an emulated automotive testbed of Raspberry PIs with Controller Area Network (CAN) interfaces and tests were performed using network traces datasets. In the same direction, the work in [13] discussed a comparative simulation-based study of these NDN deployments over Wi-Fi 6 for Internet-of-Vehicles using real vehicular traces. Both papers do not present a real NDN deployment.

The work in [14] presents an NDN solution over ZigBee that evaluated, through three different scenarios, the suitability and ease of use of NDN in the IoT context. They compared their NDN implementation over ZigBee to the basic wireless NDN over Wi-Fi communication that uses the UDP/IP protocol stack to send NDN packets. In their tests, they tried different sets of Interest exchange rates and data packet sizes. They have measured the total time spent by each exchange (from sending the Interest to receiving the data) and also the ratio of Interests satisfied in each pool. Such tests were performed in controlled environments far from what is expected in a real smart city infrastructure.

There are some solutions for the well-known multiplication of interest and data packets in NDN networks over Wi-Fi. In [15], the authors proposed an Access Point-based proxy of Interest mechanism to mitigate WLAN channel competition. In addition, they have proposed a layer-based NDN WLAN multicast data rate selection mechanism for adaptive video streaming to improve the video bit rate. It is an interesting solution but it was only tested through simulation, such as proposed by [16]. In [17,18] follow the same line of validation. In [18], the authors use a small physical testbed consisting of an Access Point and 20 Raspberry Pi-based Wi-Fi clients. Ghasemi et al. [19] present a solution for adaptive video streaming service over NDN as a public ICN/NDN service for daily Internet users. They also show their deployment on the global NDN testbed. The authors discuss results from their implementation solution that presents a reasonable Quality-of-Experience of the service over the global NDN testbed using the videos of the NDN website. The paper [20] proposes an architecture based on NDN and discusses its main functionalities for ITS in smart cities. Still, this paper is purely discussion without any implementation.

Our testbed offers a set of different alternatives in relation to the solutions discussed before. We implemented NDN in a city-scale network that operates simultaneously with the TCP/IP stack without any interference between them. Our container-based solution is scalable and allows the inclusion of other network paradigms transparently, on-the-fly and without any interruption in the networks in operation. Furthermore, to the best of our knowledge, there are no NDN implementations with multiple and heterogeneous radio access interfaces per network node of this scale.

3. Integrating NDNs in Smart Cities

Our goal was to provide a city-scale NDN-based infrastructure in the city of Aveiro, Portugal. Aveiro already has an Internet-based communication infrastructure distributed throughout the city as part of the Aveiro Tech City Living Lab (ATCLL). This metropolitan network provides various communication technologies, such as ITS-G5, Cellular (4G/5G),

IEEE 802.11 (Wi-Fi) and LoRa, which are used to offer various services to several verticals. One of these verticals is the vehicular domain, where vehicles can communicate with each other and with the infrastructure for safety and/or entertainment purposes [21]. Thus, the nodes of the vehicular network, On-Board Units (OBUs) and Road Side Units (RSUs), were modified to include the NDN stack without harming the current Internet stack in operation (TCP/IP), allowing the operation of both paradigms at the same time.

In the next sections we briefly describe the characteristics of Named Data Networks and their differences to the TCP/IP-based architectures. Then we quickly present the communication infrastructure installed in the ATCLL, and finally we detail how we have deployed our NDN solution in the ATCLL network equipments, highlighting the peaceful coexistence between the two paradigms.

3.1. Named Data Networks

Currently, the Internet communication model TCP/IP is based on a host-centric network vision paradigm, where the request for a given content begins by locating and identifying the node in which the content is stored, followed by the establishment of a connection from which the communication of the request to acquire the content will be made, which is repeated for each request made by a consumer, as seen in Figure 1.

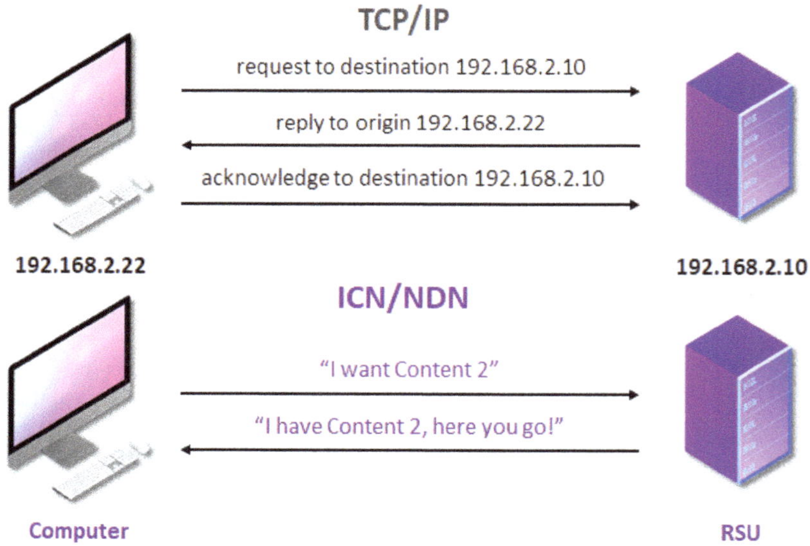

Figure 1. TCP/IP vs. NDN paradigm.

This type of model was not developed taking into account some requirements that are of interest to the IoT, such as the large number of connected devices, node mobility support, in-network caching or the security of the transmitted contents, but was rather a model that only considered the connection between a limited number of devices/computers. Some approaches have been proposed by many communities to solve these problems and needs, one of them is NDN, which places information dissemination at the centre of the network layer.

In NDN architectures, the node interested in some content (Consumer) requests it from its neighbors through its unique name without any indication of location (IP for example). The network nodes disseminate this request until it reaches the node where the desired content is located, a Producer or any other node with the Content on its cache. Thus, communication between nodes follows a Request-Response exchange model with native multicast support, which is quite efficient in mobile environments [3]. The Producer sends the requested Content using the opposite path to that taken by the Interest, following a receiver-oriented approach.

In an NDN architecture, the Consumer requests Contents without knowing the providing host, and the communication follows a receiver-driven approach, i.e., the data follow the reverse route of the request (Figure 2a). The system is then responsible for mapping the requested data and its location. An NDN architecture is characterized by the adopted approach on each of the following key functionalities: naming and name resolution, routing and forwarding, and caching [22]. NDN nodes use three structures in the Forwarding process (Figure 2b). The Forwarding Information Base (FIB) contains the names of a given Content, provided by the routing protocol, and associates them with the interfaces on which they were received. In addition to the FIB, NDN nodes use two other tables: Pending Interest Table (PIT) and Content Store (CS). Each PIT entry contains the name of the Interest packet and the respective interfaces through which it arrived at the node. On the other hand, the CS stores the data packets and works as a buffer managed by caching policies [3,23].

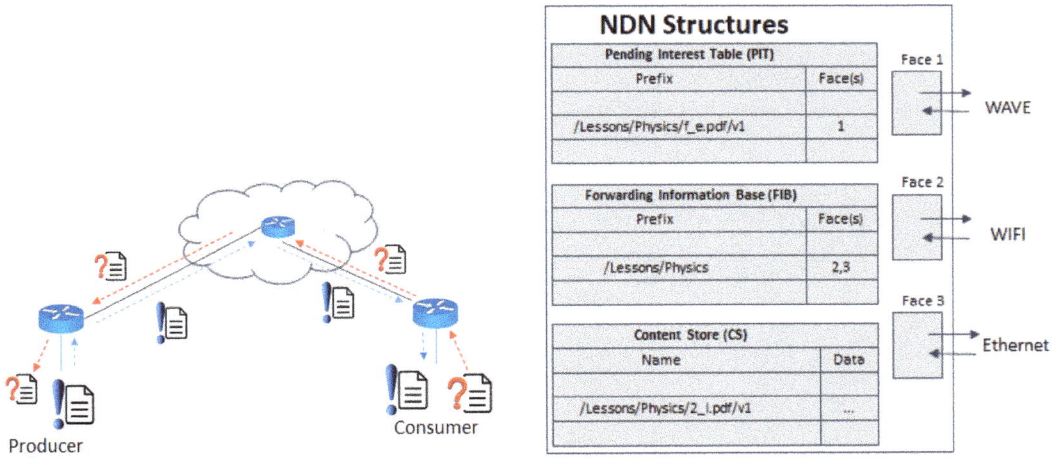

(**a**) (**b**)

Figure 2. NDN basic principle. (**a**) Consumer and Producer interaction. (**b**) Node's control structures.

In short, in Named Data Networks the contents are independent of their physical location, such as IP addresses, being accessible from anywhere on the network. In addition, caching mechanisms speed up their access by distributing and keeping copies in the network nodes. The forwarding and routing mechanisms of NDNs make the nodes completely independent, not requiring the previous formation of networks, clusters or platoons for communication to flow between them. These characteristics better meet requirements such as scalability, heterogeneity and dynamicity that are increasingly present in new applications for metropolitan networks.

Yet, despite all the expected benefits of solutions that address the limitations of Internet protocols, due to their widespread use, it is not expected to be simply phased out, but a gradual replacement over time [24]. In this direction, our option was to design an architecture that allows peaceful coexistence between the IP-based network and the NDN solution. For a better understanding of such solution for coexistence and shared use of devices, below we give an overview of the city-scale communication infrastructure in the ATCLL.

3.2. Aveiro Tech City Living Lab Infrastructure

Aveiro Tech City Living Lab (ATCLL) is an advanced, large-scale sensing, communications and computation infrastructure, spread throughout the city of Aveiro in Portugal. The ATCLL is composed of an advanced communications infrastructure and a platform for the analysis and management of urban data that, together, make it possible to provide an open and large-scale technological laboratory in the city. The access infrastructure is supported by state-of-the-art fiber technology, reconfigurable radio units, 5G-NR radio and 5G network services, aggregating and interconnecting a range of sensors and remote information collection units that span the entire area of the city. Buses and garbage collection vehicles have also been equipped with communication units with Wi-Fi, ITS-G5 and 5G technologies, mobility sensors (GPS, traffic radars, LiDARs, and video cameras) and environmental sensors (such as temperature, humidity, pollution). This set of sensors record mobility and environmental data, building a complete live map of these parameters in the city, and providing the required data for its management, such as traffic monitoring and safe driving systems.

As illustrated in the left side of Figure 3, network nodes are also installed in the vehicles and local boats to transmit mobility and environment data to the data processing centre. In the vehicular network supported by ATCLL, the vehicles are equipped with On-Board Units (OBUs) to establish a connection with the Road Side Units (RSUs) and with the other vehicles. OBUs and RSUs were implemented using PC Engines APUs (https://www.pcengines.ch/apu.htm accessed on 7 June 2022) (Photo on the right side of the Figure 4) using an Ubuntu Linux distribution, offering vehicle-to-vehicle and vehicle-to-infrastructure communication over IEEE 802.11p interfaces (WAVE with ITS-G5 support), Wi-Fi, Ethernet and cellular communication (ready to integrate 5G modules). APUs use CPU AMD GX-412TC with 4 cores, 4GB DRAM, 30GB HDD, and Debian GNU/Linux 11.

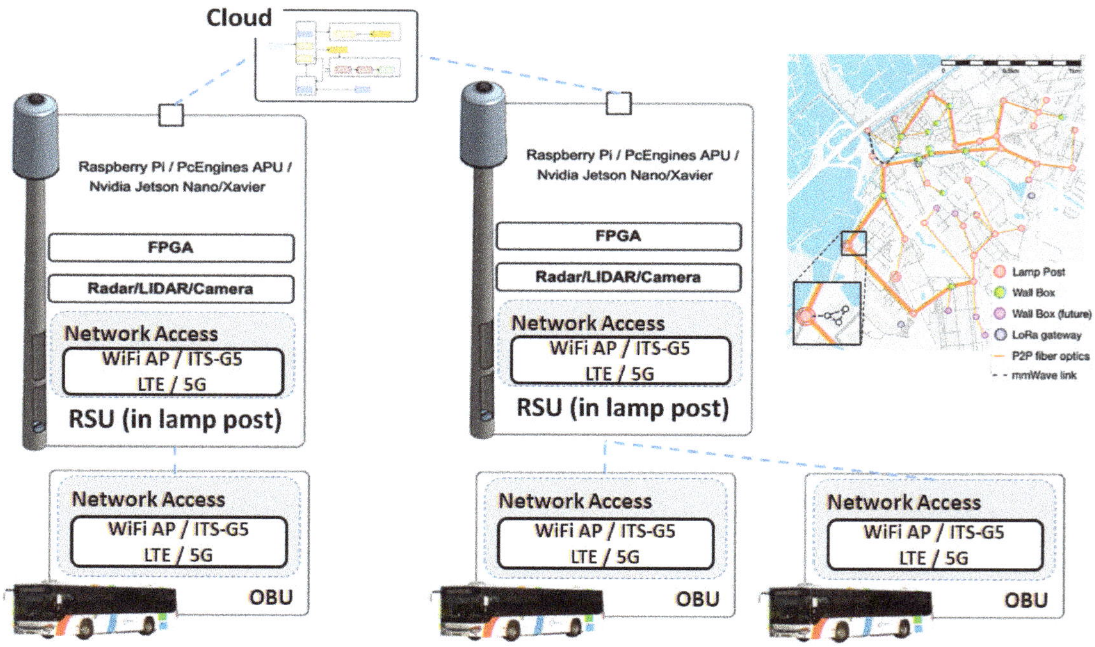

Figure 3. Aveiro Tech City Living Lab: communication architecture (on the **left**) and infrastructure's map (on the **right**).

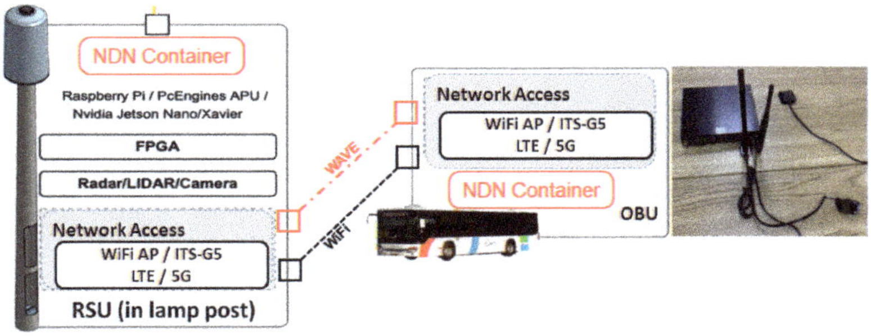

Figure 4. Deployment of NDN over ATCLL mobile communication units.

With restricted devices such as APUs, a scheme that allows dynamic publication of services on demand becomes essential. Thus, our solution for the deployment of NDN adopts a strategy of microservices made available through lightweight containers. This strategy allows us not only to deploy on demand, but also to include services migration to meet the mobility of OBUs, in addition to facilitating the configuration of devices on-the-fly as needed by the applications. Moreover, using lightweight containers, we isolate the NDN network stack from the TCP/IP network stack, ensuring complete independence from the operation of the networks while preparing the ATCLL for new network paradigms that may arise in the future.

To implement our NDN using containers we opted for Docker (https://www.docker.com/ accessed on 7 June 2022). Docker is a platform for service virtualization where containers operate in isolation using their own software, libraries and configuration files. We prepare our Docker base image with S.O. Ubuntu (https://ubuntu.com accessed on 7 June 2022) where we install our NDN implementation. From this base image, we created a container on each node (RSU/OBU), highlighted in red in Figure 4, and configured them to communicate with each other, both by IEEE 802.11 Wi-Fi and by ITS-G5, in order to constitute a parallel and independent network.

3.3. NDN in the ATCLL

Our NDN solution was specially designed for IoT infrastructures taking advantage of mobility prediction information. In our project, we developed improvements regarding the Forwarding [3], Caching and Energy-efficiency [23], as well as exploring mobility-aware prediction capabilities [4]. For all these developments, we have used the NDN implementation available in the ndnSIM network simulator, an open source NS-3-based NDN simulator [25]. The ndnSIM offers efficient features for simulating the device faces of the NDN nodes. However, these mechanisms do not work in real communication devices interfaces. In order to appraise our solution in a mobile-aware environment as is the case of the ATCLL infrastructure, first we had to port our adjustments to an adequate format to be installed in constrained mobile devices. In this way, we performed this port in a modular approach, with the aforementioned changes to Forwarding and Caching processes being implemented in a Single-Data request format at the current time, with mobility-awareness capabilities, while the publish-subscribe capabilities should be included in following versions.

3.3.1. NDN Configuration

The NDN daemon is in charge of performing all the necessary setup of the NDN stack modules that implement the NDN architecture, such as forwarding and caching tables (PIT, CS, FIB) as well as verifying which physical interfaces exist on the device and installing an interface of communication between them and the NDN stack to allow for traffic routing using NDN packets as opposed to the commonly known IP. The components of these modules can be configured via a control file which is analyzed by the daemon upon start, holding variables such as blacklisting of physical interfaces to be installed, among others.

Similarly, the NLSR routing protocol [10] behaves in analogous fashion, being in charge of calculating routing tables for each node in the infrastructure, thus, establishing their neighbourhood and guaranteeing that every Interest packet arriving at the edge of the network via the wireless medium can be properly forwarded towards their destination through a suitable NDN face, which can be controlled upon initialization via a control file. While all nodes in the network that wish to maintain NDN communication must have the NDN daemon installed and running, only fixed nodes composing the infrastructure are required to have the NLSR protocol running in order to route NDN traffic, due to their initial static neighbourhoods. Moreover, only certain nodes will act as either Producers or Consumers. This behavior is introduced via applications, which are designed specifically with a goal in mind, such as asking for a specific file, and must include proper packet creation in the NDN format and successful connection to the NDN daemon installed on the node. These applications are in charge of locally managing the content to send back to Consumers, acting as a Producer, and manage the Content received, when acting as a Consumer, using the daemon as a process to communicate with the outside world.

In its current state, the NDN platform does not intrinsically transmit traffic in the physical medium; instead, its network layer protocol makes use of the existing foundations set by IP which includes physical channels such as Ethernet wires and logical connections such as UDP or TCP tunnels over the existing Internet (Figure 5). To this extent, all communications between two direct nodes, as is the case with edge nodes and routers in the infrastructure, use UDP tunnelling to successfully route data (Interest or DAta packets) amongst them. On the other hand, when using the wireless medium, mainly between edge and mobile nodes, communications will take effect in a broadcast manner, in which all nodes will receive and transmit data without being directly influenced by its neighbours as expected from the case under study.

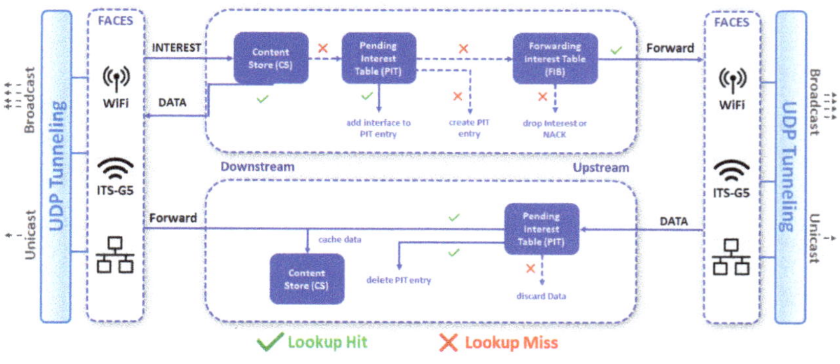

Figure 5. Forwarding process.

Figure 6 details the messages exchanged between a mobile Consumer and a mobile Producer, and the operations performed on each network element, when different radio access technologies are used.

3.3.2. Migration and Scalability

The adjustments mentioned previously result in a device capable of routing NDN traffic. In an ever-changing network environment, as is the case of our study, it is important that these capabilities can be replicated to various other devices in the network. This was achieved by the creation of a Docker image running the bare-minimum components in order to run the solution, which can be transferred seamlessly to other devices running the Docker daemon on-the-fly.

Such behavior promotes the scalability, as new devices can be prepared in a quick manner to attend to issues such as mobile node network connectivity, as well as preemptive caching to attend the efficient content delivery, among others.

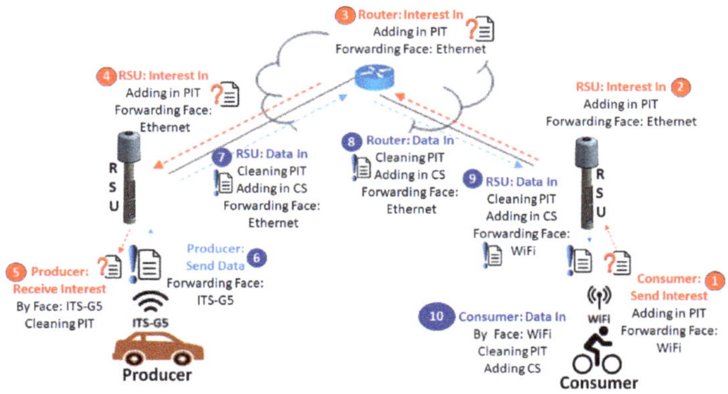

Figure 6. Messages exchanged following the NDN architecture and heterogeneous radio access technologies.

4. Experimentation Setup and Results

To evaluate the performance of our NDN implementation in a real-world wireless environment, three scenarios have been designed. The first scenario addresses real experimentation in the city of Aveiro, using the ATCLL, while the second and third scenarios occur in laboratory environment for a more fine-grain validation of a set of network characteristics. In the first one, we evaluate the content dissemination through mobile nodes traveling around the city using a single radio access technology (detailed in Section 4.1). In the second scenario we study the impact of multiple wireless interfaces in the performance of the NDN. In this case ITS-G5 and Wi-Fi are used simultaneously for the communication between OBUs and RSUs (Section 4.2). Finally, in Section 4.3 we study how the network characteristics (the Maximum Transmission Unit (MTU), and consequently, the chunk size) impact the performance of the NDN. All the scenarios consider that mobile Consumers communicate with a single Producer strategically placed in the infrastructure. Such interaction occurs using the single-Data request communication paradigm, which is traditional in NDN.

4.1. NDN in Single Wireless Radio Scenarios

To evaluate our solution in a real scenario, we have used a limited number of nodes belonging to the previously mentioned ATCLL infrastructure, consisting of four edge nodes (RSUs) capable of both wireless and wired communication, one backend router, which will act as a content Producer, and two mobile nodes, one posing as a vehicle moving through unpredictable traffic and the other as a pedestrian.

At first, the vehicle maneuver through an itinerary that connects with all four considered RSUs at some point in time, with the sole purpose of retrieving content that it is interested in, under the form of real files consisting of images and text documents, caching them for future use (Figure 7a). After 574 s, the vehicle completes its fist itinerary and move towards the pedestrian which is placed outside the coverage of any given RSU, meaning that its only point of contact with the contents is through the moving vehicle. At this moment, the pedestrian downloads the content from the vehicle's Content Store (Figure 7b). Then, after 1000 s the vehicle starts a second trip around the city to retrieve the remaining contents from the RSUs. This second trip is shorter than the previous, and the vehicle contacts only with three RSUs (Figure 7c). Finally, the vehicle ends its journey next to the pedestrian, providing another opportunity for the pedestrian to retrieve the remaining contents from the vehicle's cache (Figure 7d).

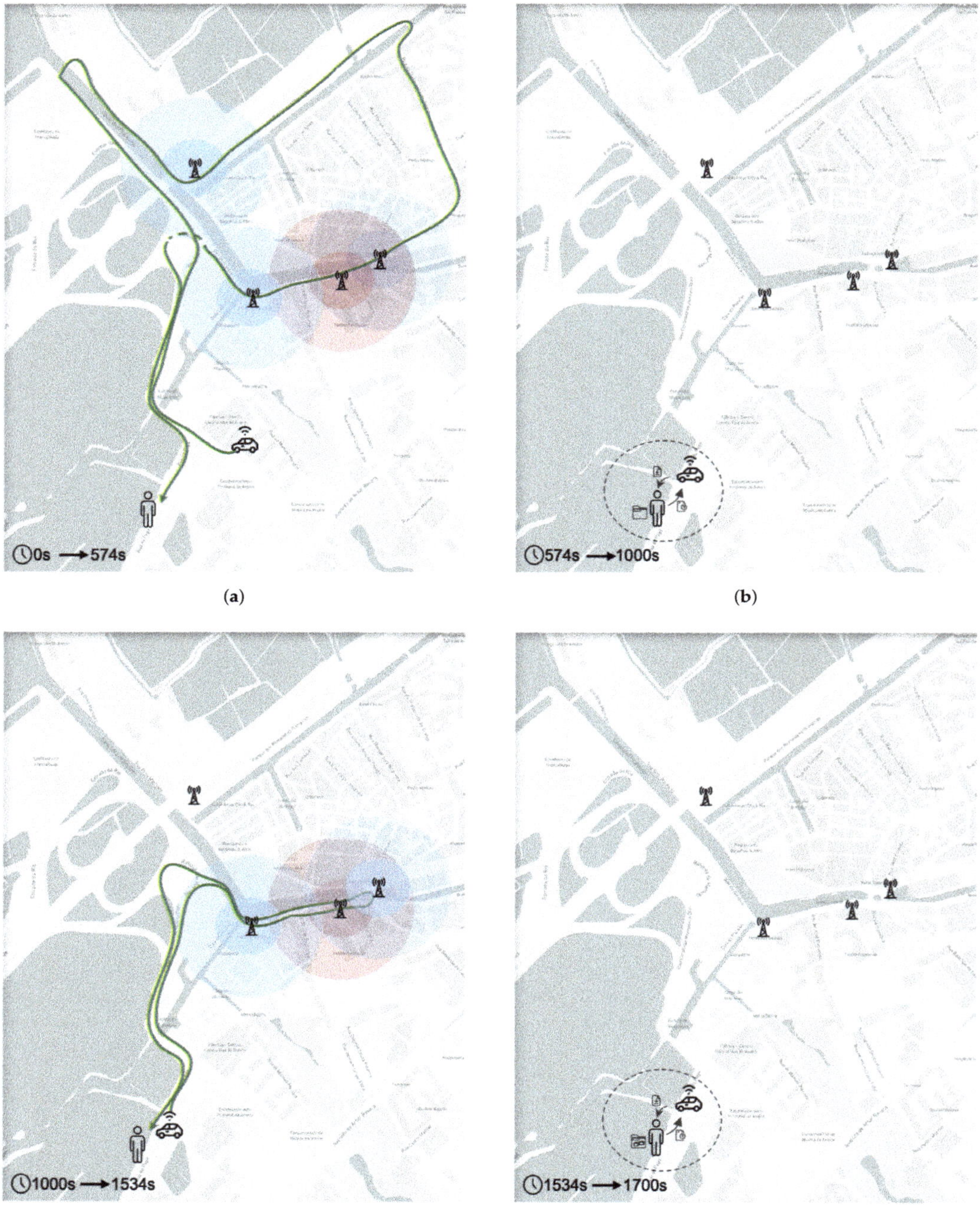

Figure 7. Route performed by the mobile node in the first experiment. (**a**) Between t = 0 s and t = 574 s the vehicle connects with four RSUs. (**b**) Between t = 574 s and t = 1000 s both vehicle and pedestrian are in communication range. (**c**) Between t = 1000 s and t = 1534 s the vehicle travels around the city, connecting with three RSUs. (**d**) Finally, the vehicle ends its journey next to the pedestrian.

Table 1 refers to the parameters considered in this scenario. Six files were considered with increasing size values which would be transmitted through ITS-G5 on the wireless section of the network. A single file will be under dissemination at a given time, meaning that only after the previous file is successfully downloaded, the next one will start its transmission. This approach aims to test two aspects: how the architecture behaves when multiple files are being delivered in consecutive manner, and the impact of the mobility in the delivery of large files, requiring the connection to multiple points of access to the infrastructure.

Table 1. City scenario parameters.

Experiment Duration	1700 s		
Interest Timeout	4s		
Chunk Size	1490		
Communication Type	ITS-G5		
Content	**Type**	**Size**	**Number of Chunks**
	.docx	16 KB	11
	.jpg	36 KB	23
	.jpg	68 KB	46
	.jpg	104 KB	71
	.pdf	8.2 MB	5760

The performance of the NDN solution in this scenario is discussed using the following indicators:

- **Interest satisfaction ratio:** measures the number of Interest packets arriving at the Producer, and how many of the Data packets, sent upon the reception of the Interest packet, arrive at the Consumer. This helps us to understand if the heuristics set on the Forwarding module, responsible for decisions about the forwarding of Interest and Data packets, are properly defined;
- **Network overhead:** a mobile network node moving at uncertain speed and at an unpredictable rate of contact with the network has a direct influence on the number of packets exchanged with the network mostly due to the inevitable packet loss that will occur, increasing the need for Interest retransmission. The number of Interests and Data packets traversing the network will be considered in this indicator;
- **Network contact time:** despite the strategic placement of RSUs, due to the unknown behavior of the vehicle traffic and weather conditions, it is inevitable that disconnections will occur. To ascertain the efficiency of content forwarding when contact is available, and the impact this may have in the results, we verify how long a mobile node is connected to the network, in seconds, throughout the experience.

Figure 8 presents the raw number of unique content chunks received by both types of mobile nodes in the network-vehicle and pedestrian. By interlacing the jumps in the amount of data received with the layout of the edge nodes in the network, we can observe the first itinerary of the vehicle (between 100 and 450 s in Figure 8a). Following this initial behavior, we see that the vehicle stops receiving content for a while, while the pedestrian starts to receive content, despite being isolated from the infrastructure, which proves that both mobile nodes made contact and the latter received the data through cache's vehicle. Then, the vehicle entering the second part of the circuit received the exclusive additional data, while the pedestrian retrieved all the additional content available in the second cycle. In general, we see that, although the vehicle could not retrieve the full extent of the available files, it was able to properly download data from RSUs placed around the city and deliver its content to other nodes via caching mechanisms.

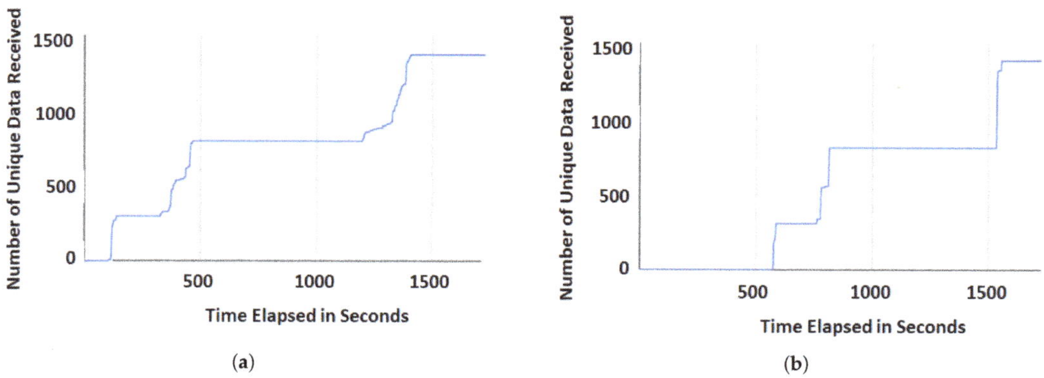

Figure 8. Number of unique data (chunks) received during the experiment. (**a**) Data downloaded by the vehicle. (**b**) Data downloaded by the pedestrian.

Figure 9a presents an overview of the number of packets (Interests and Data) transmitted and received by the vehicle. On the one hand, the number of packets shown by the red (satisfied Interests) and yellow (sent Data) bars shows that few retransmissions occurred when contact with the network was possible. On the other hand, the blue bar shows the small amount of replicated data due to the vehicle's simultaneous contact with several RSUs, mainly in the central part of the city where three edge nodes coexist. The results related to the mobile node representing the pedestrian were unconsidered as they took full advantage of the connection time to the vehicle, downloading all available cached contents with minimal retransmissions. Any Interest packets that resulted in timeouts and subsequent retransmission by the mobile nodes occur due to absence of contact with the network as opposed to issues regarding efficient Forwarding of the content through available paths. This can be addressed by adding more points of contact to the network.

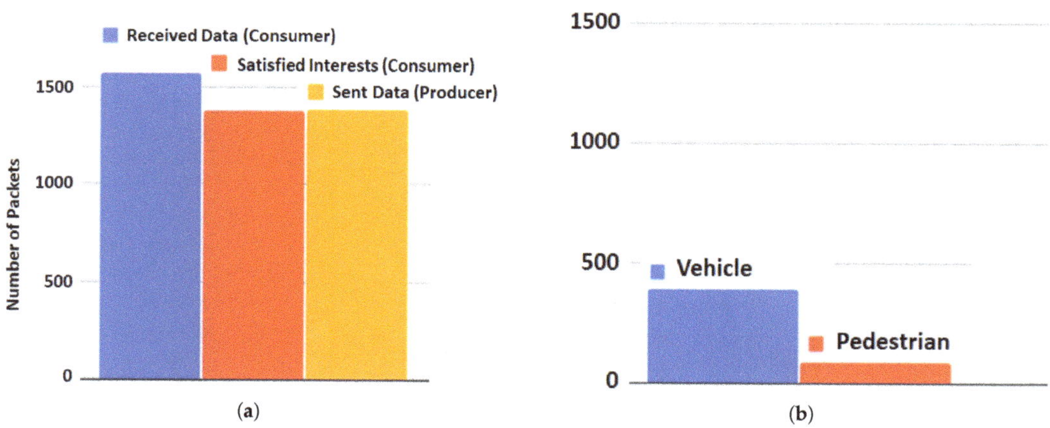

Figure 9. Network overhead. (**a**) Total amount of packets transmitted/received by the vehicle. (**b**) Total contact time between each mobile node and the infrastructure.

Figure 9b depicts the period of time each mobile node was successfully connected to the infrastructure, and thus able to send and receive content, compared to the total time elapsed. We can conclude that part of the reason the mobile nodes were not able to completely transfer all intended files was due to the reduced contact time with any cache node, totalling 26% for the vehicle (442 s over 1700 s) and 6% for the pedestrian (192 s over 1700 s), which only had the other mobile node as a source of content. Therefore, it can be implied that increasing the amount of contact time with the network could directly improve the Interest satisfaction ratio.

4.2. NDN Using Multiple Radio Wireless Interfaces

A vehicular network presents a number of connectivity challenges that can be improved by adding extra points of contact across the city. Therefore, we decided to evaluate the behavior of the NDN using ITS-G5 and Wi-Fi simultaneously in ad-hoc mode. These tests were executed in a small-scale environment, following the same network elements used in the previous experiment (OBUs and RSUs), although without mobility. The goal was to evaluate the behavior of the NDN when two or more wireless communication technologies are available. In this evaluation there was the transmission of a single file, divided in several chunks, throughout the testbed.

To further evaluate the impact of different application behaviors we explore two paradigms: pipelined and non-pipelined. In a pipelined scenario all chunks will be requested by the Consumer at once, virtually injecting a great mass of packets in the network in an effort to quickly retrieve the data. In a non-pipelined scenario the content chunks will be requested sequentially, with new chunks only being requested upon successful retrieval of the previous ones, hence guaranteeing a slower yet reliable delivery of the file requested.

The test itself was performed by splitting the request of a 6 MB sized file through both wireless communication technologies, meaning that each will be in charge of retrieving half of the content chunks of the file (Figure 10). Furthermore, we considered both Unicast and Multicast communication methods to further enrich the study of the behavior of both technologies. In this scenario the performance of the NDN solution is assessed by measuring the number of packets transmitted (Interests) and received (Data), the number of timeouts, and the average delay per chunk which refers to the time between the creation of an Interest packet for a given chunk and the successful retrieval of its corresponding Data, taking into account all necessary retransmissions.

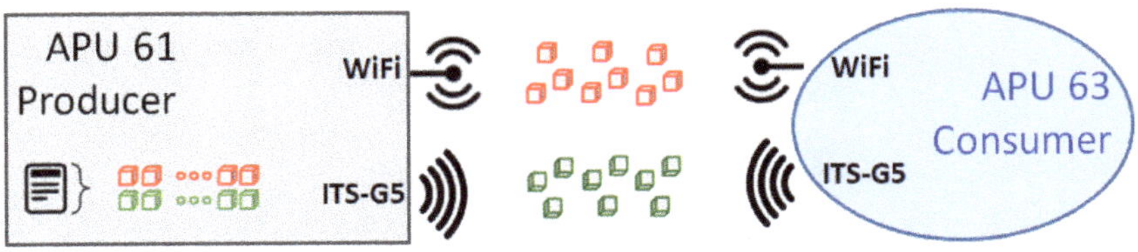

Figure 10. NDN in multiple wireless interfaces—requesting a 6 MB file using both ITS-G5 and Wi-Fi.

We started our tests by transmitting the entire file using only one of the technologies to have a baseline when managing the use of two technologies simultaneously. Table 2 shows the results on distributing the 6 MB file (1403 chunks for MTU = 4470).

Table 2. Non-pipelined results—full 6 MB file using ITS-G5 or Wi-Fi.

	ITS-G5				Wi-Fi			
	Time	**OutInts**	**InData**	**TimedOuts**	**Time**	**OutInts**	**InData**	**TimedOuts**
Multicast	94 s	1412	1403	9	303 s	1440	1403	37
	Average delay per chunk: 27.89 ms				Average delay per chunk: 57.07 ms			
	Time	**OutInts**	**InData**	**TimedOuts**	**Time**	**OutInts**	**InData**	**TimedOuts**
Unicast	25 s	1403	1403	0	82 s	1403	1403	0
	Average delay per chunk: 17.87 ms				Average delay per chunk: 58.09 ms			

Having established the performance benchmark for each technology, we split the file in half to be transmitted by one of the technologies so that we can assess the combined performance. Table 3 details the results obtained using the non-pipelined application for each technology.

Table 3. Non-Pipelined results—full 6 MB file using both ITS-G5 and Wi-Fi (3 MB to each interface).

	ITS-G5				Wi-Fi			
	Time	**OutInts**	**InData**	**TimedOuts**	**Time**	**OutInts**	**InData**	**TimedOuts**
Multicast	69 s	719	711	8	149 s	729	711	18
	Average delay per chunk: 28.16 ms				Average delay per chunk: 57.37 ms			
	Time	**OutInts**	**InData**	**TimedOuts**	**Time**	**OutInts**	**InData**	**TimedOuts**
Unicast	14 s	711	711	0	42 s	711	711	0
	Average delay per chunk: 18.41 ms				Average delay per chunk: 58.73 ms			

The summary of the results when both technologies are used simultaneously is shown in Table 4. Firstly, we can verify that using multicast brings about several delays in the delivery of the full sized file, no matter which technology is considered, in part due to the use of the shared wireless medium to forward data, which results in timeouts. Secondly, it is notable that using unicast methodology presents much better quality of results, both in terms of reliability and speed of content delivery, and allows us to conclude that ITS-G5 is much faster in delivering content as is to be expected. When we consider both technologies to share efforts in requesting a file to the network, improvements are noticeable in all aspects. On the one hand, by sharing the wireless medium in different bands it alleviates the collision problem, which implies greater reliability in the delivery of packets. On the other hand, by sharing the burden over two different physical interfaces we can minimize the occurrence of issues caused by the forwarding of several packets simultaneously, as is noticeable by the reduced number of packets attended by each of them. Ultimately, the delivery of the full sized file over two different technologies presented better results over using solely Wi-Fi and worse results over using solely ITS-G5, although these values may change when congestion starts to occur in the network due to the increased number of nodes taking part in the wireless medium communications.

Table 4. Total download time using multi-technology and non-pipelined application.

	Time	OutInts	InData	TimedOuts	Average Delay per Chunk
Multicast	149 s	1448	1422	26	42.76 ms
Unicast	42 s	1422	1422	0	38.57 ms

Tables 5 and 6 present the results regarding the use of the pipelined application in which all content requests are sent in bulk. We identify a significantly worse general performance when using a single communication technology, both under the unicast or multicast methods. Multicast methods suffered several Interest timeouts, due to the insufferable congestion introduced into the wireless medium, that resulted in increased periodicity to deliver the intended file. However, by using both Wi-Fi and ITS-G5 to forward the content, we were able to obtain much better performance under the form of reduced interest timeouts which ultimately resulted in lower times to retrieved the file successfully.

Table 5. Pipelined results—full 6 MB file using both ITS-G5 and Wi-Fi (3 MB to each interface).

	ITS-G5				Wi-Fi			
	Time	**OutInts**	**InData**	**TimedOuts**	**Time**	**OutInts**	**InData**	**TimedOuts**
Multicast	331 s	27,441	711	26,730	212 s	11,857	711	11,146
	Average delay per chunk: 516.93 ms				Average delay per chunk: 2957.79 ms			
	Time	**OutInts**	**InData**	**TimedOuts**	**Time**	**OutInts**	**InData**	**TimedOuts**
Unicast	19 s	1551	711	840	37 s	2650	711	1939
	Average delay per chunk: 1644.91 ms				Average delay per chunk: 3019.83 ms			

Despite this increase in performance, it is clear that trying to retrieve content in a bulk, as is the case of the pipelined application, can be detrimental to the behavior of the network, and thus presents significant challenges to overcome which are not entirely related to the architecture itself, but instead are affected by it.

Table 6. Total download time using multi-technology and pipelined application.

	Time	OutInts	InData	TimedOuts	Average Delay per Chunk
Multicast	331 s	39,298	1422	37,876	516.93 ms
Unicast	37 s	4201	4266	2779	2197.48 ms

The results show that there are grounds to believe that the use of several communication technologies to forward content simultaneously is beneficial to the well fare of architectures using NDN communications in mobile networks.

4.3. NDN According to the Network Characteristics

Over the course of our tests we have noticed that certain network conditions could have a significant impact in the NDN protocol operation, specially in the packet forwarding over the wireless medium. Such conditions can, in some cases, lead to a higher number of congestion situations. To this extent, we have performed a study on how different network characteristics affect the behavior of the NDN in an effort to further ascertain how better they can be utilized to minimize any negative consequences. The network parameters under study in this scenario are: (i) the Interest timeout, which represents the amount of time that a given node will wait for the corresponding Data packet to arrive; and (ii) the Maximum Transmission Unit (MTU), and consequently, the chunk size. We ran this experiment in the same laboratory environment used in the previous scenario. The main goal was the transmission of a 6 MB file from the infrastructure to the mobile node using wireless technologies, in an effort to obtain the lowest download time possible in a reliable manner. Finally, the indicators used to assess the performance of the NDN in this scenario are as follows:

- **Download duration:** measures the total time needed to download the Content starting from the moment that the first chunk is requested;
- **Number of timeout events:** every time a Consumer sends and Interest a timer is set, indicating the maximum amount of time available for the Data packet to arrive. If the timer expires, the Consumer retransmits the Interest packet, therefore increasing the network overhead.

Figure 11a shows that increasing the timeout value slightly can bring benefits in the time needed to download the file, but can also be detrimental or present no significant results as this value is increased to higher values. It is also clear that the NDN packet size has notable improvements: by simply doubling the size of the packet we were able to download the file 60% faster. Improvements to download time stagnate as we approach higher values to this metric, due to the fact that it is easier for longer sized packet to be lost in the wireless channel. A safe value to assume as a *sweet-spot* for our scenario would be an Interest timeout value of 6 seconds and a 4470 chunk-sized NDN packet, achieving one of the smallest download times while still leaving a window for any transmission errors to occur.

These values can be further attested by verifying the number of occurred timeouts for the different network characteristics, as depicted in Figure 11b. As the number of Interest timeouts increases, the number of timeouts naturally decreases because more time is given for the data packet to be delivered. However, it also needs to be taken into account that higher Interest timeout values will result in delayed Interest retransmissions, if needed, and consequently longer download times. To this extent, it is emphasized that a middle term must be achieved for a proper and reliable behavior of the network altogether.

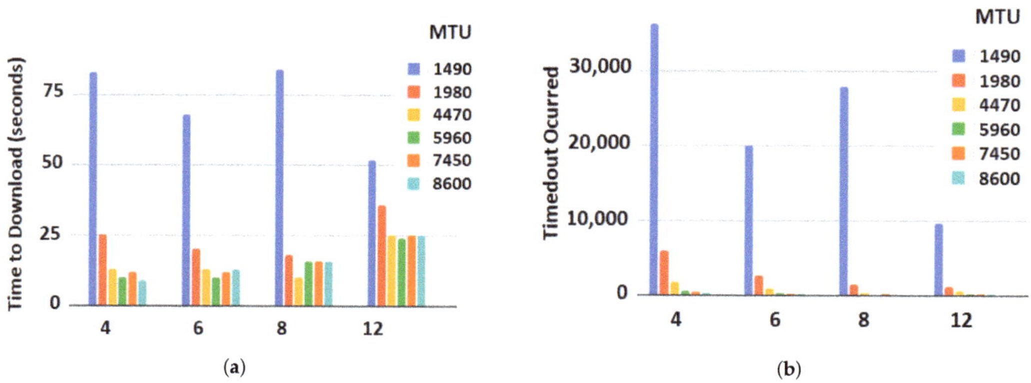

Figure 11. Download durations and number of timeout events for different MTUs and Interest timeouts. (**a**) Time to download file. (**b**) Timeouts.

5. Conclusions

In this article, we have discussed the implementation and performance of an NDN solution over a real smart city communication infrastructure. Our implementation, a container-based solution to be deployed in the network nodes, either mobile nodes or those belonging to the non-mobile infrastructure, was designed to allow the simultaneous operation of NDN and TCP/IP communications, offering facilities for on-the-fly network upgrades without compromising the network operation, and it allows the use and the performance evaluation of both communication paradigms simultaneously.

Evaluation tests, using real hardware equipment and mobility, have shown that our NDN solution, a modified version of the original NDN implementation with new heuristics about the forwarding, caching and support for publish-subscribe communications, is able to deliver the Content to the requested mobile users, directly from the infrastructure, or from the cache of other mobile nodes. In addition, we have shown that the NDN architecture can successfully explore the use of multiple communication interfaces for a faster and reliable Content retrieval. In the end, we analyzed how the performance of the NDN architecture is impacted by the network characteristics, experimenting different MTUs, that consequently lead to different chunk sizes.

As a future work, we aim to develop new heuristics for the management of the multiple communication interfaces by selecting the best interface for a given application according to the status of each interface. Another line of research includes the adaptation of the NDN solution to SDN-based environments and its assessment in the SDN-capable ATCLL communication infrastructure. Finally, to improve security in packet exchange, authentication mechanisms for NDN should be explored.

Author Contributions: Data curation, L.G.; Investigation, L.G., C.S. and M.L.; Supervision, C.S. and M.L.; Validation, L.G., C.S. and M.L.; Visualization, C.S. and M.L.; Writing—original draft preparation, L.G.; Writing—review and editing, C.S. and M.L. All authors have read and agreed to the published version of the manuscript.

Funding: This work was supported by the European Regional Development Fund (FEDER), through the Competitiveness and Internationalization Operational Programme (COMPETE 2020) of the Portugal 2020, Regional Operational Program of Lisbon (FEDER) and Foundation for Science and Technology, project InfoCent-IoT (POCI-01-0145-FEDER-030433), and by FCT/MEC through national funds and when applicable co-funded by FEDER – PT2020 partnership agreement under the project UIDB/50008/2020-UIDP/50008/2020.

Institutional Review Board Statement: Not applicable.

Informed Consent Statement: Not applicable.

Data Availability Statement: Not applicable.

Acknowledgments: We would like to thank Gustavo Amaral for the support on the experimentation.

Conflicts of Interest: The authors declare no conflict of interest

References

1. Ahlgren, B.; Dannewitz, C.; Imbrenda, C.; Kutscher, D.; Ohlman, B. A survey of information-centric networking. *IEEE Commun. Mag.* **2012**, *50*, 26–36. [CrossRef]
2. Zhang, L.; Afanasyev, A.; Burke, J.; Jacobson, V.; Claffy, k.; Crowley, P.; Papadopoulos, C.; Wang, L.; Zhang, B. Named Data Networking. *SIGCOMM Comput. Commun. Rev.* **2014**, *44*, 66–73. [CrossRef]
3. Gameiro, L.; Senna, C.; Luís, M. ndnIoT-FC: IoT Devices as First-Class Traffic in Name Data Networks. *Future Internet* **2020**, *12*, 207. [CrossRef]
4. Hernandez, D.; Gameiro, L.; Senna, C.; Luís, M.; Sargento, S. Handling Producer and Consumer Mobility in IoT Publish-Subscribe Named Data Networks. *IEEE Internet Things J.* **2022**, *9*, 868–884. [CrossRef]
5. Aboodi, A.; Wan, T.C.; Sodhy, G.C. Survey on the Incorporation of NDN/CCN in IoT. *IEEE Access* **2019**, *7*, 71827–71858. [CrossRef]
6. Jacobson, V.; Smetters, D.K.; Thornton, J.D.; Plass, M.F.; Briggs, N.H.; Braynard, R.L. Networking Named Content. In Proceedings of the 5th International Conference on Emerging Networking Experiments and Technologies, Rome, Italy, 1–4 December 2009; ACM: New York, NY, USA, 2009; pp. 1–12. [CrossRef]
7. Jacobson, V.; Smetters, D.K.; Thornton, J.D.; Plass, M.; Briggs, N.; Braynard, R. Networking Named Content. *Commun. ACM* **2012**, *55*, 117–124. [CrossRef]
8. Ioannou, A.; Coileain, F.O.; Collins, D.; Zhang, Y.; Chen, B. CLONE: An NDN Architecture for Content Distribution at Remote Tourist Sites—A TCP/IP and NDN Comparison. In Proceedings of the Proceedings of the 5th ACM Conference on Information-Centric Networking, Boston, MA, USA, 21–23 September 2018; Association for Computing Machinery: New York, NY, USA, 2018; pp. 212–213. [CrossRef]
9. Guimarães, P.H.V.; Ferraz, L.H.G.; Torres, J.V.; Mattos, D.M.F.; Andrés F., M.P.; Andreoni, L.M.E.; Alvarenga, I.D.; Rodrigues, C.S.C.; Duarte, O.C.M.B. Experimenting Content-Centric Networks in the future internet testbed environment. In Proceedings of the 2013 IEEE International Conference on Communications Workshops (ICC), Budapest, Hungary, 9–13 June 2013; pp. 1383–1387. [CrossRef]
10. Wang, L.; Lehman, V.; Mahmudul Hoque, A.K.M.; Zhang, B.; Yu, Y.; Zhang, L. A Secure Link State Routing Protocol for NDN. *IEEE Access* **2018**, *6*, 10470–10482. [CrossRef]
11. Mick, T.; Tourani, R.; Misra, S. LASeR: Lightweight Authentication and Secured Routing for NDN IoT in Smart Cities. *IEEE Internet Things J.* **2018**, *5*, 755–764. [CrossRef]
12. Papadopoulos, C.; Shannigrahi, S.; Afanaseyv, A. In-Vehicle Networking with NDN. In Proceedings of the 8th ACM Conference on Information-Centric Networking, Paris, France, 22–24 September 2021; Association for Computing Machinery: New York, NY, USA, 2021; pp. 127–129.
13. De Sena, Y.A.B.L.; Dias, K.L. Native versus Overlay-based NDN over Wi-Fi 6 for the Internet of Vehicles. In *Lecture Notes of the Institute for Computer Sciences, Social Informatics and Telecommunications Engineering*; Springer: Cham, Switzerland, 2021; Volume 424. [CrossRef]
14. Abane, A.; Daoui, M.; Bouzefrane, S.; Muhlethaler, P. NDN-over-ZigBee: A ZigBee support for Named Data Networking. *Future Gener. Comput. Syst.* **2019**, *93*, 792–798. [CrossRef]
15. Yang, W.; Wu, F.; Tian, K. High Performance Adaptive Video Streaming Using NDN WLAN Multicast. In Proceedings of the 8th ACM Conference on Information-Centric Networking, Paris, France, 22–24 September 2021; Association for Computing Machinery: New York, NY, USA, 2021; pp. 42–51.
16. Wu, F.; Yang, W.; Ren, J.; Lyu, F.; Ding, X.; Zhang, Y. Adaptive Video Streaming Using Dynamic NDN Multicast in WLAN. In Proceedings of the IEEE INFOCOM 2020—IEEE Conference on Computer Communications Workshops (INFOCOM WKSHPS), Toronto, ON, Canada, 6–9 July 2020; pp. 97–102. [CrossRef]
17. Li, M.; Pei, D.; Zhang, X.; Zhang, B.; Xu, K. NDN Live Video Broadcasting over Wireless LAN. In Proceedings of the 2015 24th International Conference on Computer Communication and Networks (ICCCN), Las Vegas, NV, USA, 3–6 August 2015; pp. 1–7. [CrossRef]
18. Li, M.; Pei, D.; Zhang, X.; Zhang, B.; Xu, K. Interest-suppression-based NDN live video broadcasting over wireless LAN. *Front. Comput. Sci.* **2017**, *11*, 675–687. [CrossRef]
19. Ghasemi, C.; Yousefi, H.; Zhang, B. Internet-Scale Video Streaming over NDN. *IEEE Netw.* **2021**, *35*, 174–180. [CrossRef]
20. Bouk, S.H.; Ahmed, S.H.; Kim, D.; Song, H. Named-Data-Networking-Based ITS for Smart Cities. *IEEE Commun. Mag.* **2017**, *55*, 105–111. [CrossRef]
21. Rainer, B.; Petscharnig, S. Challenges and Opportunities of Named Data Networking in Vehicle-To-Everything Communication: A Review. *Information* **2018**, *9*, 264. [CrossRef]
22. Nour, B.; Sharif, K.; Li, F.; Biswas, S.; Moungla, H.; Guizani, M.; Wang, Y. A survey of Internet of Things communication using ICN: A use case perspective. *Comput. Commun.* **2019**, *142–143*, 95–123. [CrossRef]

23. Marques, D.; Senna, C.; Luís, M. Forwarding in Energy-Constrained Wireless Information Centric Networks. *Sensors* **2022**, *22*, 1438. [CrossRef] [PubMed]
24. Conti, M.; Gangwal, A.; Hassan, M.; Lal, C.; Losiouk, E. The Road Ahead for Networking: A Survey on ICN-IP Coexistence Solutions. *IEEE Commun. Surv. Tutor.* **2020**, *22*, 2104–2129. [CrossRef]
25. Mastorakis, S.; Afanasyev, A.; Zhang, L. On the Evolution of NdnSIM: An Open-Source Simulator for NDN Experimentation. *SIGCOMM Comput. Commun. Rev.* **2017**, *47*, 19–33. [CrossRef]

future internet

MDPI

Article

An Efficient Location-Based Forwarding Strategy for Named Data Networking and LEO Satellite Communications

Pablo Iglesias-Sanuy [1], José Carlos López-Ardao [2,*], Miguel Rodríguez-Pérez [2], Sergio Herrería-Alonso [2], Andrés Suárez-González [2] and Raúl F. Rodríguez-Rubio [2]

1 WINLAB, Rutgers University, North Brunswick, NJ 08902, USA
2 atlanTTic Research Center for Telecommunication Technologies, Universidade de Vigo, 36270 Vigo, Spain
* Correspondence: jardao@det.uvigo.es

Abstract: Low Earth orbit (LEO) satellite constellations are increasingly gaining attention as future global Internet providers. At the same time, named data networking (NDN) is a new data-centric architecture that has been recently proposed to replace the classic TCP/IP architecture since it is particularly well suited to the most common usage of the Internet nowadays as a content delivery network. Certainly, the use of NDN is especially convenient in highly dynamic network environments, such as those of next LEO constellations incorporating inter-satellite links (ISL). Among other native facilities, such as inbuilt security, NDN readily supports the mobility of clients, thus helping to overcome one of the main problems raised in LEO satellite networks. Moreover, thanks to a stateful forwarding plane with support for multicast transmission and inbuilt data caches, NDN is also able to provide a more efficient usage of the installed transmission capacity. In this paper, we propose a new location-based forwarding strategy for LEO satellite networks that takes advantage of the knowledge of the relative position of the satellites and the grid structure formed by the ISLs to perform the forwarding of NDN packets. So, forwarding at each node is done using only local information (node and destination locations), without the need of interchanging information between nodes, as is the case with conventional routing protocols. Using simulation, we show that the proposed forwarding strategy is a good candidate to promote the efficient and effective future use of the NDN architecture in LEO satellite networks.

Keywords: NDN; LEO; ICN; routing; forwarding

Citation: Iglesias-Sanuy, P.; López-Ardao, J.C.; Rodríguez-Pérez, M.; Herrería-Alonso, S.; Suárez-González, A.; Rodríguez-Rubio, R.F. An Efficient Location-Based Forwarding Strategy for Named Data Networking and LEO Satellite Communications. *Future Internet* **2022**, *14*, 285. https://doi.org/10.3390/fi14100285

Academic Editor: Luis Javier Garcia Villalba

Received: 5 August 2022
Accepted: 27 September 2022
Published: 29 September 2022

1. Introduction

The TCP/IP architecture, originally proposed to provide reliable communication between pairs of nodes in the Internet [1], is a host-centric architecture in which packets only identify communication endpoints using IP addresses. However, the general purpose of the Internet has changed significantly since its origin, and today's applications and services use it mainly as a distribution network [2]. Thus, the need to identify the ends of communication in IP networks has become nothing but a shortcoming of the system. Information-centric networks (ICNs) adopt a different paradigm for communication on the Internet. In ICNs, the current host-centric TCP/IP architecture is replaced with a new data-centric architecture that brings data into focus. One of the most popular ICN proposals is the named data networking (NDN) project, proposed in 2010 and funded by the NSF Future Internet Architecture Program. This architecture changes the dynamics of network interaction from sending a packet to a given destination to fetching data by addressing its name, regardless of its location. NDN has some relevant advantages over IP. Firstly, the data identity is decoupled from its location, and hence, the data can be easily moved. Additionally, any node can reply to a data request if it has a copy of the desired data. This means that the NDN architecture natively provides transparent in-network caching for content providers. In addition, it separates consumer mobility from data transmission due to its consumer-driven interaction.

At the same time, there has been renewed interest in low Earth orbit (LEO) satellite networks to provide global Internet connectivity with low latency and high bandwidth. These new satellite networks can play an important role in multiple applications, such as fleet management, maritime or aeronautical tracking systems, and even in disaster relief, since they guarantee connectivity in remote areas where it is difficult or expensive to install ground infrastructure. Indeed, they can also be a complement to fifth generation (5G) networks, either as an additional radio access technology (RAT) or as a 5G core (5GC) backhaul; they will surely compete with large distance connectivity providers, such as submarine cable. SpaceX Starlink is currently the most advanced satellite Internet provider. As of July 2022, SpaceX Starlink has a fully operational constellation of nearly 2600 satellites and more than 500,000 estimated subscribers, and it has applied for permission to deploy up to 42,000 satellites [3]. Its strongest competitors are Amazon's Project Kuiper, OneWeb, Telesat's Lightspeed, Boeing and Kepler Communications, but SpaceX is the only corporation that already has a fully operational network.

The possibility of interconnecting LEO satellites through inter-satellite links (ISLs) will be one of the strongest advantages of these constellations, though this feature has not been used by commercial deployments yet. ISLs are laser links, also known as free space optical (FSO) links, that take advantage of vacuum conditions for lower latency, higher data rates, smaller antenna sizes, narrower beams and lower power requirements. The use of ISLs between neighboring satellites results in a grid topology that is particularly convenient for maintaining a well-structured satellite network, the core idea of the work in this paper.

At present, it is not clear which will be the networking design of these large constellations of LEO satellites, but the idea of using NDN in them is gradually gaining attention since this architecture is especially convenient for highly dynamic network environments [4,5]. Its strongest advantage over IP is the mobility management, which is one of the biggest challenges of LEO constellations, given the orbital speed of these satellites (LEO satellites need about 90 min to orbit the Earth [6]). Additionally, NDN provides a stateful forwarding plane with support for multicast transmission and inbuilt data caches, thus resulting in a more efficient usage of the installed transmission capacity.

The main goal of this paper is to contribute to unveil the possibilities that emerge from applying the NDN architecture to large LEO satellite constellations using ISLs. Some open problems in these networks are related with forwarding and routing. In NDN, the routing plane is in charge of obtaining available routes, while the forwarding plane and the *strategy* layer make decisions about the preference and usage of routes based on their performance/status. However, routing tables in NDN may consume more memory space and bandwidth in comparison to common IP routing tables [7].

In this paper, we propose a new location-based forwarding strategy for LEO satellite networks that takes advantage of the knowledge of the relative position of the satellites and the grid structure formed by the ISLs to perform the forwarding of NDN packets. So, forwarding at each node is done using only local information (node and destination locations), without the need of interchanging information between nodes, as is the case with conventional routing protocols.

Simulation results show that this forwarding strategy results in high network utilization when we try to use the highest available bandwidth for multiple flows between several consumers and a producer, and so, it is a good candidate to promote the adoption of the NDN architecture in LEO satellite networks.

This paper is organized as follows. Section 1 introduces the work in this paper. Section 2 is a review of the related work found in the literature. Section 3 describes the general architecture of an NDN network. Section 4 introduces the forwarding problem and presents our new GeoTag-based forwarding strategy. Section 5 evaluates the proposed solution. Finally, Section 6 presents some conclusions and future lines of research.

2. Related Work

The problem of routing in LEO satellite networks with ISLs has been widely addressed in the literature, and it is now commonly accepted that adaptive routing is an absolute requirement to optimize network utilization [8]. Thus, the work in [9] presents and evaluates different adaptive routing schemes in a constellation of LEO satellites, also in comparison to static routing. The most important conclusion in this work is that, although taking network load into account can improve both network utilization and QoS, it also makes the routing architecture more complex. Another relevant conclusion of this work is that isolated traffic adaptive routing (i.e., using only locally available information) is an interesting and efficient alternative to more complex distributed routing schemes.

However, the use of traffic adaptive routing schemes in regularly meshed networks with many alternative paths, as it is our case, may cause traffic load oscillations between alternative paths that are particularly inconvenient under heavy traffic load conditions [10]. For this reason, many works propose adaptive multipath routing mechanisms that implement load balancing [11–14]. Finally, Ref. [15] surveys other advanced routing algorithms, such as virtual topology routing and virtual node routing.

In the case of NDN, although there are many works about routing in NDN networks [16–18], to the best of our knowledge, there are only a few works addressing the problems related to NDN routing in LEO satellite networks. In particular, we have found only two works about this type of networks, and they are mainly focused on mobility management due to satellite handovers [19,20].

In this paper, we present a multipath location-based forwarding strategy that benefits from some NDN features to completely avoid the need of interchanging information between nodes, as is the case with conventional adaptive routing protocols in large LEO satellite networks (and so the scalability and load oscillation problems related to them). In particular, our forwarding strategy takes advantage of the knowledge of the relative positions of the satellites and the grid structure formed by the ISLs to choose the next-hop along a minimum-hop path. When several minimum-hop paths are possible at an NDN node, the next-hop is locally chosen randomly in order to provide the desired load balancing.

The work in [21] also proposes a location-based mechanism for IP-based LEO satellite networks that uses the geographical position of the destination to search for the satellite node serving it. However, this algorithm just chooses a single path to minimize the physical distance between the source and the destination, instead of choosing any path minimizing the number of hops between the edge nodes, as in our proposal. In relation to this, it is very important to note that a very recent work [22] demonstrated that all the shortest-distance paths belong to the set of minimum-hop paths in LEO constellations as long as the inclination is not too large (i.e., less than 68°) or, alternatively, the phasing offset is large enough.

3. NDN Architecture

The most distinctive feature of NDN is the use of names to identify data rather than locations. While IP names the network location of the host to send the message to, NDN names the content of the message itself, without specifying any location. In the NDN architecture, every data chunk has a name, and the network layer uses data names to communicate. NDN communication consists of a pull-based model, i.e., the receiver sends an *Interest* packet that contains the name of the requested data chunk, and fetches one *Data* packet back from either the original data producer, or from a node cache. A clear advantage of this architecture is that any node that has the requested data packet can reply to the interest, without needing an overlay solution that changes the addresses of the remote hosts, like in TCP/IP.

With respect to the forwarding of NDN packets, each NDN node maintains three data structures: the forwarding information base (FIB), the pending interest table (PIT) and the content store (CS). Thus, once an interest packet is received at an interface of an NDN node, the following procedure is executed:

1. The NDN node checks the CS for a copy of the requested data. If there are matching data in the CS, they are sent following the reverse path of the interest packet.
2. Otherwise, the NDN node searches the PIT to find any prior interest with the same name from a different input interface. If there is a matching PIT entry, the incoming interface is added to the list of incoming interfaces for that entry.
3. In the case of no match, neither in the CS nor in the PIT, the NDN node inserts a new entry for the name of the interest packet into the PIT, and the interest forwarding strategy decides whether, when and where to forward the incoming interest, among all the possible outgoing interfaces, according to the FIB for a given name or, alternatively, using a *forwarding hint*. (When a name cannot be announced globally, the interest packet can carry the ISP that hosts the producer. That is referred to as a *forwarding hint*. So, when an NDN node does not find a matching prefix for the name in an interest packet, it uses the forwarding hint to forward it.)

Data packets follow the reverse path of their corresponding interest packets, hop by hop, to reach all the requesters. When a data packet arrives at an NDN node, it checks the PIT to get the incoming interfaces for that name, and sends a copy through every one of them. Then, that PIT entry is removed, and the NDN node optionally saves a copy of the data in its CS. The CS is optional and works like a cache, that is, it is used to store the most popular data chunks traversing the node. If no matching entry exists in the PIT, the data packet is dropped.

NDN networks can run routing protocols to announce the reachability of data names, like IP does with IP addresses. The FIB also works just like the traditional IP forwarding table, but using NDN names instead of IP prefixes. Routing protocols are usually responsible for populating the FIB with entries relating names to output interfaces.

4. GeoTag-Based Forwarding Strategy

As previously explained, an interest forwarding strategy is the logic that decides in each NDN node whether, when and where to forward an incoming interest, among all the possible outgoing interfaces, according to the FIB for a given name or, alternatively, a forwarding hint. Already existing strategies are the best route strategy, that follows the longest prefix match principle; the multicast strategy; and the self-learning strategy, which broadcasts interests for the first time to learn a single path toward data. In [23], the authors surveyed the main forwarding strategies proposed in the literature for NDN-based wireless networks.

In this paper, we propose a new location-based forwarding strategy for NDN-based LEO satellite networks that takes advantage of the knowledge of the relative position of the satellite nodes and the grid structure formed by the ISLs to perform the forwarding of the interest packets without the need of interchanging information between nodes, as is the case with conventional routing protocols. We consider the most common scenario where each satellite keeps four ISLs connected to its four immediate neighbors: two intra-plane ISLs with the immediate previous and next satellites in the same orbital plane, and two inter-plane ISLs with the closest satellites of both immediate adjacent planes, as shown in Figure 1. That is, for our purposes, we can consider the satellite network as a grid covering the Earth and, therefore, we propose to identify each satellite by two Cartesian coordinates. The horizontal coordinate represents the ordinal number of the orbital plane while the vertical one represents the number of the satellite within the orbital plane, as shown in Figure 2.

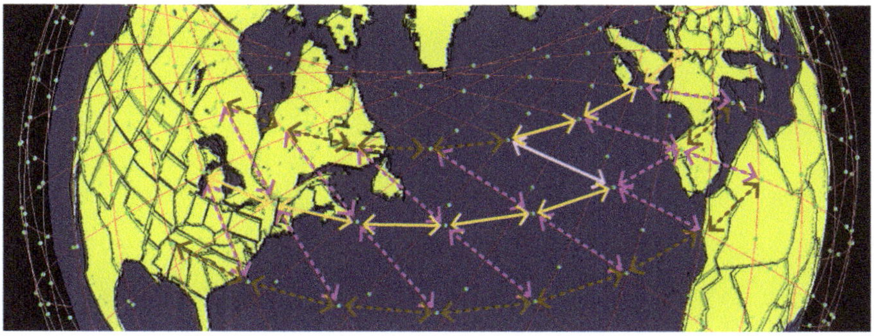

Figure 1. Part of a constellation showing the ISLs between neighboring satellites. Orange arrows show links between adjacent planes and pink arrows show links between consecutive satellites in the same orbital plane. The non-dashed links show a possible path between Western Europe and the Great Lakes region in the USA.

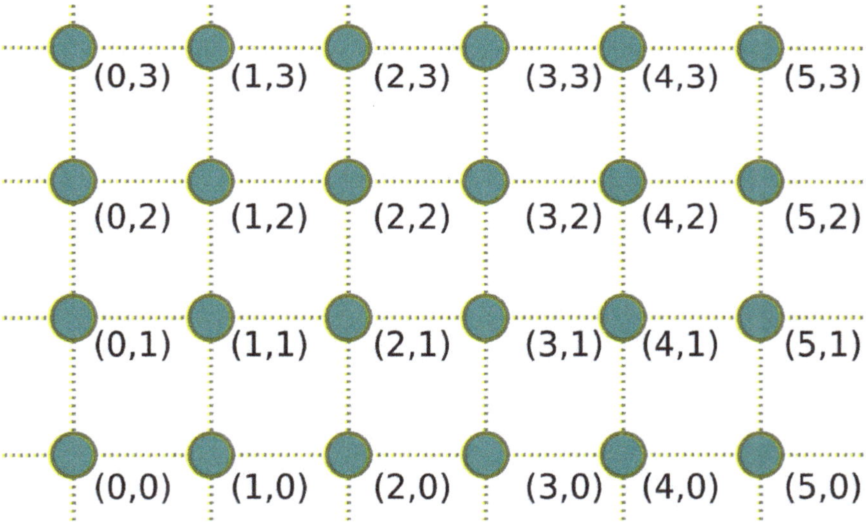

Figure 2. Representation of the grid structure showing both the ISL links and the satellite notation.

Ground nodes that generate interest packets are referred to as *consumers*, and ground nodes that have the original data are called *producers*, although recall that it is also possible to maintain copies at intermediate NDN node caches. Both consumer and producer ground nodes must choose periodically the *best* (closest) access satellite to connect to so that they direct their beams to only one satellite for upstream communication, whereas downstream messages come from only that satellite. However, the constant mobility of LEO satellites forces the ground nodes to perform a handover every five minutes approximately. As pointed out in [24], there are two alternatives to solve the producer mobility problem in NDN: either the mobile producer makes the data available at a reachable named location, or it reports constantly its point of attachment (PoA), that is, the address of the satellite to which it is connected to, similarly to how the DNS system works for IP. In this paper, we choose this second option and, thus, the PoA will be the two Cartesian coordinates of this satellite.

The NDNLPv2 (NDN Link Adaptation Protocol version 2) protocol [25] offers the possibility to send a GeoTag into the interest packets, that is, a predefined 3-byte header that is meant to carry Cartesian 3D coordinates (GPS coordinates, usually). We propose using the first two bytes to encode the orbital plane and the ordinal position within the orbital plane as two Cartesian coordinates, while the third remaining byte is reserved for possible future expansion. In this way, the consumer, using the name of the desired data, obtains the PoA of the satellite connected to the producer (i.e., its two Cartesian coordinates), and sends these coordinates into the GeoTag field of the corresponding interest packet. Then, it forwards this packet to its own access satellite.

When an incoming interest arrives at a satellite node and it must be forwarded, the forwarding strategy must choose one of the five possible next-hop interfaces contained in the FIB: the four ISL links to the neighboring satellites (port, starboard, bow and stern) and the link to the ground. For that, the GeoTag of the interest is analyzed as follows:

- If the GeoTag matches with the address of the satellite, then the interest is forwarded to ground.
- Otherwise, the interest must be forwarded to another satellite. The next-hop satellite is obtained based on its own coordinates and the coordinates included in the GeoTag, choosing as the *best* next hop the one with the closest index of the bi-dimensional address relative to its own address. That is, if the GeoTag of the consumer is (x_c, y_c) and the GeoTag of the output satellite (connected to the producer) is (x_p, y_p):
  - When $|x_p - x_c| > |y_p - y_c|$, the next hop is in the horizontal direction $\Longrightarrow$ The next-hop is the satellite on the starboard side if $x_p > x_c$, or the satellite on the port side if $x_p < x_c$.
  - When $|x_p - x_c| < |y_p - y_c|$, the next hop is in the vertical direction $\Longrightarrow$ The next-hop is the satellite on the bow if $y_p > y_c$, or the satellite on the stern if $y_p < y_c$.
  - When $|x_p - x_c| = |y_p - y_c| \neq 0 \Longrightarrow$ The direction of the next-hop (horizontal or vertical) is chosen randomly in order to provide load balancing.

Once the forwarder satellite has obtained the next hop satellite, it forwards the interest packet to it. Eventually, the interest packet will reach the producer, and this will deliver the data packet without the GeoTag, just following the PIT entries back to the consumers, as usual. It is obvious that if caches are used at the satellite nodes, data can be sent back to the consumers from an intermediate node without the need to reach the producer. It is also possible that there are several producers for a given name. In this case, the consumer will choose the exit satellite closest to its access satellite.

Note that, if we have a consumer with coordinates (x_c, y_c) that obtains data from a producer with coordinates (x_p, y_p), the problem is equivalent to that of routing interest packets along a grid from a source edge at $(0,0)$ to the destination edge at (W, H), where $W = |x_p - x_c|$ and $H = |y_p - y_c|$. Clearly, any path between these edges chosen by our forwarding strategy will comprise a minimum number of hops, and this number is $W + H$. Without loss of generality, let us assume that $W \geq H$ $(H \geq W)$. In this case, the path followed by the interest packet between the edges consists of $W - H$ movements to the right ($H - W$ upward movements) and then a random sequence of H (W) rightward movements and other H (W) upward movements in an arbitrary order.

As an example, Figure 3 shows the minimum-hop paths, represented with red (yellowish) lines, between a consumer at $(0,0)$ and two producers at $(5,3)$ and $(2,3)$.

We could slightly modify our strategy to use any other two disjoint minimum-hop paths, but as the number of consumers is much higher than the number of producers, load balancing is much more necessary close to the producers due to the aggregated data traffic from them. Moreover, close to the consumers, we prefer to use a single path in order to benefit from the eventual use of caches at the NDN nodes. In this way, our mechanism performs load balancing from the producer between the two paths around the diagonal, and then uses a single path to the consumers.

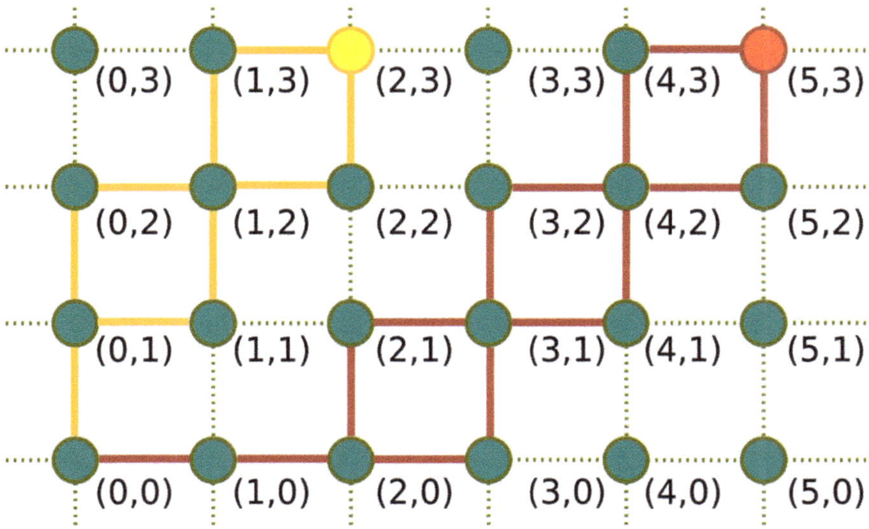

Figure 3. Minimum-hop paths between edge nodes.

5. Results

We tested the proposed GeoTag-based forwarding strategy using ndnSIM [26], the open-source NS-3 module that implements the NDN communication model, and a new GeoTag-based forwarding strategy module we developed for it [27].

The LEO constellation was configured with 30 orbital planes of 30 satellites, an inclination angle of 60° and an altitude of 400 km. Regarding the positions of the ground nodes, we chose different longitudes between latitudes 35° and 45° since this area corresponds to a highly populated region in the Northern Hemisphere, where the proposed constellation should provide good coverage. However, for the longitude, we avoided differencing between land and sea masses to keep the scenario simple. Additionally, the tracking period was established to 20 s.

We simulated a variable number of consumers executing the *ConsumerPcon* application. With this application (analogous to iPerf [28] in TCP/IP), we can measure the network utilization since it implements a TCP-like congestion control algorithm that allows the flows between the consumers and the producer to use all the available bandwidth. Note that, in this scenario, the rate of interest packets issued per second is not fixed, but is constantly changed by the chosen window adaptation algorithm. We also configured the consumers to start sending the interest packets at different instants to prevent the synchronization of the window adaptation algorithms. Regarding the producer, it is configured to always reply to the interest packets with data packets of a fixed size (1050 bytes). Finally, to keep things simple, we configured all the links in the simulation experiments with the same capacity.

For testing the performance of our forwarding strategy, we firstly measured the percentage of bandwidth usage at the link from the producer to its satellite node. Each test comprises 10 random executions of the same scenario, applying a random position displacement to the ground nodes. Initially, we simulated a scenario where all the consumers send interests for the same data name from the producer.

Figure 4 shows the average link usage at the producer (with 95% confidence intervals) for 4, 8, 16 and 32 consumers. We can see that our forwarding strategy achieves an average link usage from the producer of 90% with 4 consumers, but this percentage decreases as the number of consumers increases, falling to 70% for 32 consumers. The reason for this decrease is that, as the number of consumers increases, the probability that a NDN node receives multiple interests for the same data increases. In this case, the satellite node will only forward one interest packet, thus saving some bandwidth at the producer since all these pending interests will be fulfilled with just a single data packet. Note that, under these conditions, the links to the consumers from their corresponding satellites will get fully utilized, while there is still spare capacity at the producer.

To illustrate this effect, we repeat the experiment, but this time, each consumer will request different data.

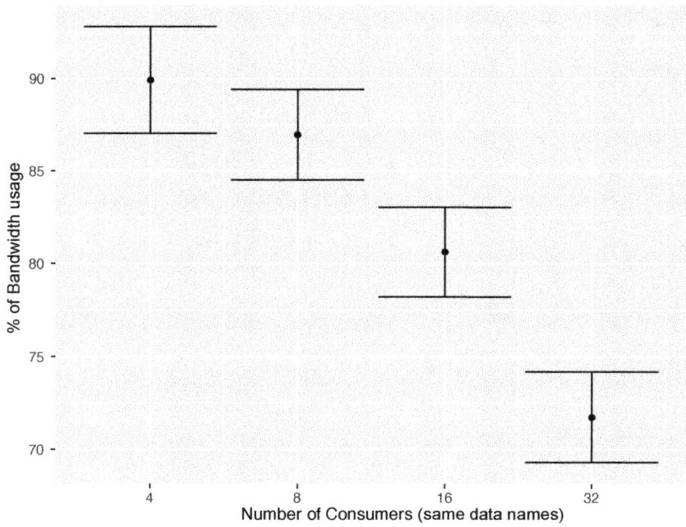

Figure 4. Percentage of bandwidth usage at the producer when consumers require the same data names.

As expected, the results in Figure 5 show that the average link usage for 4 consumers is now slightly higher (94%). In addition, in this case, the link usage slightly increases with the number of consumers, reaching a value of 95% for 16 and 32 consumers, which is an excellent result for our Forwarding Strategy, despite the short periods of unavailability caused by the satellite handovers. (When a handover occurs at the consumer in the time between the original interest transmission and the reception of the corresponding data, the data do not reach the new satellite serving the consumer but the old one. This is because in an NDN network, data packets follow the reverse path recorded in the PIT tables by interest packets. For the purpose of this paper, we relied on the retransmission mechanism of the consumer application to overcome this issue with good enough results, although it is possible to devise a mechanism to overcome this problem as in [20].) The reason for the increase in the bandwidth usage with the number of consumers is caused by the sawtooth behavior of the TCP-like congestion avoidance algorithm. As the number of unsynchronized consumers increases, their aggregated demand grows smoother than their individual demands. This effect is also present when using regular TCP transmissions.

To take a closer look into the effects of interest aggregation at the forwarders, we represent in Figure 6 the normalized total number of bytes exchanged in the network (the total number of bytes was normalized by the maximum value obtained in all the simulation experiments) in the previous experiments, both for the case of consumers requesting the *same* named content and for *different* contents.

We can observe that more of the network capacity is used when consumers request the same content. This is expected since some traffic in shared links serves several consumers and, therefore, downstream links can carry more capacity. That is, a single packet at a shared link results in multiple copies when it reaches the consumers. We can also appreciate that the total traffic increases with the number of consumers in the *same* content case, as the opportunities for aggregating interest packets at the forwarders increase.

Finally, the effects of caching when requesting the same content can be seen in Figure 7.

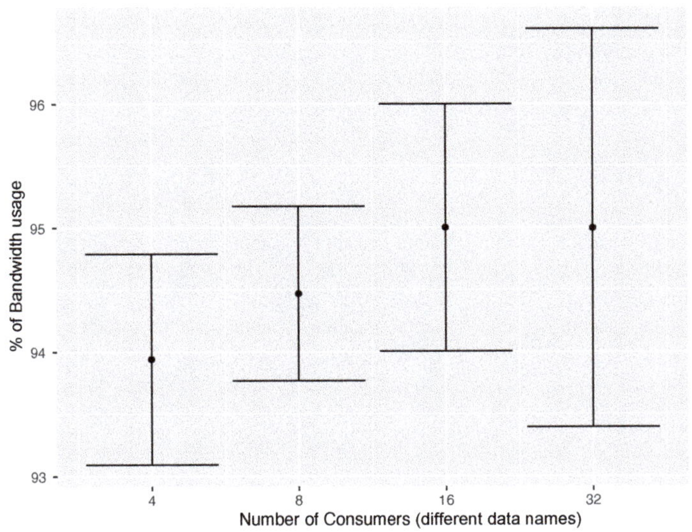

Figure 5. Percentage of bandwidth usage at the producer when consumers require different data names.

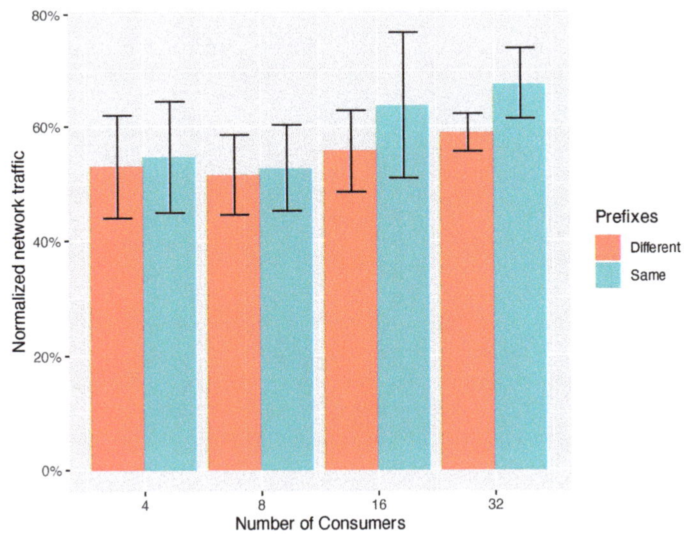

Figure 6. Total amount of data exchanged (normalized) through the network when a set of consumers requests content from a producer at their maximum attainable rate.

As before, as the number of consumers increases, the amount of total data transmitted also augments. Even with the link from the producer saturated, the consumers can still get high throughput because their data copies do not have to travel from the producer itself. With no caches, this sharing is only due to the coalescing of interest packets. However, with active caching, the sharing also comes from copies of data previously forwarded by individual satellite nodes.

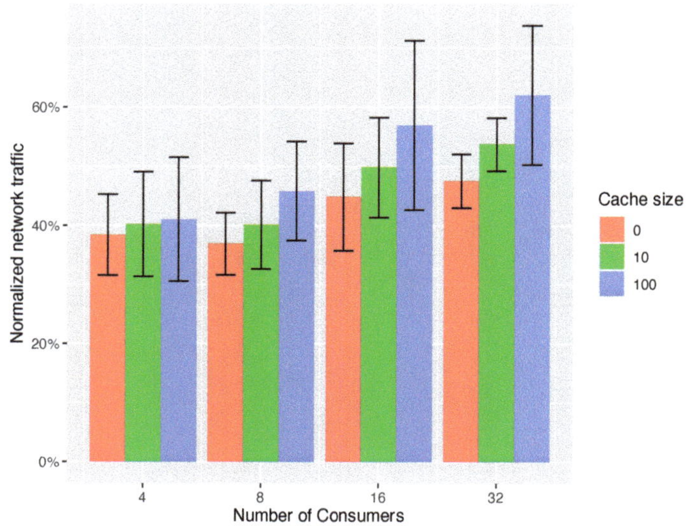

Figure 7. Total amount of data exchanged (normalized) through the network when a set of consumers requests the same content from a producer at their maximum attainable rate for different cache sizes.

6. Conclusions

In this paper, we proposed a new location-based forwarding strategy that can be a very good candidate to enable the efficient and effective future use of the NDN architecture in LEO satellite networks. The proposed forwarding strategy takes advantage of the knowledge of the relative positions of the satellite nodes, together with the grid structure formed by the inter-satellite links, to perform the forwarding of the NDN interest packets without the need of interchanging information between nodes, as is the case with conventional adaptive routing protocols, thus avoiding the scalability problems related to them. In addition, our forwarding strategy performs load balancing from the producer between the two paths around the diagonal toward each consumer, and then uses a single path toward it. The reason to use this alternative is that load balancing is much more necessary close to the producers, while using a single path close to the consumers could benefit from the eventual use of caches at the NDN nodes.

The simulation experiments we have conducted show that the proposed forwarding strategy works properly, and it is able to achieve an average link usage at the producer of 95%. Moreover, our experiments show that an NDN-based constellation with our proposal makes even better use of network capacity when consumers request the same content from the producer. The usage improves even more if caches are deployed in the NDN nodes.

Author Contributions: Conceptualization, S.H.-A. and M.R.-P.; methodology, R.F.R.-R.; software, P.I.-S.; validation, S.H.-A. and A.S.-G.; investigation, M.R.-P. and P.I.-S.; writing—original draft preparation, J.C.L.-A.; writing—review and editing, J.C.L.-A., R.F.R.-R. and A.S.-G.; supervision, R.F.R.-R. and A.S.-G.; funding acquisition, M.R.-P. and S.H.-A. All authors have read and agreed to the published version of the manuscript.

Funding: This work has received financial support from grant PID2020-113240RB-I00, financed by MCIN/AEI/10.13039/501100011033, and by the Xunta de Galicia (Centro singular de investigación de Galicia accreditation 2019–2022) and the European Union (European Regional Development Fund—ERDF).

Data Availability Statement: Not applicable, the study does not report any data.

Conflicts of Interest: The authors declare no conflict of interest.

References

1. Jacobson, V.; Smetters, D.K.; Thornton, J.D.; Plass, M.F.; Briggs, N.H.; Braynard, R.L. Networking named content. In Proceedings of the 5th International Conference on Emerging Networking Experiments and Technologies, CoNEXT '09, Rome, Italy, 1–4 December 2009; Association for Computing Machinery: New York, NY, USA, 2009; pp. 1–12. [CrossRef]
2. Zhang, L.; Afanasyev, A.; Burke, J.; Jacobson, V.; Claffy, K.; Crowley, P.; Papadopoulos, C.; Wang, L.; Zhang, B. Named data networking. *ACM SIGCOMM Comput. Commun. Rev.* **2014**, *44*, 66–73. [CrossRef]
3. SpaceX Submits Paperwork for 30,000 More Starlink Satellites. Available online: https://spacenews.com/spacex-submits-paperwork-for-30000-more-starlink-satellites/ (accessed on 2 May 2022).
4. Bhattacherjee, D.; Singla, A. Network topology design at 27,000 km/h. In Proceedings of the 15th International Conference on Emerging Networking Experiments and Technologies, Orlando, FL, USA, 9–12 December 2019; pp. 341–354. [CrossRef]
5. Pan, H.; Yao, H.; Mai, T.; Zhang, N.; Liu, Y. Scalable Traffic Control Using Programmable Data Planes in a Space Information Network. *IEEE Netw.* **2021**, *35*, 35–41. [CrossRef]
6. Darwish, T.; Kurt, G.; Yanikomeroglu, H.; Lamontagne, G.; Bellemare, M. Location Management in IP-based Future LEO Satellite Networks: A Review. *arXiv* **2021**, arXiv:2101.08336.
7. Saxena, D.; Raychoudhury, V.; Suri, N.; Becker, C.; Cao, J. Named Data Networking: A survey. *Comput. Sci. Rev.* **2016**, *19*, 15–55. [CrossRef]
8. Mohorcic, M.; Werner, M.; Svigelj, A.; Kandus, G. Adaptive routing for packet-oriented intersatellite link networks: Performance in various traffic scenarios. *IEEE Trans. Wirel. Commun.* **2002**, *1*, 808–818. [CrossRef]
9. Franck, L.; Maral, G. Static and adaptive routing in ISL networks from a constellation perspective. *Int. J. Satell. Commun.* **2002**, *20*, 455–475. [CrossRef]
10. Svigelj, A.; Mohorcic, M.; Kandus, G. Oscillation suppression for traffic class dependent routing in ISL network. *IEEE Trans. Aerosp. Electron. Syst.* **2007**, *43*, 187–196. [CrossRef]
11. Zhao, N.; Long, X.; Wang, J. A multi-constraint optimal routing algorithm in LEO satellite networks. *Wirel. Netw.* **2021**. [CrossRef]
12. Ma, X. Adaptive distributed load balancing routing mechanism for LEO satellite IP networks. *J. Netw.* **2014**, *9*, 816–821. [CrossRef]
13. Zhu, J.; Rao, Y.; Fu, L.Y.; Chen, W.; Shao, X. Load balancing routing based on agent for polar-orbit LEO satellite networks. *J. Inf. Comput. Sci.* **2012**, *9*, 1373–1384.
14. Rao, Y.; Wang, R.C. Agent-based load balancing routing for LEO satellite networks. *Comput. Netw.* **2010**, *54*, 3187–3195. [CrossRef]
15. Markovitz, O.; Segal, M. Advanced Routing Algorithms for Low Orbit Satellite Constellations. In Proceedings of the ICC 2021—IEEE International Conference on Communications, Montreal, QC, Canada, 14–23 June 2021. [CrossRef]
16. Wang, L.; Hoque, A.; Yi, C.; Alyyan, A.; Zhang, B. *OSPFN: An OSPF Based Routing Protocol for Named Data Networking*; Technical Report NDN-003; NDN: Needham, MA, USA, 2012. Available online: https://www.named-data.net/techreport/TR003-OSPFN.pdf (accessed on 4 August 2022).
17. Mahmudul Hoque, A.; Amin, S.O.; Alyyan, A.; Zhang, B.; Zhang, L.; Wang, L. NLSR: Named-data link state routing protocol. In Proceedings of the ICN 2013—ACM SIGCOMM Workshop on Information-Centric Networking, Hong Kong, China, 12 August 2013; pp. 15–20. [CrossRef]
18. Dai, H.; Lu, J.; Wang, Y.; Liu, B. A two-layer intra-domain routing scheme for named data networking. In Proceedings of the IEEE Global Telecommunications Conference (GLOBECOM), Anaheim, CA, USA, 3–7 December 2012; pp. 2815–2820. [CrossRef]
19. Liang, T.; Xia, Z.; Tang, G.; Zhang, Y.; Zhang, B. NDN in large LEO satellite constellations: A case of consumer mobility support. In Proceedings of the 8th ACM Conference on Information-Centric Networking, ICN'21, Paris, France, 22–24 September 2021; Association for Computing Machinery: New York, NY, USA, 2021; pp. 1–12. [CrossRef]
20. Xia, Z.; Zhang, Y.; Liang, T.; Zhang, X.; Fang, B. Adapting Named Data Networking (NDN) for Better Consumer Mobility Support in LEO Satellite Networks. In Proceedings of the 24th International ACM Conference on Modeling, Analysis and Simulation of Wireless and Mobile Systems, Alicante, Spain, 22–26 November 2021; Association for Computing Machinery: New York, NY, USA, 2021; pp. 207–216.
21. Henderson, T.; Katz, R. On distributed, geographic-based packet routing for LEO satellite networks. In Proceedings of the Globecom '00-IEEE, Global Telecommunications Conference, Conference Record (Cat. No.00CH37137), San Francisco, CA, USA, 27 November–1 December 2000; Volume 2, pp. 1119–1123. [CrossRef]
22. Chen, Q.; Yang, L.; Guo, D.; Ren, B.; Guo, J.; Chen, X. LEO Satellite Networks: When Do All Shortest Distance Paths Belong to Minimum Hop Path Set? *IEEE Trans. Aerosp. Electron. Syst.* **2022**, *58*, 3730–3734. [CrossRef]
23. Tariq, A.; Rehman, R.A.; Kim, B.S. Forwarding Strategies in NDN-Based Wireless Networks: A Survey. *IEEE Commun. Surv. Tutor.* **2020**, *22*, 68–95. [CrossRef]
24. Zhang, Y.; Afanasyev, A.; Burke, J.; Zhang, L. A survey of mobility support in Named Data Networking. In Proceedings of the 2016 IEEE Conference on Computer Communications Workshops (INFOCOM WKSHPS), San Francisco, CA, USA, 10–14 April 2016; pp. 83–88. [CrossRef]
25. NDNLPv2-NFD. Available online: https://redmine.named-data.net/projects/nfd/wiki/NDNLPv2 (accessed on 31 May 2022).
26. Afanasyev, A.; Moiseenko, I.; Zhang, L. *ndnSIM: NDN Simulator for NS-3*; Technical Report NDN-0005; NDN: Needham, MA, USA, 2012. Available online: https://named-data.net/publications/techreports/trndnsim/ (accessed on 4 August 2022).

27. Rodríguez Pérez, M.; Herrería Alonso, S.; Iglesias Sanuy, P. *A LEO Constellation Simulator Module for ndnSIM*; Universidade de Vigo: Vigo, Spain, 2022. Available online: https://github.com/ICARUS-ICN/icarus-ndnsim (accessed on 4 August 2022).
28. iPerf—The TCP, UDP and SCTP Network Bandwidth Measurement Tool. Available online: https://iperf.fr/ (accessed on 20 June 2022).

future internet

MDPI

Article

A Novel Multipath Transmission Scheme for Information-Centric Networking

Yong Xu [1,2], Hong Ni [1,2] and Xiaoyong Zhu [1,2,*]

1 National Network New Media Engineering Research Center, Institute of Acoustics, Chinese Academy of Sciences, No. 21, North Fourth Ring Road, Haidian District, Beijing 100190, China
2 School of Electronic, Electrical and Communication Engineering, University of Chinese Academy of Sciences, No. 19(A), Yuquan Road, Shijingshan District, Beijing 100049, China
* Correspondence: zhuxy@dsp.ac.cn; Tel.: +86-131-2116-8320

Abstract: Due to the overload of IP semantics, the traditional TCP/IP network has a number of problems in scalability, mobility, and security. In this context, information-centric networking (ICN) is proposed to solve these problems. To reduce the cost of deployment and smoothly evolve, the ICN architecture needs to be compatible with existing IP infrastructure. However, the rigid underlying IP routing regulation limits the data transmission efficiency of ICN. In this paper, we propose a novel multipath transmission scheme by utilizing the characteristics and functions of ICN to enhance data transmission. The process of multipath transmission can be regarded as a service, and a multipath transmission service ID (MPSID) is assigned. By using the ICN routers bound to the MPSID as relay nodes, multiple parallel paths between the data source and the receiver are constructed. Moreover, we design a path management mechanism, including path selection and path switching. It can determine the initial path based on historical transmission information and switch to other optimal paths according to the congestion degree during transmission. The experimental results show that our proposed method can improve the average throughput and reduce the average flow completion time and the average chunk completion time.

Keywords: multipath transmission; path selection; path switching; ICN

Citation: Xu, Y.; Ni, H.; Zhu, X. A Novel Multipath Transmission Scheme for Information-Centric Networking. *Future Internet* **2023**, *15*, 80. https://doi.org/10.3390/fi15020080

Academic Editors: José Carlos Lopez-Ardao, Miguel Rodríguez Pérez and Sergio Herrería Alonso

Received: 9 January 2023
Revised: 3 February 2023
Accepted: 15 February 2023
Published: 17 February 2023

1. Introduction

According to Cisco, internet users and traffic have grown rapidly over the past five years. By 2023, sixty-six percent of the global population will be internet users and average global fixed broadband speeds will be 110 Mbps [1]. The development of emerging network applications such as virtual reality (VR), ultra-high-definition video, unmanned driving, and industrial automation has put forward new requirements for the existing TCP/IP networks in terms of high bandwidth, massive communication, and low latency. Due to the lack of a native mechanism to support efficient content distribution, the host-centric traditional network architecture has been unable to meet the needs of current network services. With this background, a new network paradigm, information-centric networking (ICN) [2,3], has been proposed to solve the problems faced by existing TCP/IP networks.

Unlike traditional TCP/IP, ICN decouples the locator and identifier of resources, thus avoiding the problems caused by IP semantic overload, such as mobility, security, and scalability [4]. ICN uses a globally unique ID to name network entities and deploys in-network caching, which naturally supports massive content distribution. The development of ICN has led to many related projects. According to the difference in routing and forwarding methods, the existing ICN paradigms can be mainly divided into two types, the ICN paradigm of name-based routing and the ICN paradigm of stand-alone name resolution. The former mainly uses hierarchical and aggregated names, and the name resolution process is coupled with message routing. Typical examples of this paradigm include Content Centric Networking (CCN) [4], Named Data Networking (NDN) [5], and

Content Network (CONET) [6]. However, the deployment of such an ICN paradigm is costly, as significant upgrades to the infrastructure of traditional IP networks are required. The latter mainly uses flat names, and the name resolution process and the message routing process are decoupled, such as MobilityFirst [7], Network of Information (NetInf) [8], and On-Site, Elastic, Autonomous Network (SEANet) [9]. In this type of ICN paradigm, the Name Resolution System (NRS) plays an important role in maintaining the relationship between network address (NA) and identifiers (ID). Publishers of content or service can register their NAs and the corresponding ID with the NRS. Then, users can obtain the NAs of the publishers from the NRS according to the content or service ID, and transmit the data through the NA routing function. Since it is not practical to override the existing IP network, using IP addresses as NAs is a feasible solution [10]. This type of ICN paradigm is more compatible with the existing IP infrastructure, thus enabling smooth evolution. Note that our proposed multipath transmission scheme is based on the latter ICN paradigm.

However, due to compatibility with the existing IP infrastructure, ICN packets are still transmitted along the default "IP path" given by the underlying IP routing regulation. The single path IP routing protocols stubbornly believe that all ICN packets travel from source to destination by the same default shortest path. Therefore, it is difficult to guarantee user demands for high reliability and bandwidth when applied with the IP single path routing protocol, especially in the scenario of massive content distribution in ICN.

In order to solve the problems of the low resource utilization and poor fault recovery capability of single-path methods, some multi-path transmission protocols have been proposed [11,12]. Although Equal-Cost Multipath Routing (ECMP) [11] has been used in data center networks, its improvement effect in asymmetric network topology is not significant. Segment Routing IPv6 (SRv6) [12] has been proposed in recent years to solve the problem of IP routing rigidity, but the excessive header overhead will reduce the data carrying efficiency. In addition, some researchers have focused on how to utilize the characteristics and resources of the underlying IP network at the application layer [13–17]. These overlay methods build relay paths to avoid failure or congestion by detecting information such as delay and packet loss rate of the underlying network in real time. Although the overlay methods complement traditional network protocols, their performance is still unable to break through the bottleneck of the protocol stack due to the deployment at the application layer, and they cannot be directly applied to ICN. Moreover, there have been some attempts to implement multipath forwarding based on multiple addresses of terminal devices identified by globally unique identifiers in the ICN paradigm of stand-alone name resolution [18,19]. However, these methods can only be applied in the scenario of multi-homed terminals.

In this context, we propose a novel multipath transmission scheme applied in an IP-compatible ICN architecture. Our goal is to make full use of the characteristics and functions of ICN and the diversity of IP paths to improve the data transmission efficiency and robustness. The main contents of this paper are as follows:

- We introduce the overall layout of our ICN protocol stack and propose a novel multipath transmission scheme in an IP-compatible ICN architecture. We regard the multipath transmission process as a kind of service and the multipath transmission service ID (MPSID) is assigned. Based on the MPSID, multiple parallel paths can be constructed between data sources and receivers by utilizing ICN routers as relay nodes.
- We propose a path management mechanism to make full use of multipath resources. To reduce the overhead and avoid a poor selection, the initial path can be determined according to the historical transmission information. Besides, by measuring the congestion degree of the selected path during transmission, path switching can be performed to avoid bad paths.
- We conduct a series of experiments to verify the performance of our multipath transmission scheme. The experimental results show that our proposed method has a significant improvement in terms of average throughput, average flow completion time, and average chunk completion time.

The rest of this paper is organized as follows. We review the related research on multipath transmission mechanisms of IP networks and ICN networks, respectively, in Section 2. In Section 3, we describe the overall layout of our ICN protocol stack and the principle and process of ID-based multipath transmission scheme. In Section 4, the path management mechanism is described in detail. Then, we conduct simulation experiments of the multipath transmission scheme and discuss the experimental results in Section 5. Finally, we conclude our research and look forward to future work in Section 6.

2. Related Work

Multipath transmission techniques have been extensively studied due to their improvement in transmission robustness and efficiency. Since the multipath transmission scheme we propose is applied in an IP-compatible ICN architecture, we present a comprehensive study on the multipath transmission mechanism of IP networks and ICN, respectively, in this section.

2.1. Multipath Transmission Mechanism of IP Networks

According to the difference in design and deployment, the multipath transmission mechanism of the existing IP networks can be classified into the overlay method and the underlay method.

The overlay method creates a virtual overlay topology on the underlying network and forms multiple end-to-end paths through logical links. The Resilient Overlay Network (RON) [13] is a typical example of the overlay method. RON periodically measures the status information of virtual links between overlay nodes, then distributes topology and path information to each neighbor node, and finally forwards traffic by building IP tunnels to avoid congested links. However, the overhead of periodic detection and information distribution is not conducive to large-scale deployment of RON. Based on RONs, some studies have proposed a multi-homing overlay network (MON) [14], which introduces the multi-homing characteristics of devices into overlay routing. Although the MON improves transmission benefits, the high overhead is still its fatal drawback. In order to improve the scalability of the overlay method, the authors in [15] proposed the One-Hop Source Routing (OHSR), which attempts to recover from congestion or failure by randomly selecting optional relay nodes for indirect routing. However, the random selection manner does not guarantee the status of the indirect path. In addition, Path Probing Relay Routing (PPRR) [16] uses a random search method to discover alternative detour paths and determines the path state by probing on demand. When serious congestion or failure occurs on the direct path of the underlying IP network, PPRR can quickly switch to an alternative path. Moreover, the Topology-Aware Reliable Overlay Multipath (TAROM) [17] can accurately find relay nodes based on global topology and routing information, but it is difficult to obtain dynamic internet topology and routing information timely and accurately.

The underlay method uses a layer-2 or layer-3 routing and forwarding mechanism to ensure end-to-end multipath connectivity. ECMP is a classic multi-path routing protocol, which is widely used in Data Center Networking (DCN). ECMP distributes traffic equally over multiple available forwarding ports based on the diversity of the underlying physical paths. However, the traffic is equally distributed, which also determines that ECMP is difficult to apply to asymmetric networks. Therefore, researchers proposed Weighted Cost Multi-Path (WCMP) [20] based on ECMP. WCMP distributes traffic proportionally across paths based on link state. However, WCMP is inflexible and cannot reroute flows based on congestion information. In [21], the authors proposed a dynamic acyclic multipath routing algorithm, which keeps multiple possible next hops and weights for one destination address, and the router assigns packets of the same destination address to multiple next hops according to the estimated weight ratio. Compared with the traditional IP routing protocol, Multi-Protocol Label Switching (MPLS) [22] provides a new network switching method, which maps IP addresses into short and fixed-length labels. MPLS uses label switching instead of IP routing table lookup to significantly improve forwarding efficiency.

As a result of its high price and poor security, MPLS has been stretched under the condition of high bandwidth demand. In addition, researchers have proposed segment routing in the past few years. Segment routing [23] is a packet routing and forwarding mechanism based on source routing. The forwarding path is encoded in the packet header before sending. During the forwarding process, routers can determine who the next hop is according to the header address field. Segment routing has attracted much attention due to its flexible path selection and forwarding. For instance, the authors in [24] proposed an online and offline algorithm for path selection based on segment routing, and the experimental result showed that both methods are effective. SRv6 [12], proposed in recent years, combines the advantages of IPv6 and segment routing, and has been widely deployed on Wide Area Networks (WANs). Although these segment routing technologies can specify the transmission path, the excessive header overhead will reduce the data carrying efficiency. In wireless networks, how to implement self-organizing multipath routing through relay devices has been extensively studied [25,26], but these technologies can only be applied in wireless scenarios.

2.2. Multipath Transmission Mechanism of ICN

With the development of ICN, there have been many studies on ICN multipath transmission. In this subsection, we introduce the multipath transmission techniques of the ICN paradigm of name-based routing and stand-alone name resolution, respectively.

In the ICN paradigm of name-based routing, the data packets do not loop since they take the reverse path of request packets, so the multi-path transmission mechanism relies on its flexible forwarding plane in this ICN paradigm. In [27], the authors used three colors to mark the performance of the forwarding port, and the router can preferentially select the port with the best performance to forward the data packet, thus avoiding the congested path. In [28], the authors proposed an on-demand multi-path forwarding mechanism based on the principle of minimum RTT priority, which forwards user requests from different interfaces in proportion according to RTT. In [29], the authors regard the process of request packet forwarding as a multiple attribute decision making problem, and proposed an Entropy-based Probabilistic Forwarding (EPF) strategy. EPF integrates multiple network state metrics to accurately calculate the state of interfaces by objectively assigning weights to the metrics. In [30], the Markov Decision Process (MDP)-based forwarding strategy was introduced, which uses queuing theory to estimate the real-time network state, and builds an MDP model to guide request packets for probabilistic forwarding. In [31], the authors regarded the problem of multi-path congestion control and request forwarding as a global objective optimization problem for maximum throughput and minimum network cost, and derived a set of optimal distributed algorithms for dynamic request forwarding. In addition, some researchers have used reinforcement learning algorithms to design forwarding strategies [32,33]. ICN routers perform probabilistic forwarding to explore potential cache replicas in the network, and then use the learned results to guide request packet forwarding. However, these methods are only suitable for the innovative ICN architecture, so it is difficult to apply them in practice.

In the ICN paradigm of stand-alone name resolution, there have been some attempts to implement multi-path forwarding based on multiple addresses of terminal devices identified by globally unique identifiers. In [18], the authors proposed a novel forwarding scheme in the multi-address scenario in this ICN paradigm. In this scheme, multiple addresses of the device are carried into the destination address group field of the ICN data packet header, which enables each data packet to be matched to multiple interfaces in the hop-by-hop forwarding process, and the router can select an optimal interface for each data packet according to the state information of the interfaces. In [19], the authors proposed a network-assisted multipath transmission mechanism, in which the in-network nodes can split data into different paths according to locally generated policies. Specifically, multiple network addresses can be obtained by querying the NRS according to the globally unique identifier (GUID), so as to establish multiple paths to destination mobile devices.

3. Design

We first outline a practical protocol stack of an IP-compatible ICN architecture in this section. Then, we analyze the disadvantages of the single-path transmission method to clarify the motivation of our study. Finally, we describe the principle and process of the proposed multipath transmission scheme in detail.

3.1. Protcol Stack

In order to reduce deployment costs, it is reasonable to deploy the new ICN architecture on top of an existing IP network, so as to fully utilize the IP infrastructure. Figure 1 shows a practical ICN protocol stack, in which the ID layer is incrementally deployed on top of the IP layer, thereby extending the functions of the network layer. At the network layer, a protocol called the identifier protocol (IDP) [34] is running, which defines a set of regulations specifying how to operate on the NA according to the ID of the data packet, including adding, deleting, and modifying. Above the ID layer is a transport layer, on which we design a transport protocol [35], including the ICN packet format, transmission process, congestion control mechanism, and retransmission mechanism. The transport layer protocol can provide an efficient and reliable data transmission service, and it belongs to the scope of the transport layer.

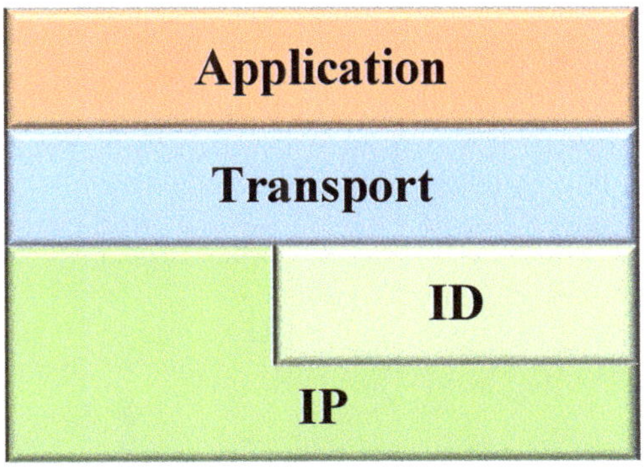

Figure 1. The protocol stack of ICN.

In ICN, the chunk is the basic data unit for transmission and caching, which is bound to an addressable globally unique name. Most of the existing switches or routers are two-layer or three-layer structures; therefore, the ICN packet can be normally routed and forwarded on the existing network. In order to achieve smooth evolution, we consider the deployment of hybrid networking, where the entire network consists of IP routers and ICN routers. The ICN router has a complete ICN protocol stack, as well as computing and storage capabilities. In addition, we apply the concept of late binding in ICN [36], which means that the NA of the packets can be modified according to the policy generated by the local computing module during the forwarding process. Here are some typical examples: In the chunk retrieval scenario, the ICN routers can select the optimal replica by modifying the destination NA of the request packets according to the NAs of content providers obtained by querying the NRS via a chunk ID [37]. In mobility scenarios, the mobile device, whose NA is changed, can re-register its NA and ID relationship with the NRS; then, the ICN routers can modify the destination NA of data packets according to the NA obtained by querying the NRS via mobile device ID, so as to ensure transmission continuity [38].

3.2. Motivation

However, due to the compatibility with the existing IP infrastructure, ICN packets are still transmitted in a best-effort approach employed by IP routing rules. With the increase in the number of access devices, the congestion probability of the default path gradually increases. Once congestion or failure occurs, it often takes seconds or even minutes for routing protocols to converge to a normal state [39]. The end-to-end connections may experience outages for seconds or minutes during this process. Therefore, rigid IP single path routing has been unable to meet the demand of massive data distribution. A reasonable data transmission scheme is necessary to improve the transmission reliability and efficiency of such IP-compatible ICN architectures.

With a complete protocol stack, the ICN routers can resolve an ID to network address(es) by initiating a query operation to the NRS. In addition, the ICN routers can also use the IDP to process the network address(es) of packets, which may include adding, deleting, and modifying the network address. Therefore, it is a feasible solution to utilize the ICN routers to act as relay nodes to enhance transmission. In addition, researchers have proven through experiments that the single-hop indirect path formed by one relay node could achieve significant gains in terms of round-trip delay, packet loss rate, and throughput [13]. Based on this finding, an ICN router can form a relay path between the sender and the receiver, so multiple ICN routers can be used for multipath transmission. In the next subsection, we will introduce the design of our proposed multipath transmission scheme in detail.

3.3. Overview of Transmission Process

In ICN, network services are treated as a kind of network entity and are labeled with identifiers, such as multicast services [40] and storage services [34]. Inspired by this, we consider multipath transmission as a kind of service and assign the corresponding MPSID. Some ICN in-network routers need to register the relationship of their NAs to the MPSID with the NRS so that they can be addressed by other network devices. These ICN routers are regarded as relay nodes for end-to-end transmission, which can reroute traffic sent by the data source to the destination. Therefore, in addition to the default shortest path, the data source can use multiple suitable relay ICN routers to construct multiple parallel single-hop relay paths between the data source and the receiver.

Figure 2 shows the process of the proposed multipath transmission scheme.

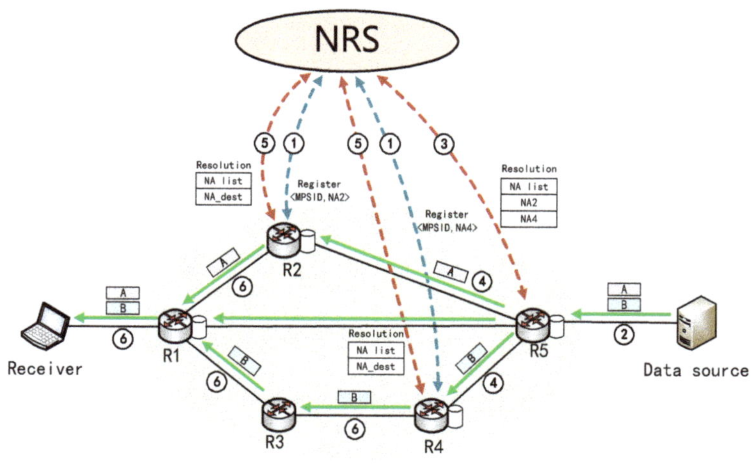

Figure 2. An overview of the multipath transmission process.

When the ICN routers (R2 and R4) supporting the multipath transmission service go online, they will register the mapping between its NA and MPSID with the NRS (step 1).

After the connection between the two sides is established, the data source starts transferring local data chunks. Each chunk to be transferred is segmented according to the Maximum Transmission Unit (MTU) and an ICN header is added to form an ICN data packet. The data source can transmit these packets using the default path; in addition, if a higher bandwidth is required, the data source can mark a multipath transmission label in the preference field of these ICN data packets according to the application requirements, and then send packets to the corresponding edge router (here, R5) (step 2).

For security reasons, the NAs of in-network ICN routers should not be exposed to devices outside the ICN network area, so the query operation to the NRS is initiated by the edge node. When receiving the first marked data packet of the first chunk during this transmission, R5 initiates a query operation to the NRS according to the MPSID to obtain the NA list (here, NA2 and NA4). During the whole end-to-end transmission, the operation of initiating a query to the NRS by the edge router only happens once and the subsequent query operation is not needed (step 3).

According to the node selection strategy, R5 selects several suitable NAs from the NA list and adds them to the selected node set. Then, R5 changes the destination NA of the data packet to one of the selected NAs and caches the network address translation information locally. It is worth noting that the granularity of data scheduling is chunk, so segmented packets of the same chunk will be processed in the same way. Since IPs can be used as NAs in this paper, these data packets will be transmitted to the corresponding relay node (R2 or R4) through an IP routing function (step 4).

When receiving the first data packet of a chunk, the relay ICN routers (R2 or R4) will send a name resolution request to the NRS according to the destination ID to obtain the NA of the receiver device. It should be noted that this query operation only occurs when the first data packet of a chunk is received. The relay router will form NA translation rules locally. When subsequent packets of the same chunk arrive, their destination NAs are automatically changed according the translation rules (step 5).

Finally, the relay ICN router (R2 or R4) modifies the destination NA of the data packet to the NA of the receiver device and forwards it (step 6).

In contrast to some overlay routing methods, our multipath transmission scheme is implemented based on the MPSID at the network layer rather than the application layer. In the process of multipath transmission, three kinds of IDs are mainly involved, which are the IDs of chunks to be transmitted, the ID of the receiver device, and the MPSID. The source ID and destination ID of the data packet are the chunk ID and the receiver device ID, respectively. The MPSID does not appear in the header of the packet, and the edge node can decide whether to use the MPSID according to the preference field. Locators and identifiers are separated, which is one of the characteristics of ICN. In the process of multipath transmission, the MPSID acts as an identifier, and each NA bound to an MPSID acts as a locator, which conforms to the basic characteristics of ICN. In this paper, we refer to the design pattern of flat IDs from MobilityFirst [7], and use a 20-byte flat ID to name network entities, including the MPSID. In addition, the NRS plays a vital role. Firstly, the data source can obtain the NAs of these relay ICN routers by querying the NRS with an MPSID to discover multiple paths. Secondly, the relay ICN routers can query the NRS through the destination ID of the data packets to obtain the NA of the receiver device. Existing ICN multipath transmission mechanisms either rely on its flexible forwarding plane or on the multi-homing feature of devices. To the best of our knowledge, unlike existing mechanisms, we are the first to regard multipath transmission as a service and use an ID to identify it. At the transport layer, we use the previously proposed transport protocol [35], which has congestion control and a retransmission mechanism, to ensure efficient and reliable transmission by adjusting the request sending rate within a single chunk.

4. Path Management

In our scheme, an ICN router bound with an MPSID can constitute a one-hop indirect path, which can provide a data relay service for end-to-end transmission. Therefore, making full use of path resources is vital. To this end, we propose in this section a path management mechanism in detail. Path management consists of two phases: path selection and path switching. In the path selection phase, we select several relay nodes as initial temporary solutions based on historical transmission information. In the path switching phase, we measure the congestion degree of these selected paths. If any of these paths experience heavy congestion, we discard this congested path and find an alternative.

4.1. Path Selection

Some studies [41] have shown that only a small number of relay nodes can provide optimal relay paths for most end-to-end transmission pairs, and these nodes usually have high betweenness centrality. Therefore, we make some ICN routers with high betweenness centrality register their NA-MPSID relationship with the NRS. Betweenness centrality [42] of a node i is the sum of the fraction of all-pairs shortest paths that pass through i, which is denoted as follows:

$$BC(i) = \sum_{s,t \in V} \frac{\sigma_{st}(i)}{\sigma_{st}} \tag{1}$$

where V is the set of nodes in topology, σ_{st} denotes the number of shortest paths from s to t, and $\sigma_{st}(i)$ is the number of shortest paths from s to t that go through i.

After obtaining the NA list of the relay nodes by querying the NRS with the MPSID, the edge router needs to select several appropriate NAs from the NA list to form the initial multipath. In [43], the authors point out that relay node selection is an NP-hard problem. Although some node selection algorithms have been proposed in the overlay network [13,15,16], their performance is poor in terms of robustness or scalability. The RON [13] and other similar systems require the relay nodes to probe the entire network and periodically exchange probe information, and are hence not scalable. OHSR [15] adopts a random selection strategy and thus is scalable, but it is hard to avoid poor relay nodes. PPRR [16] can maintain a top set of optional relay nodes for each destination, thus reducing the probing overhead. However, the probe overhead is still not negligible as the number of relay nodes increases. Therefore, we need a more scalable way to discover available relay paths.

To reduce the overhead and avoid a poor selection, we let the edge router maintain an alternate path state table (*PST*). The *PST* records the NAs of the relay nodes and the corresponding transmission success rate (*TSR*). The *TSR* is obtained according to the success rate of a relay node providing data relay service in the past period of time, so it can reflect the relay node's ability to provide multipath transmission service to a certain extent. The *TSR* can be calculated as follows:

$$TSR_i = \frac{s_i}{s_i + f_i} \tag{2}$$

where s_i and f_i represent the number of times that the path formed by relay node i is congested and not congested, respectively. s_i and f_i are the results of long-term historical observations. If the path formed by the selected relay node i is not congested during one transmission, the corresponding s_i is increased by one; that is,

$$s_i = s_i + 1 \tag{3}$$

Conversely, if the path is congested during transmission, the corresponding f_i is decreased by 1; that is,

$$f_i = f_i - 1 \tag{4}$$

In this way, the node's *TSR* can be calculated and updated during every transmission. The larger the *TSR* of a relay node, the higher the probability of it being selected. The rationale for this method is that a relay node that previously provided a sufficiently well-conditioned path to various destinations is likely be selected again with higher probability in the future [15]. There may be some ICN routers registering or deregistering with the MPSID on the NRS; therefore, the edge routers need to update the NA entries of the *PST* according to the NA list of the relay nodes obtained from the NRS. In addition, if a relay node has not been selected for a long time or a new relay node is added to the *PST*, its *TSR* will be automatically set to the default value. The *PST* has a simple structure and its update does not require additional probe overhead, so it is scalable.

The *PST* can provide prior knowledge, and thus can reduce the difficulty and overhead of node selection. Suppose that P is the set of alternative nodes obtained from the NRS and the set of selected nodes is denoted as S. Our goal is to pick k optimal relay nodes from P according to the *PST* and add them to S to form initial multiple paths. If all the relay nodes' *TSRs* in the *PST* are the same, the edge routers will randomly select k. In this paper, we set the value of k to 3. The reason is that it has been shown in [44] that most of the benefits can be obtained by using three to four relay paths. By selecting more than one relay routers based on the *PST*, the edge router can ensure that a single unlucky selection is not fatal. The path selection strategy based on the *PST* cannot ensure that all selected paths are in a good state, but it can improve the availability of the initial path to a certain extent. The path selection strategy needs to cooperate with the path switching strategy, which will be introduced in the next subsection. The detailed path selection algorithm is shown in Algorithm 1.

Algorithm 1 Path Selection Algorithm

Input: *MPSID, PST, k*
Output: *S*
Initialization: *S = NULL, P = NULL*
At the beginning of transmission:
  P = querybyNRS(MPSID)
  update the NA entries of *PST* according to *P*
  S = SelectTop_K(P, PST, k)
During transmission:
  execute the path switching algorithm //describe in Algorithm 2
At the end of transmission:
  for every NA_i in *S* **do**
    $TSR_i = TSR_i + 1$
    release *P, S*
end for

4.2. Path Switching

As described above, we determine the initial paths based on prior knowledge of the *PST*. However, these ICN relay nodes usually need to carry more traffic and are a frequent location of congestion. During the transmission, once a path experiences heavy congestion, it is necessary to re-examine its availability. To alleviate congestion, a common method is to allocate fewer data chunks to congested paths. This method is feasible when the path is slightly congested for a short time. Once a path is heavily congested for a long time, it means that the path cannot provide high-quality transmission services. In this case, it is wise to delete the congested path from S and look for another alternative.

In this paper, we use the packet loss rate to judge whether a sub-path is congested. Considering that a high sampling frequency may cause redundant calculation overhead, we set a time interval, Δt, during which the receiver samples the instantaneous packet loss rate. According to the experimental results from [45], we set Δt to 0.2 s. As shown in Equation (4), the packet loss rate is calculated in a smooth manner to avoid erroneous estimates caused by short-term burst traffic.

$$LOSS_i(t) = \alpha \times loss_i(t) + (1-\alpha) \times LOSS_i(t-\Delta t) \tag{5}$$

where $LOSS_i(t)$ is the average packet loss rate of path i, and $loss_i(t)$ is the ratio of the number of lost packets to the number of transmitted packets during Δt. The receiver calculates the packet loss rate of each path and feeds the congestion information back to the edge router through request packets, and the latter performs path switching. We introduce a congestion threshold, L_{thresh}. According to the relationship between $LOSS_i(t)$ and L_{thresh}, we can judge whether the path is in the heavy congestion state. The path switching algorithm pseudocode is shown in Algorithm 2.

$$congestion\_flag = \begin{cases} 0, 0 < LOSS_i(t) < L_{thresh} \\ 1, L_{thresh} < LOSS_i(t) \end{cases} \quad (6)$$

The main overhead of multipath transmission is determined by the maintenance cost of the path state. The higher the maintenance cost, the higher the complexity. Assume that the number of optional relay nodes is N. In a RON, each node needs to maintain the path state with other $N - 1$ nodes, so the maintenance cost is $O(N^2)$. PPRR selects M from N optional nodes to form the top set and maintains the path state between M nodes according to the probing results, so its complexity is $O(M^2)$. The method adopted in this paper is based on prior information of historical transmission, and does not require any probing operation, so the complexity is also O(1). This shows that our method is scalable. Moreover, we use a practical data scheduling strategy in this paper. To avoid out-of-order arrival of packets of the same chunk, we set the granularity of data scheduling to chunk level instead of packets. The edge node adopts the round robin method based on the packet loss rate, and preferentially allocates the chunks to be transmitted to the path with the smallest packet loss rate.

Algorithm 2 Path Switcing Algorithm

```
1: Input: loss(t), LOSS(t-Δt), PST, P
2: Output: S
3: for every Δt during transmission do
4:   count the packet loss rate loss_i(t) of selected path
5:   LOSS_i(t) = α × loss_i(t) + (1-α) × LOSS_i(t-Δt)
6:   for every NA_i in S do
7:     if LOSS_i(t) > L_thresh then
8:       S = S\{NA_i}
9:       TSR_i = TSR_i − 1
10:      S = S ∪ SelectTop_K(P\S, PST,1)
11:    end if
12:  end for
13: end for
```

5. Performance Evaluation

In this section, we conduct simulation experiments and analyze the experimental results. First, the experiment setup is introduced. Then, we compare the proposed multipath transmission scheme with other related works from different metrics. Finally, we analyze and discuss the experimental results.

5.1. Experiment Setup

We implement our proposed multipath transmission scheme based on NS-3 [46], which is designed to meet the needs of academic research and teaching. As shown in Figure 3, we use a real-world topology, the European Academic Network (GEANT2) [47], to comprehensively evaluate the performance of our multipath transport scheme. The default latency and bandwidth between backbone nodes in the topology are 100 Mbps and 1 ms, respectively. In addition, we added a data source node (S) and some receiver nodes (C1 to C5) at the edge of the network, marked with rectangles and triangles, respectively. The default bandwidth between the data source node (or receiver nodes) and the corresponding edge nodes is 1000 Mbps, and the default latency is 5 ms. In our scenario, chunk is the basic data unit, and the default size of chunk is set to 2 MB. A complete chunk will be divided into smaller segments for transmission, and the segment size is set to 1250 bytes. Additionally, each router has a maximum queue length of 1000 packets. To make a comprehensive comparison, we use three transmission methods at the network layer, which are single path methods, i.e., Open Shortest Path First (OSPF) [48], ECMP, and our proposed multipath transmission scheme. OSPF, ECMP, and the proposed multipath transmission scheme belong to the category of network layer. At the transport layer, we use the transmission protocol proposed in previous work [35] to ensure reliable and stable data transmission.

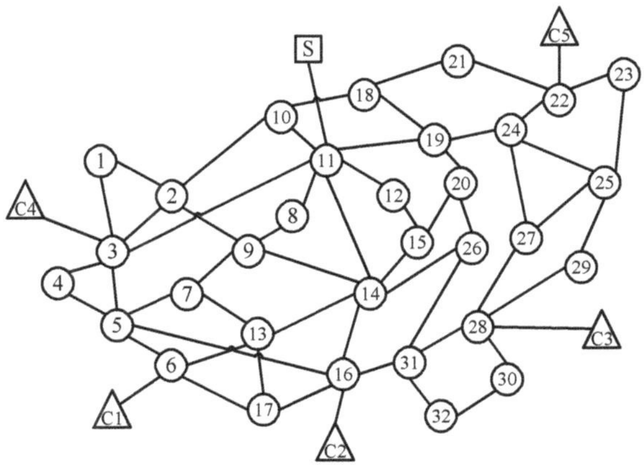

Figure 3. Simulation topology (GEANT2).

We evaluate the performance of our scheme based on three metrics: average throughput, average flow completion time (FCT), and average chunk completion time (CCT). The average CCT can reflect the response speed of each transmission method. The shorter the chunk completion time, the faster it can be delivered to the upper layer application. This is among the most important metrics for data transmission.

5.2. Throughput

Throughput is among the most important indicators of data transmission performance. In this subsection, we simulate the throughput of the multipath transmission scheme under different numbers of relay nodes and different link bandwidths.

Firstly, we set the number of relay nodes of GEANT2 to 5, 10, 15, 20, 25, and 30, respectively, to analyze the influence of the number of relay nodes on the throughput. The binding relationship between the MPSID and the NAs of these relay nodes has been registered on NRS. Then, we conducted five rounds of simulation. We let the data source node S continuously send data to a receiver node (C1 to C5), respectively, in each round and measured the steady-state throughput. As shown in Figure 4, it can be observed that when the number of relay nodes is ten, the average throughput can reach a maximum of about 169 Mbps. When the number of relay nodes is five, the average throughput is close to the maximum. This means that using only a few relay nodes can significantly improve the efficiency of multipath transmission. However, as the number of relay nodes increases, the average throughput decreases instead. The reason is that too many candidates increase the complexity of path selection and switching algorithms, so that it cannot quickly converge to the optimal relay. According to the experimental results, we set the number of relay nodes to ten in the following section.

Secondly, we simulated the average throughput under different bandwidths by changing the link bandwidth between the backbone nodes of the topology to 50 Mbps, 100 Mbps, 150 Mbps, and 200 Mbps, respectively. The experimental results are shown in Figure 5. It can be seen that the average throughputs of the single path method under different bandwidths were roughly 47 Mbps, 91 Mbps, 134 Mbps, and 179 Mbps, which is the lowest among the three methods. This is because the single path method cannot take advantage of other paths other than the default shortest path. It can also be observed that under different bandwidths, ECMP can improve the average throughput to about 62 Mbps, 113 Mbps,

158 Mbps, and 201 Mbps, respectively, but the improvement is not obvious. The reason is that ECMP is not always effective, since it only works on symmetric links. The symmetric links generally exist in data center networks [49], such as Fat-tree topology and Clos topology, but rarely exist in carrier networks or WANs. Compared with the other two methods, our proposed multipath transmission scheme can achieve the highest steady-state throughput under different link bandwidths, which are about 88 Mbps, 169 Mbps, 247 Mbps, and 302 Mbps, respectively. This also shows that our proposed method can achieve a significant improvement in average throughput.

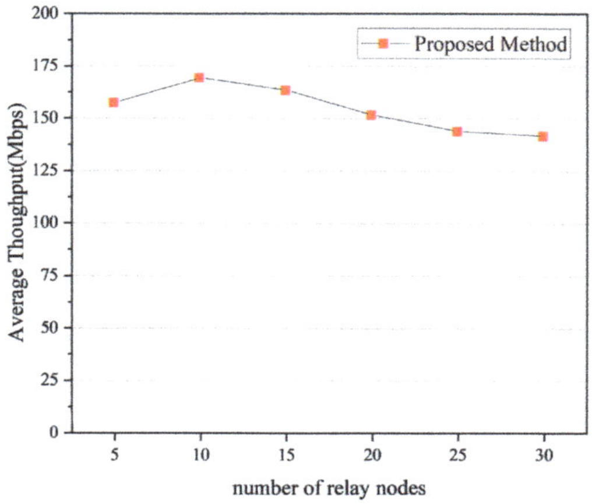

Figure 4. The average throughput under different number of relay nodes.

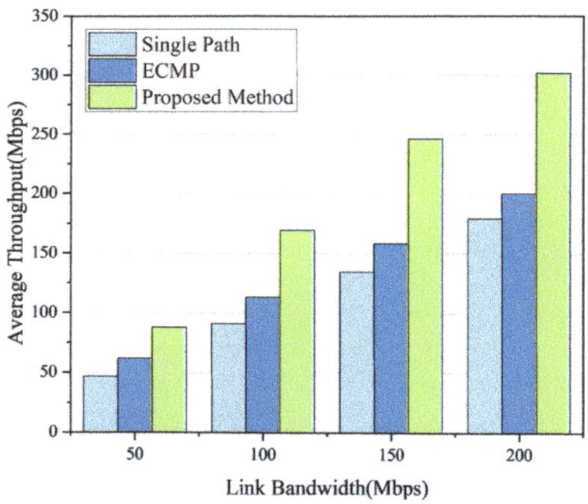

Figure 5. The average throughput under different link bandwidths.

5.3. Flow Completion Time

The average FCT can reflect the speed at which data transmission is completed, so we use this metric to evaluate our scheme. In this subsection, we simulated the average FCT of each transmission method under different link bandwidths by setting the link bandwidth between backbone nodes to 50 Mbps, 100 Mbps, 150 Mbps, and 200 Mbps, respectively. We performed five rounds of simulations. In each round, the data source node S connects with one of the receiver nodes (C1 to C5) and transmits 100 chunks. Figure 6 shows the average FCT of all connections transmitting 100 chunks under different bandwidths. As we expected, the average FCT of the single path method is the longest, about 36.5 s, 18.8 s, 12.8 s, 9.5 s, respectively, because it can only use the bandwidth resources on the shortest path. It can also be observed that under different bandwidths, ECMP can reduce the average FCT to about 27.7 s, 15.2 s, 10.8 s, and 8.6 s, respectively. The data packets will be transmitted to the destination along multiple equal-cost paths, thus reducing the average FCT. Compared with the other two methods, our proposed method can significantly reduce the average FCT to 19.5 s, 10.1 s, 6.9 s, and 5.7 s, respectively. This illustrates the effectiveness of our proposed method.

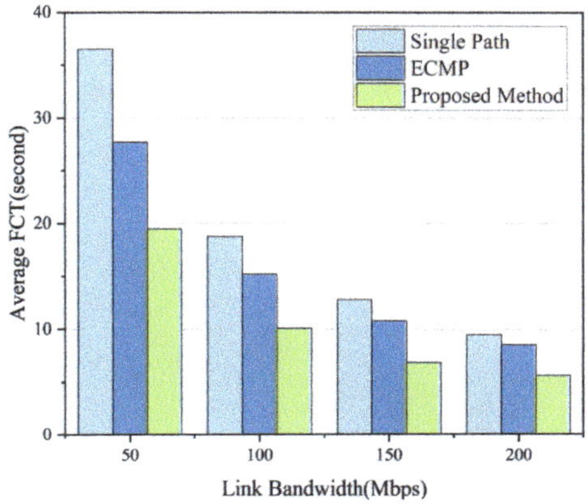

Figure 6. The average FCT under different link bandwidths.

5.4. Chunk Completion Time

In the ICN scenario, a chunk needs to be completely received before it can be delivered to the application layer. Therefore, the average chunk completion time is an important indicator showing the performance of the transmission method. The simulation steps are the same as those described in Section 5.3, and we measured the average chunk completion time under different bandwidths. Figure 7 shows the experimental results. It can be seen that as the bandwidth increases, the average CCT of each method decreases. The average CCT of the single path method is the highest, roughly 2.16 s, 1.75 s, 1.41 s, and 1.28 s under different bandwidths. It can also be observed that both ECMP and our proposed method can reduce the average CCT, and our proposed method is significantly more effective. Compared with the other two methods, our proposed method can reduce the average CCT to 1.44 s, 1.1 s, 0.74 s, and 0.56 s, respectively, under different bandwidths. This also shows that our proposed method can speed up the transmission, thus ensuring timely delivery of data to upper-layer applications.

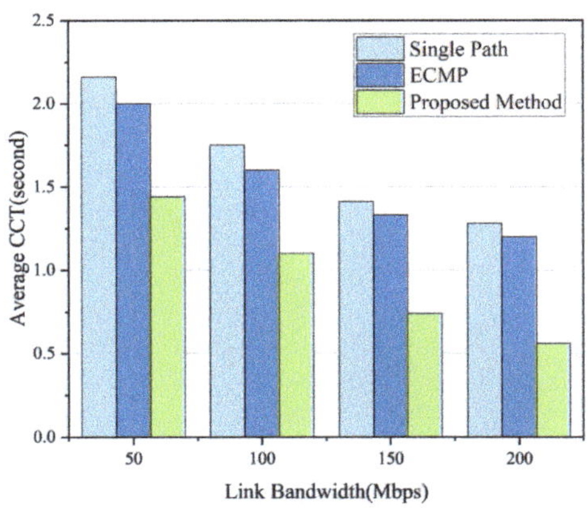

Figure 7. The average CCT under different link bandwidths.

In addition, as shown in Figure 8, we obtained the CCT cumulative distribution of data source node S and receiver node C1 when transmitting 100 chunks under 200 Mbps bandwidth. We can see that the proportions of CCT less than 1 s for the single path method, ECMP, and our proposed method are 0.74, 0.67, and 0.91, respectively, and the proportions of less than 2 s are 0.86, 0.98, and 0.98, respectively. Therefore, we can conclude that the single path method does not perform as well as the multipath method in reducing the CCT. Moreover, compared with ECMP, our proposed method can maintain the CCT at a lower level.

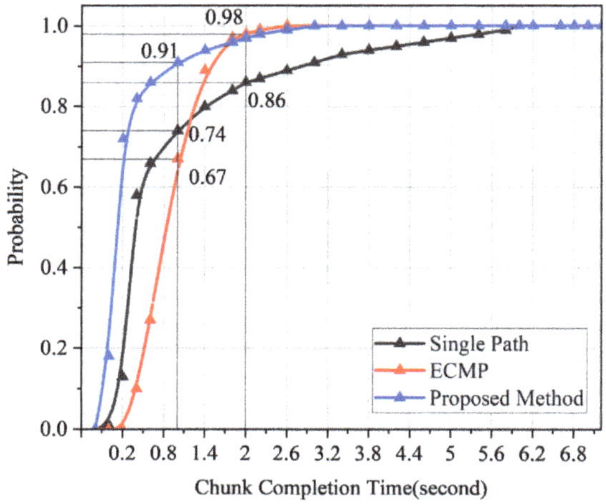

Figure 8. The cumulative distribution of the CCT.

6. Conclusions

In this paper, we propose a novel multipath transmission scheme which is applied in an IP-compatible ICN architecture. First, we introduce the overall layout of our ICN protocol stack and describe the principle and process of multipath transmission in detail. Then, we focus on how to make full use of multipath resources and propose a path management mechanism. In the path selection phase, we select several relay nodes as initial temporary solutions based on historical transmission information. In the path switching phase, path reselection can be performed during transmission by measuring the congestion degree of the selected sub-paths. Finally, we conduct simulation experiments to evaluate the performance of our proposed method. The experimental results show that our proposed method has an excellent performance in terms of average throughput, average flow completion time, and average chunk completion time.

In future research, we will focus on the following two aspects. Firstly, a reasonable data scheduling strategy is required to allocate the data ratio of each sub-path to ensure load balance. Secondly, the mathematical apparatus used in this paper is valid, but quite simple. Considering that the path selection method based on historical information cannot avoid the underlying overlapping paths, we will redesign the path selection strategy according to the network topology information obtained from the controller to eliminate the correlation of paths.

Author Contributions: Conceptualization, Y.X., H.N. and X.Z.; methodology, Y.X., H.N. and X.Z.; software, Y.X.; writing—original draft preparation, Y.X.; writing—review and editing, H.N. and X.Z.; supervision, X.Z.; project administration, X.Z.; funding acquisition, H.N. All authors have read and agreed to the published version of the manuscript.

Funding: This work was funded by the Strategic Leadership Project of the Chinese Academy of Sciences: SEANET Technology Standardization Research System Development (project no. XDC02070100).

Institutional Review Board Statement: Not applicable.

Informed Consent Statement: Not applicable.

Data Availability Statement: Not applicable.

Acknowledgments: We would like to express our gratitude to Jinlin Wang, Rui Han, and Zhiyuan Wang for their meaningful support of this work.

Conflicts of Interest: The authors declare no conflict of interest.

References

1. Cisco Annual Internet Report (2018–2023) White Paper. Available online: https://www.cisco.com/c/en/us/solutions/collateral/executive-perspectives/annual-internet-report/white-paper-c11-741490.html (accessed on 30 December 2022).
2. Ahlgren, B.; Dannewitz, C.; Imbrenda, C.; Kutscher, D.; Ohlman, B. A survey of information-centric networking. *IEEE Commun. Mag.* **2012**, *50*, 26–36. [CrossRef]
3. Jiang, X.; Bi, J.; Nan, G.; Li, Z. A survey on Information-centric Networking: Rationales, designs and debates. *China Commun.* **2015**, *12*, 1–12. [CrossRef]
4. Jacobson, V.; Smetters, D.K.; Thornton, J.D.; Plass, M.F.; Briggs, N.H.; Braynard, R.L. Networking named content. In Proceedings of the 5th International Conference on Emerging Networking Experiments and Technologies, Rome, Italy, 1–4 December 2009; pp. 1–12.
5. Zhang, L.; Afanasyev, A.; Burke, J.; Jacobson, V.; Claffy, K.; Crowley, P.; Papadopoulos, C.; Wang, L.; Zhang, B. Named data networking. *ACM SIGCOMM Comput. Commun. Rev.* **2014**, *44*, 66–73. [CrossRef]
6. Detti, A.; Melazzi, N.B.; Salsano, S.; Pomposini, M. CONET: A content centric inter-networking architecture. In Proceedings of the ACM SIGCOMM Workshop on Information-Centric Networking, Toronto, Canada, 19 August 2011; pp. 50–55.
7. Raychaudhuri, D.; Nagaraja, K.; Venkataramani, A. MobilityFirst: A robust and trustworthy mobility-centric architecture for the future internet. *ACM SIGMOBILE Mob. Comput. Commun. Rev.* **2012**, *16*, 2–13. [CrossRef]
8. Dannewitz, C.; Kutscher, D.; Ohlman, B.; Farrell, S.; Ahlgren, B.; Karl, H. Network of Information (NetInf)—An informationcentric networking architecture. *Comput. Commun.* **2013**, *36*, 721–735. [CrossRef]
9. Wang, J.; Chen, G.; You, J.; Sun, P. SEANet: Architecture and Technologies of an On-site, Elastic, Autonomous Network. *J. Netw. New Media* **2020**, *9*, 1–8.
10. Zeng, L.; Ni, H.; Han, R. An Incrementally Deployable IP-Compatible-Information-Centric Networking Hierarchical Cache System. *Appl. Sci.* **2020**, *10*, 6228. [CrossRef]
11. Christian, E. Analysis of an Equal-Cost Multi-Path Algorithm. Available online: https://datatracker.ietf.org/doc/html/rfc2992 (accessed on 30 December 2022).
12. Wu, Y.; Zhou, J. Dynamic Service Function Chaining Orchestration in a Multi-Domain: A Heuristic Approach Based on SRv6. *Sensors* **2021**, *21*, 6563. [CrossRef]
13. Andersen, D.; Balakrishnan, H.; Kaashoek, F.; Morris, R. Resilient overlay networks. In Proceedings of the Eighteenth ACM Symposium on Operating Systems Principles, Banff, AB, Canada, 21–24 October 2001; pp. 131–145.
14. Zhu, Y.; Dovrolis, C.; Ammar, M. Combining multihoming with overlay routing (or, how to be a better ISP without owning a network). In Proceedings of the IEEE INFOCOM 2007-26th IEEE International Conference on Computer Communications, Anchorage, AK, USA, 6–12 May 2007; pp. 839–847.
15. Gummadi, P.; Madhyastha, H.; Gribble, S.; Levy, H.; Wetherall, D. Improving the Reliability of Internet Paths with One-hop Source Routing. In Proceedings of the OSDI, San Francisco, CA, USA, 6–8 December 2004; p. 13.
16. Cheng, C.; Huan, Y.; Kung, H.; Wu, C. Path probing relay routing for achieving high end-to-end performance. In Proceedings of the IEEE Global Telecommunications Conference, Dallas, TX, USA, 29 November–3 December 2004; pp. 1359–1365.
17. Tang, C.; McKinley, P. Improving multipath reliability in topology-aware overlay networks. In Proceedings of the 25th IEEE International Conference on Distributed Computing Systems Workshops, Columbus, OH, USA, 6–10 June 2005; pp. 82–88.
18. Ma, P.; You, J.; Wang, J. An efficient multipath routing schema in multi-homing scenario based on protocol-oblivious forwarding. *Front. Comput. SCI-CHI.* **2020**, *14*, 1–12. [CrossRef]
19. Mukherjee, S.; Baid, A.; Seskar, I.; Raychaudhuri, D. Network-assisted multihoming for emerging heterogeneous wireless access scenarios. In Proceedings of the IEEE 25th Annual International Symposium on Personal, Indoor, and Mobile Radio Communication, Washington, DC, USA, 2–5 September 2014; pp. 1520–1524.
20. Zhou, J.; Tewari, M.; Zhu, M.; Kabbani, A.; Poutievski, L.; Singh, A.; Vahdat, A. WCMP: Weighted cost multipathing for improved fairness in data centers. In Proceedings of the Ninth European Conference on Computer Systems, Amsterdam, The Netherlands, 14–16 April 2014; pp. 1–14.
21. Raju, J.; Garcia-Luna-Aceves, J. A new approach to on-demand loop-free multipath routing. In Proceedings of the Eight International Conference on Computer Communications and Networks, Boston, MA, USA, 11–13 October 1999; pp. 522–527.
22. Mannie, E. Generalized Multi-Protocol Label Switching (GMPLS) Architecture. Available online: https://www.rfc-editor.org/rfc/rfc3945 (accessed on 30 December 2022).
23. Filsfils, C.; Previdi, S.; Ginsberg, L.; Decraene, B.; Litkowski, S.; Shakir, R. Segment Routing Architecture. Available online: https://www.rfc-editor.org/rfc/rfc8402 (accessed on 30 December 2022).

24. Bhatia, R.; Hao, F.; Kodialam, M.; Lakshman, T. Optimized network traffic engineering using segment routing. In Proceedings of the 2015 IEEE Conference on Computer Communications (INFOCOM), Hong Kong, China, 26 April–1 May 2015; pp. 657–665.
25. Lim, H.; Lim, C.; Hou, J.C. A coordinate-based approach for exploiting temporal-spatial diversity in wireless mesh networks. In Proceedings of the 12th annual international conference on Mobile computing and networking, Los Angeles, CA, USA, 23–29 September 2006; pp. 14–25.
26. Tsai, M.; Yang, H.; Huang, W. Axis-based virtual coordinate assignment protocol and delivery-guaranteed routing protocol in wireless sensor networks. In Proceedings of the 2015 IEEE Conference on Computer Communications (INFOCOM), Anchorage, AK, USA, 6–12 May 2007; pp. 2234–2242.
27. Yi, C.; Afanasyev, A.; Wang, L.; Zhang, B.; Zhang, L. Adaptive forwarding in named data networking. *ACM SIGCOMM Comput. Commun. Rev.* **2012**, *42*, 62–67. [CrossRef]
28. Udugama, A.; Zhang, X.; Kuladinithi, K.; Goerg, C. An on-demand multi-path interest forwarding strategy for content retrievals in ccn. In Proceedings of the 2014 IEEE network operations and management symposium (NOMS), Krakow, Poland, 5–9 May 2014; pp. 1–6.
29. Lei, K.; Wang, J.; Yuan, J. An entropy-based probabilistic forwarding strategy in named data networking. In Proceedings of the 2015 IEEE international conference on communications (ICC), London, UK, 8–12 June 2015; pp. 5665–5671.
30. Su, J.; Tan, X.; Zhao, Z.; Yan, P. MDP-based forwarding in named data networking. In Proceedings of the 2016 35th Chinese Control Conference (CCC), Chengdu, China, 27–29 July 2016; pp. 2459–2464.
31. Carofiglio, G.; Gallo, M.; Muscariello, L. Optimal multipath congestion control and request forwarding in information-centric networks: Protocol design and experimentation. *Comput Netw.* **2016**, *110*, 104–117. [CrossRef]
32. Chiocchetti, R.; Perino, D.; Carofiglio, G.; Rossi, D.; Rossini, G. Inform: A dynamic interest forwarding mechanism for information centric networking. In Proceedings of the 3rd ACM SIGCOMM Workshop on Information-Centric Networking, Hong Kong, China, 12 August 2013; pp. 9–14.
33. Bastos, I.; Moraes, I. A forwarding strategy based on reinforcement learning for Content-Centric Networking. In Proceedings of the 2016 7th International Conference on the Network of the Future (NOF), Búzios, Brazil, 16–18 November 2016; pp. 1–5.
34. Dang, S.; Han, R. An In-Network Cooperative Storage Schema Based on Neighbor Offloading in a Programmable Data Plane. *Future Internet* **2021**, *14*, 18. [CrossRef]
35. Xu, Y.; Ni, H.; Zhu, X. An Effective Transmission Scheme Based on Early Congestion Detection for Information-Centric Network. *Electronics* **2021**, *10*, 2205. [CrossRef]
36. Liao, Y.; Sheng, Y.; Wang, J. A deterministic latency name resolution framework using network partitioning for 5G-ICN integration. *Int. J. Innov. Comput. Inf. Control* **2019**, *15*, 1865–1880.
37. Song, Y.; Ni, H.; Zhu, X. An enhanced replica selection approach based on distance constraint in ICN. *Electronics* **2021**, *10*, 490. [CrossRef]
38. Zhou, T.; Sun, P.; Han, R. An Active Path-Associated Cache Scheme for Mobile Scenes. *Future Internet* **2022**, *14*, 33. [CrossRef]
39. Boutremans, C.; Iannaccone, G.; Diot, C. Impact of link failures on VoIP performance. In Proceedings of the 12th International Workshop on Network and Operating Systems Support for Digital Audio and Video, Miami, FL, USA, 12–14 May 2002; pp. 63–71.
40. Li, B.; Wang, J. An Identifier and Locator Decoupled Multicast Approach (ILDM) Based on ICN. *Appl. Sci.* **2021**, *11*, 578. [CrossRef]
41. Liao, J.; Tian, S.; Wang, J.; Li, T.; Qi, Q. Load-balanced one-hop overlay multipath routing with path diversity. *Ksii. T Internet Inf.* **2014**, *8*, 443–461.
42. Brandes, U. On variants of shortest-path betweenness centrality and their generic computation. *Soc. Netw.* **2008**, *30*, 136–145. [CrossRef]
43. Ekici, F.; Gözüpek, D. Joint overlay routing and relay assignment for green networks. *Comput. Netw.* **2015**, *79*, 323–344. [CrossRef]
44. Bui, V.; Zhu, W.; Bui, L. Optimal relay placement for maximizing path diversity in multipath overlay networks. In Proceedings of the 2008 IEEE Global Telecommunications Conference, New Orleans, LA, USA, 30 November–4 December 2008; pp. 1–6.
45. Wu, F.; Yang, W.; Sun, M.; Ren, J.; Lyu, F. Multi-path selection and congestion control for ndn: An online learning approach. *IEEE Trans. Netw. Serv.* **2020**, *18*, 1977–1989. [CrossRef]
46. ns-3 Network Simulator. Available online: https://www.nsnam.org (accessed on 1 February 2023).
47. Yampolskiy, M.; Hamm, M. Management of multidomain end-to-end links-a federated approach for the pan-european research network geant 2. In Proceedings of the 10th IFIP/IEEE International Symposium on Integrated Network Management, Munich, Germany, 21–25 May 2007; pp. 189–198.
48. Moy, J. OSPF Version 2. Available online: https://www.rfc-editor.org/rfc/rfc2178 (accessed on 1 February 2023).
49. Bari, M.; Boutaba, R.; Esteves, R.; Granville, L.; Podlesny, M.; Rabbani, M.; Zhani, M. Data center network virtualization: A survey. *IEEE Commun. Surv. Tut.* **2012**, *15*, 909–928. [CrossRef]

future internet

Article

A Replica-Selection Algorithm Based on Transmission Completion Time Estimation in ICN

Zhiyuan Wang [1,2], Hong Ni [1,2] and Rui Han [1,2,*]

[1] National Network New Media Engineering Research Center, Institute of Acoustics, Chinese Academy of Sciences, No. 21 North Fourth Ring Road, Haidian District, Beijing 100190, China
[2] School of Electronic, Electrical and Communication Engineering, University of Chinese Academy of Sciences, No. 19(A) Yuquan Road, Shijingshan District, Beijing 100049, China
* Correspondence: hanr@dsp.ac.cn

Abstract: As the Internet communication model changes from host-centric to content-centric, information-centric networking (ICN) as a new network architecture has received increasing attention. There are often multiple replicas of content in ICN, and how to reasonably utilize the characteristics of multiple replicas to further improve user experience is an important issue. In this paper, we propose a replica-selection algorithm, called the transmission completion time estimation (TCTE) algorithm. TCTE maintains the state of replica nodes in the domain with passive measurements in a limited domain of an enhanced name resolution system (ENRS), then estimates the transmission completion time of different replica nodes and selects the smallest one. When no replica is found in the ENRS domain, the nearest-replica algorithm will be used, so TCTE will not increase the traffic in the core network. Experiments show that TCTE not only effectively improves the user's download rate and edge node throughput, reduces download rate fluctuations, reduces user download delay, and improves fairness, but also has universal applicability.

Keywords: information-centric networking; replica selection; name resolution

Citation: Wang, Z.; Ni, H.; Han, R. A Replica-Selection Algorithm Based on Transmission Completion Time Estimation in ICN. *Future Internet* **2023**, *15*, 120. https://doi.org/10.3390/fi15040120

Academic Editors: José Carlos Lopez-Ardao, Miguel Rodríguez Pérez, Sergio Herrería Alonso and Gyu Myoung Lee

Received: 6 February 2023
Revised: 13 March 2023
Accepted: 23 March 2023
Published: 25 March 2023

1. Introduction

As the number of users increases, the huge demand for higher data rates and bandwidth puts the current Internet under increasing pressure. The communication model of the Internet is changing from a host-centric to a content-centric paradigm [1]. The user's motivation is to obtain the requested content, not to establish a connection with the service provider's device. In the current end-to-end communication model, users need to know the IP of the service provider to establish a connection to transmit content. When a large number of users obtain the same content, all users need to communicate with the server provider's device. Duplicate transmission of content causes a lot of wasted bandwidth. Many ICN architectures have been proposed, such as CCN [2], NDN [3], Mobilityfirst [4], NetInf [5], DONA [6], and SEANet [7]. In recent years, the feasibility of combining ICN with software-defined network (SDN) [8], intent-based networking (IBN) [9], and other technologies has also attracted new attention to ICN.

ICN separates content from its producer's location [10]. Users can complete content acquisition from any node in the network that caches that replica based on the content's name. The caching of replica nodes in the network puts the content closer to the user. Therefore, the traffic in the core network can be effectively alleviated. In ICN, the content is a named data chunk (NDC) with a globally unique name. As the routers in the ICN network can cache the content, there are multiple replicas of content in the network. When a user requests content, how to find and select a suitable replica node to improve the user download rate and network throughput is an important issue, which has been studied by many researchers [11,12]. According to the approach of name resolution, ICN can be divided into two types: the approach of name-based routing (NBR), in which the process of name routing is the process of replica selection, and the approach of standalone name resolution (SNR). NDN belongs to the former. It forwards the request to the content producer. The replica node can only be selected when it is on the forwarding path, and the replica node outside the path cannot be used. Although many works have been studied, the NBR-based approach is more complex for the process of discovering replica nodes. These approaches either require a cluster head for decision making, or forwarding based on ICN router local information [11]. SNR-based approaches such as [4] can easily obtain a list of network addresses (NA) of replica nodes in the name resolution system (NRS), and then select a replica node in the list. However, the existing replica-selection algorithm does not consider the expected transmission time of NDC. What users need most is to obtain a high and stable transmission rate, so as to have a low transmission completion time.

In this paper, we proposed the novel replica-selection algorithm based on transmission completion time estimation (TCTE). When the replica node is in the ENRS [13] domain, TCTE selects the replica node with the shortest transmission completion time; when the replica node is outside the domain, TCTE selects the nearest replica node. Therefore, TCTE mainly affects the traffic distribution within the ENRS domain without increasing the traffic in the core network.

The main contributions of this paper are as follows:

- To solve the problem of replica node selection in ICN network, we propose the replica-selection algorithm based on the transmission completion time estimation. The transmission completion time of NDC is related to the size of NDC, the expected transmission rate, and the round-trip time (RTT). Therefore, we designed a method to obtain the expected transmission rate, RTT, and NDC size.
- To solve the problem that the replica nodes may not be selected again due to bad performance in a short time, we design an "activation" mechanism.
- We conducted experiments on the proposed algorithm. Experiments show that TCTE not only effectively improves the user's download rate and edge node throughput, reduces download rate fluctuations, reduces user download delay, and improves fairness, but also has universal applicability.

The remainder of this paper is organized as follows. Section 2 discusses the existing literature on the topic. Section 3 depicts the NDC transmission process and the name resolution service. Section 4 describes our proposed algorithm. Section 5 evaluates the effectiveness of our proposed algorithm. Section 6 concludes our work.

2. Related Works

In ICN networks, content can be cached in different locations, so each piece of content may have multiple replicas, and it is an important issue how to use multiple replicas of ICN effectively. Table 1 shows the characteristics of the related work.

A name resolution service is used to help clients access content, hosts, or services using names, and there are two main approaches, the name-based routing approach (NBR) and the standalone name resolution approach (SNR) [14]. The ICN architecture has different approaches to name resolution services and different issues when performing replica selection.

In the ICN architecture using the name-based routing approach, such as NDN [3], the ICN router needs to select the outgoing interface to forward the content request according to the information it maintains. The process of content request forwarding is the process of replica selection. The shortest-path routing algorithm forwards content requests to content providers, such as OSPFN [15] and NLSR [16], so only replicas on the routing path can be selected. The algorithm has lower overheads, but users have fewer replicas of the content available to them. Many algorithms have been proposed to efficiently utilize content replicas on non-shortest-routing paths. The cluster-based method [17,18], through the cluster head node centralized caching decision and forwarding decision when receiving a content request, can realize the replica selection in the cluster. iNRR [19] explores the network in a scope-flooding approach using the first content request. In SEARCH-CNG [20], link bandwidth is defined as the ratio of the link's capacity to the number of flows over this link, and the algorithm makes forwarding decisions based on link bandwidth to avoid congested links. Both iNRR and SEARCH-CNG are flood-based methods that explore in-network replicas, thereby reducing traffic on the default forwarding path. INFORM [21] is a dynamic forwarding mechanism based on the Q-learning algorithm [22]. It discovers content replicas through flooding, and then selects forwarding paths based on indicators such as RTT. Stateful forwarding [23] is also a learning-based forwarding mechanism. The ranking value is the basis for the forwarding decision of the algorithm. It represents the priority of the forwarding path, which depends on various performance indicators, such as RTT and link state, etc. In SCAN [24], the single-hop neighbor nodes exchange the information of their cached content regularly through the Bloom filter [25]. When the content request arrives at the node, if the content exists, it will be forwarded directly to the client; if it does not exist, the content will be searched for whether it is cached in a neighbor. RFW [26] is based on random walks. A layered ICN architecture is proposed in RFW, the publisher is located in the first layer, and the client is located in the bottom layer. When the content request arrives at the layer N, the maximum random walking time is T. When the walking time exceeds T, it will be forwarded to the $N - 1$ layer until a replica of the content is found.

In the ICN architectures that use the standalone name resolution approach, such as Mobilityfirst [4] and NetInf [5], the client first obtains the locator of the content from the name resolution system (NRS), then selects a replica, and uses the content locator for routing to complete the content acquisition [27,28]. Compared with the ICN architecture using NBR, the ICN architecture using SNR can easily obtain the locators of multiple content replicas because of the existence of NRS, and then obtain the content. ERS [12] uses the distance-constrained-based name resolution system [13] to discover nearby replicas, and then calculates a node status value based on the maintained replica node queuing delay, outstanding requests, RTT, and the network distance for replica selection. Mobilityfirst uses a global name resolution system (GNRS) to obtain the locator of the replica, and then sends the content request to the nearest replica node according to the routing table. NetInf obtains the locators of all replica nodes, and selects the replica node according to factors such as network distance and delay.

In a distributed system, the problem of replica node selection also exists. Algorithms can be divided into information-independent algorithms and information-aware algorithms. Information-independent algorithms mainly include fixed strategies, random strategies, and cyclic strategies. The advantage of this type of algorithm is that it does not require perceptual information, so the overhead is low, but with the static replica-selection strategy it is often difficult to achieve the optimal effect [29]. The information-aware algorithm actively or passively measures information such as RTT, bandwidth, and loads of different replica nodes, and selects replicas based on the measurement information. Carter et al. [30] measure the available bandwidth and round-trip time (RTT) of different replica nodes, and estimate the time required for content download to guide the selection of replica nodes. However, since the algorithm needs to actively perform network detection, it will occupy additional network bandwidth. Ping-random [31] selects a set containing the five best-performing replica nodes based on ping times, and the client randomly selects a replica node in the set when making replica selection. The two random choices (2RC) [32] algorithm randomly selects two replica nodes and chooses the less-loaded node to process the request.

The approach of NBR-based ICN architecture is not suitable for NRS-based ICN architecture. MF's nearest-replica approach and ERS are suitable for the ICN architecture of NRS, but they do not consider the NDC download rate and NDC download delay that users are most concerned about. The approaches in traditional networks are often not directly applicable to ICN. This paper proposes a new replica-selection approach suitable for the ICN architecture of NRS. The main idea of TCTE is similar to the literature [30], which is based on the estimated value of the transmission completion time for replica selection. However, their replica node information maintenance and replica-selection steps are completely different.

Table 1. Characteristics of the related work.

Application Scenarios	Approaches	Characteristics
NBR-based ICN architecture	Shortest-path routing	Only replicas on the path can be selected.
	Cluster-based method	Cluster head node decision.
	iNRR	Select the nearest replica by flooding.
	SEARCH-CNG	Avoid selecting replicas on congested paths.
	INFORM	Use flooding to discover replicas, and complete replica selection based on indicators such as RTT.
	Stateful forwarding	Calculate the ranking value of the link and select the replica on the better link.
	SCAN	The information exchange of the cache node can select the neighbor replica of the node on the path.
	RFW	Applicable to a hierarchical ICN architecture, the lower-level replica is selected by random walk.
NRS-based ICN architecture	MF	Select the closest replica based on a routing table.
	ERS	Based on weighted values of congestion, RTT, and hops.
	NetInf TP	Based on latency and network distance. (No approach details provided by author.)
Traditional network	Reference [30]	Proactively detects the network and selects the replica with the lowest flow completion time.
	Ping-random	Randomly selects one of the five smallest replicas of RTT.
	2RC	Randomly takes two replica nodes and chooses the one with the smaller load.

3. NDC Transmission and NRS Overview

Due to the existence of a large number of IP infrastructures, the evolutionary deployment of ICN networks is a realistic consideration [33]. The proposed replica-selection algorithm is based on the ICN architecture compatible with IP infrastructure. In this section, we introduce the named data chunk (NDC) transmission process and name resolution system (NRS) of the model.

3.1. Overview of the NDC Transmission Process

In ICN, each piece of content is an NDC whose size is reasonable, from hundreds of KBs to a few MBs [34,35]. The ICN router caches the NDC to serve as a replica node. The identifier (ID) and the locator in the ICN are separated, the identifier can identify the NDC and the device in the network, and the locator is the network address (NA). The client sends a content request (REQ) packet to the network, and the replica node responds to the REQ packet and replies with a content data (DATA) packet. Both the REQ packet and the DATA packet will carry a source ID and a destination ID. In the REQ packet, the source ID identifies the client device, and the destination ID identifies the NDC. In the DATA packet, the source ID identifies the NDC, and the destination ID identifies the client device. An NDC transmission process can be identified according to the device ID and the NDC ID.

Due to the consideration of flow and congestion control, the replica node cannot send all the data of NDC at once, otherwise it will cause traffic bursts in the network, and even cause the network to crash. The REQ packet will carry the segmentation information of the data, indicating the starting offset and length of the requested data in the NDC. Clients can perform receiver-driven congestion control by controlling the size of the requested data per REQ packet and the frequency at which REQ packets are sent. In this paper, we use Copa-ICN [36] as the congestion control algorithm of the client, which has a good performance in terms of algorithm convergence speed and fairness. In addition, packet loss is hard to avoid during transmission. When data loss occurs, the client will re-request the lost data to ensure reliable transmission of NDC.

When the replica node receives the REQ packet, it will read the NDC data from the cache according to the request data segmentation information in the REQ packet and assemble it into the DATA packet to send. Due to the limitation of the maximum transmission unit (MTU), the replica node may split the data requested by an REQ into several DATA packets and send them. In addition, the ICN router can use the late-binding technology [13], and if the current replica node cannot retrieve the cache of the NDC, the replica node can initiate a query to the NRS and forward its REQ packet to another replica node until the REQ packet reaches the replica node that caches its requested NDC.

Figure 1 is a schematic diagram of the NDC transmission process. Due to the support of the late-binding technology, when the client acquires an NDC, it can directly send an REQ packet to the edge node without performing name resolution. After receiving the first REQ packet, the edge node will request a network address (NA) list of replica nodes from the NRS, and use the replica-selection algorithm to select a replica node RN3 (in Figure 1) to forward the REQ packet. After receiving the REQ, RN3 replies with the DATA packet according to the REQ. The DATA packet will be forwarded to the client. It should be noted that the client can obtain the size of the NDC only after receiving the first DATA packet.

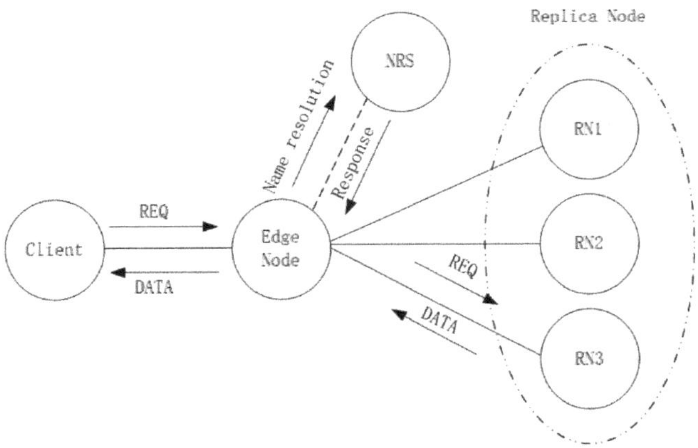

Figure 1. Diagram of NDC transmission process.

3.2. Name Resolution System

Name resolution services are provided by the name resolution system (NRS). NRS types include the enhanced name resolution system (ENRS) [13] and global name resolution (GNR). ENRS is a name resolution service of the SNR approach. It partitions space into hierarchical domains based on distance constraints. No two domains at the same level overlap. ENRS can implement low-latency name resolution services within a limited domain, while inter-domain name resolution services are provided by global name resolution (GNR). The number of layers of ENRS has no effect on our study of replica-selection algorithms, so in this paper, we use a single layer of ENRS to implement our algorithm.

Figure 2 is a schematic diagram of a single-layer ENRS. The ICN router will register its network address (NA) and cached NDC ID in the ENRS and GNR of its domain. When obtaining content, the content-requesting entity will first query ENRS for a node NA list with the content according to the ID of the content. If the query result is empty, it will continue to query GNR. Finally, the replica-selection algorithm is used to select a replica node for data transmission.

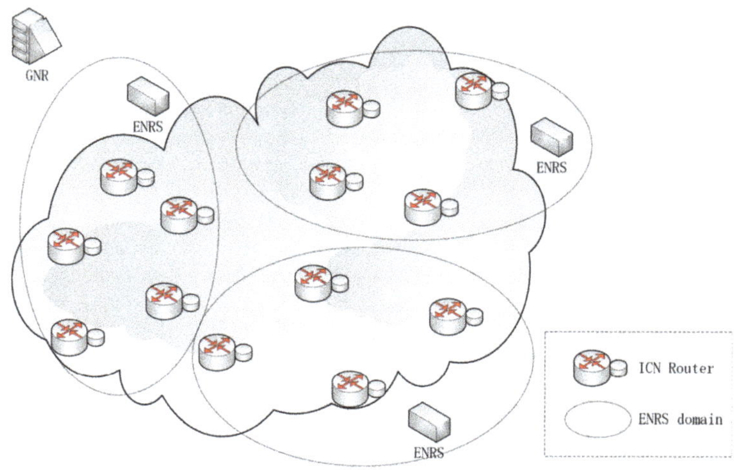

Figure 2. ENRS schematic diagram.

4. Algorithm Description

In this section, we describe the TCTE algorithm in detail and analyze the algorithm overheads.

4.1. Motivation

In an ICN, the replica-selection algorithm will directly affect the user experience. A better replica-selection algorithm can effectively increase the user's download rate and ensure a relatively stable rate, which is conducive to improving the overall throughput of the network.

Figure 3 is a schematic diagram of replica selection. Assuming that the user to R1 node is not the bottleneck, if RN1 and RN2 can be used reasonably, the speed of users' downloading NDC can be improved. In addition, users and replica nodes (RN1, RN2) are all in the same ENRS domain, so it will not increase the traffic in the core network. Considering the replica-selection algorithm based on the estimation of transmission completion time, if the estimated value of transmission completion time of the replica node of RN2 is lower than that of RN1, RN2 will be selected to provide services. After RN2 is selected once, when the replica selection is performed again, because some resources of RN2 are occupied, the estimated value of the transmission completion time may be higher than that of RN1, and RN1 will be selected at this time. By continuously selecting replica nodes with smaller transmission completion time estimates, a reasonable distribution of traffic among different replica nodes can be achieved. Therefore, we will propose a replica-selection algorithm based on transmission completion time estimation, called TCTE. TCTE enables better replica selection based on transmission completion time estimates of replica nodes in the domain.

4.2. Overview of TCTE Algorithm

Compared with the traditional IP network, ICN enables users to obtain content from a closer location through the cache replica in the network, which not only improves the user experience, but also effectively reduces the traffic of the core network, which is conducive to improving the overall performance of the network. Implementing a replica-selection algorithm based on transmission completion time estimation can improve user experience, but the replica node with the smallest estimated transmission completion time may also be a far-away replica node, which will increase traffic in the core network. In addition, it is extremely difficult and expensive to maintain replica node information within the entire network. Therefore, we implement the approach based on transmission completion time estimation within the ENRS [13] domain. On the one hand, a set of replica nodes returned by ENRS is in the same ENRS domain as the user, without obtaining a replica node that is far away. On the other hand, the overhead of maintaining replica node information in the ENRS domain is controllable. TCTE performs more complex replica-selection operations within the ENRS domain. If no replicas can be found within the ENRS domain, the TCTE will take the replica nodes obtained from the GNR. In this case, we use the nearest-replica selection algorithm as in Mobilityfirst to avoid increasing the traffic in the core network. Performing replica selection requires better awareness of the network, but the information that the client can perceive is limited. Edge routers [37] can aggregate user requests, while, due to the large number of data passing through they can better perform passive measurements on the network. Furthermore, because of the support of late-binding technology, it is a reasonable choice to run the replica-selection algorithm on the edge nodes. Inspired by the work of [12], we maintain the status of the replica nodes in the ENRS domain at the edge nodes.

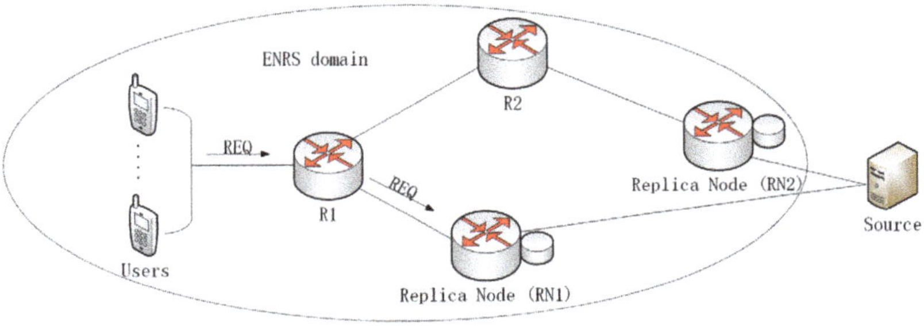

Figure 3. Schematic diagram of replica selection.

When a replica node is selected, we try to estimate the transmission completion time T for the NDC.

$$T = a * RTT + b * \frac{ChunkSize}{Rate}, \quad (1)$$

where RTT is the round-trip time between the edge node and the replica node. For a replica node in the ENRS domain, we set a timeout value of 10 s (or other reasonable values) for its RTT. When the RTT of a replica node expires, five REQ packets forwarded to the replica node will be marked for re-measurement. Take the minimum value among the measured values as the RTT of the replica node. *ChunkSize* is the size of NDC, and *Rate* is the expected transmission rate of the replica node. Parameters a and b are adjustable coefficients. *Rate* is calculated based on passive measurements of edge nodes and will be described in detail in Section 4.3. *ChunkSize* needs to complete the transmission of at least one DATA packet before it can be obtained. Therefore, before estimating the T value, it is necessary to first select a replica node to obtain a DATA packet. In order to obtain *ChunkSize* as soon as possible, the first REQ packet of NDC will be forwarded to the replica node with the smallest RTT. After the edge node receives the first DATA packet and obtains the *ChunkSize* of the NDC, it reselects the replica node with the smallest T value for NDC transmission, and saves the selection result in the replica-selection completion list. When receiving the REQ packet of the transmission process again, the edge node will directly obtain the replica node NA in the replica-selection completion list to forward the REQ packet.

The TCTE algorithm runs on the edge nodes. Algorithm 1 is the operation when TCTE receives the REQ packet. When receiving the REQ packet, TCTE first checks whether there is information about the completion of the replica selection of the NDC transmission process. If it exists, it means that the replica selection has been completed, and it then obtains the selected replica node NA and forwards the REQ. If it does not exist, the replica selection is performed. First, it retrieves the ID of the NDC from the REQ packet, and then obtains the NA list *NAList* of the replica node from the ENRS according to the ID. If the *NAList* is not empty, it selects the replica node with the smallest RTT to forward the REQ, saves the *NAList*, and marks that replica selection needs to be re-selected. If *NAList* is empty, it continues to obtain the *NAList* of the replica node from GNR. Then, it selects the nearest replica node to forward REQ, and saves the selection result in the replica-selection completed list.

Algorithm 1: Operation of the TCTE algorithm when receiving an REQ packet

```
Input: REQ
Output: NA
NA = GetNAFromReplicaSelectionCompleteList(REQ)
if NA != NULL then
    return NA
else
    EID = GetEID(REQ)
    NAList = GetNameResolutionFromENRS(EID)
    if NAList != NULL then
        Reselection = true
        NA = FindMinRTTNA(NAList)
        SaveNAList(NAList, REQ, Reselection)
    else
        NAList = GetNameResolutionFromGNR(EID)
        if NAList != NULL then
            NA = FindNearestNA(NAList)
        end if
        SaveToReplicaSelectionCompleteList(REQ, NA)
    end if
end if
return NA
```

Algorithm 2 is the operation when TCTE receives the DATA packet. When receiving the DATA packet, TCTE will check whether it needs to re-select the replica node. If it is necessary to re-select the replica node, TCTE will take out the previously saved *NAList* and obtain the *ChunkSize* from the DATA packet, then calculate the estimated value of the transmission completion time of the replica node in the *NAList* according to the status information of the replica node, and select the replica node with the smallest value and save it in the replica-selection completed list.

Algorithm 2: Operation of the TCTE algorithm when receiving a DATA packet

```
Input: DATA
Output: NULL
if is Reselection (DATA) then
    NAList = GetNAList(DATA)
    ChunkSize = GetChunkSize(DATA)
    T_min = MaxVal
    for NA_tmp in NAList then
        T_tmp = EstimatedTransmissionCompletionTime(NA_tmp, ChunkSize)
        if T_min > T_tmp then
            NA = NA_tmp
            T_min = T_tmp
        end if
    end for
    SaveToReplicaSelectionCompleteList(DATA, NA)
end if
return
```

4.3. Estimated Transmission Rate of Replica Nodes

We perceive the changes in the transmission rate of the replica nodes by passive measurements. For the measurements of the transmission rate, we smooth based on the time interval between two rate measurements, to prevent large jitters in the rate estimate. In addition, in order to solve the problem that the replica node may not be selected again due to poor performance in a short period of time, we also designed an "activation" mechanism.

We maintain the *RateRN* value of the replica node, which represents the total rate of DATA packets transmitted between the edge node and the replica node.

Therefore, the transmission rate *Rate* of the replica node is:

$$Rate = \frac{RateRN}{\max(N_c, 1)}, \tag{2}$$

where N_c is the number of NDC transmission processes that the current edge node uses the replica node.

RateRN consists of two parts:

$$RateRN = RateRN_{real} + RateRN_{extra}, \tag{3}$$

where we update $RateRN_{real}$ based on the measurement of the transmission rate of DATA packets between the replica node and the edge node.

$$RateRN_{real} = RateRN_{real} * (1 - X(t)) + \frac{DataSum}{t} * X(t), \tag{4}$$

where t is the time since the last $RateRN_{real}$ update. *DataSum* is the total size of DATA packets received within time t. $X(t)$ is a function about $t : X(t) = c * t/(1 + c * t)$ (c is an adjustable constant). $X(t)$ can dynamically adjust the weight of the historical value and the current measurement value according to the time t. When t is larger, the validity of the historical value is lower, so the weight of the current measurement value is larger. Conversely, historical values have a larger weight to smooth changes in the $RateRN_{real}$ value.

When the *Rate* calculated by the replica node is too small to be selected again, its $RateRN_{real}$ will become 0, because there is no new data transmission. Even if its link condition improves, we cannot perceive it. Therefore, we add $RateRN_{extra}$:

$$RateRN_{extra} = \begin{cases} RateRN_{extra} + t * d, & if\ RateRN_{real} = 0 \\ 0, & if\ RateRN_{real} != 0 \end{cases}, \tag{5}$$

where t is the time since the last $RateRN_{extra}$ update, and d is an adjustable constant.

In this way, even if there is no new data transmission, since $RateRN_{extra}$ increases over time, the replica node will be reselected after a period of time, thus being "activated".

4.4. Overhead Analysis

We will analyze the communication overhead, memory overhead, and computation overhead of running the TCTE algorithm on edge nodes. TCTE uses passive measurement to perceive the link status between the replica node and the edge node, and does not send additional packets into the network, so the TCTE algorithm does not bring additional communication overhead. The memory overhead of TCTE is O(n). The more replica nodes in the ENRS domain, the greater the memory overhead of TCTE. In our implementation, TCTE needs to occupy about 200 bytes of memory for a replica node. However, even maintaining the state of 10,000 replica nodes only requires a few MB of overhead, and within a reasonable ENRS domain, it will not cause a large burden on memory. The number of replica nodes in the domain is n. The maximum value of TCTE computation overhead for completing a replica selection is O(n). When a certain NDC has caches in all replica nodes, the computational overhead of TCTE reaches the maximum. However, most of the time only a few replica nodes have the same replica, so the maximum cost of a replica selection will be far less than O(n). Therefore, the additional overhead of TCTE is acceptable.

5. Performance Evaluation

In order to verify the effectiveness of the TCTE algorithm, in this section, we use the NS-3 simulator for simulation experiments. In previous work, the ICN transport protocol has been implemented in NS-3 [38]. The transmission protocol is receiver-driven and the congestion control algorithm is Copa-ICN, which can complete the data transmission from a replica node in the network based on the ID of the chunk [36]. We implemented TCTE in the edge node (router). In addition, to illustrate the effect of TCTE, we also implemented the intra-domain random selection algorithm (IDRS) with the nearest-replica algorithm (NR) [4,12].

5.1. Experimental Setup

In the experiments, we randomly generate 1000 named data chunks (NDC) between 100 KB and 10 MB, and each chunk has a globally unique ID. These chunks are randomly placed among replica nodes within the ENRS domain. The replica node registers the mapping between the cached chunk ID and its NA in the NRS. There are several users under the edge node to continuously obtain named data chunks. Each time the user acquires a chunk, a chunk will be acquired again after the transmission of the previous chunk is completed. After several tests, a is set to 2 and b is set to 1 in expression (1). The value of c in function $X(t)$ is set to 0.005. The user's request follows the Zipf distribution [37,39]. The probability that the NDC i is requested is $q_i = e * i^{-\alpha}$, $e = 1/(\sum_{i \in O} i^{-\alpha})$, and the value of α indicates the degree of concentration of the request. In the experiment, the value of α is 1. We set up two different experimental scenarios to evaluate the performance of the TCTE algorithm. The replica nodes and edge nodes in the experiments are both in the same ENRS domain.

Figure 4 is the topology of Experimental Scenario 1. In this scenario, different replica nodes do not share bottleneck links (the link between the user and the edge node is not the bottleneck). It contains fifteen replica nodes and one edge node. Between the edge node and each replica node, between zero and five routing nodes are randomly generated. The link bandwidth between every two nodes on the path from the edge node to the replica node is a random number between 20 Mbps and 100 Mbps, and the link delay is set to 2 ms. A total of 60 users are set under edge nodes, and users continue to send requests within the experimental time. The simulation experiment time is 100 s.

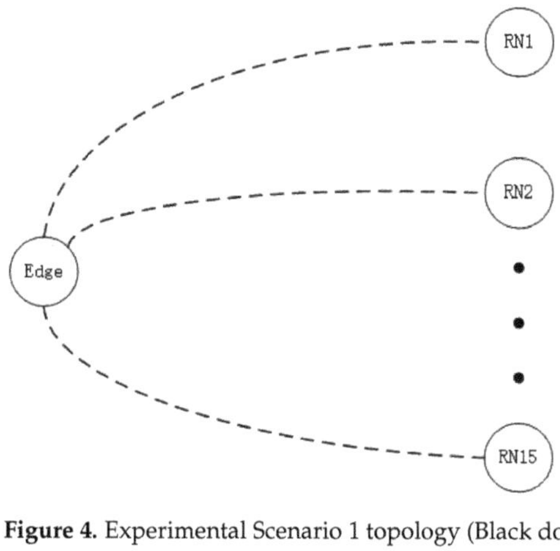

Figure 4. Experimental Scenario 1 topology (Black dots represent replica nodes RN3 to RN14).

Figure 5 is the topology of Experimental Scenario 2. In this scenario, some replica nodes share the bottleneck link, and some replica nodes do not share the bottleneck link. It is a four-layer network topology containing fourteen replica nodes and one edge node. The bandwidth of each link is a random value, as shown in Table 2. The link between the user and the edge node is not the bottleneck. The latency of each link is 2 ms. A total of 60 users are set under edge nodes, and users continue to send requests within the experimental time. The simulation experiment time is 100 s.

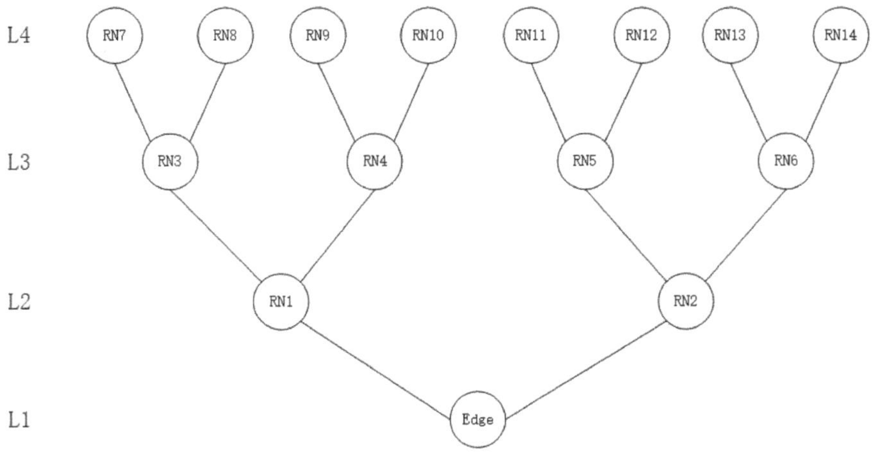

Figure 5. Experimental Scenario 2 topology.

Table 2. Bandwidth settings for Experiment Scenario 2 topology.

Link	Bandwidth (Mbps)
Between L1 and L2	Random values (20 to 100)
Between L2 and L3	Random values (100 to 200)
Between L3 and L4	Random values (200 to 300)

5.2. *Performance Comparison*

5.2.1. User's Download Rate

Figure 6a shows the variation in the download rate over time for a user in Experiment Scenario 1. We can see that the TCTE algorithm has the highest user download rate most of the time, while the IDRS and NR algorithms only have the highest user download rate very few times. In addition, the TCTE algorithm has a more stable download rate than the other two algorithms, and the other two algorithms have large fluctuations in the user download rate during the experimental time. Figure 6b shows the variation in a user's download rate over time in Experimental scenario 2. We can see that the TCTE algorithm has the highest download rate most of the time, while IDRS and NR perform poorly. Moreover, the download rate of the TCTE algorithm still has the most stable download rate. Therefore, the TCTE algorithm can enable users to obtain a higher and more stable download rate. In addition, in order to illustrate the reliability of the experimental results, we repeated the experiment 10 times, and took the average download rate of each experiment for analysis. Figure 7 shows the mean and confidence interval (95% confidence level) for the download rate.

5.2.2. Edge Node Throughput

We measured throughput at the edge node. Figure 8a shows the variation in throughput over time in Experimental Scenario 1. TCTE has much higher throughput than the other two algorithms, followed by the IDRS algorithm, and the NR algorithm is the worst. This is because it is difficult for the IDRS and NR algorithms to reasonably distribute traffic to several replica nodes that do not share bottleneck links. The NR algorithm cannot select a farther replica node, even if the farther replica node can obtain a higher transmission rate. IDRS randomly selects replica nodes, which can somehow achieve dynamic distribution of traffic on different paths, so IDRS performs better than NR. However, it is difficult to achieve the best effect by random selection.

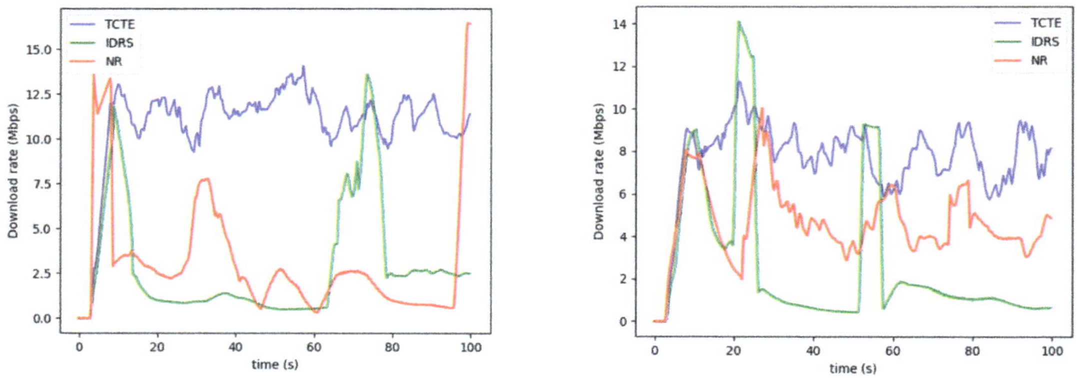

(**a**) Experiment Scenario 1. (**b**) Experiment Scenario 2.

Figure 6. User download rate over time.

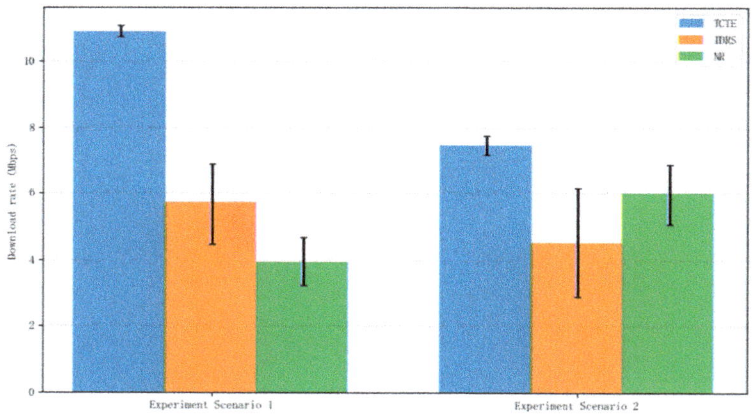

Figure 7. Mean and confidence intervals for download rates.

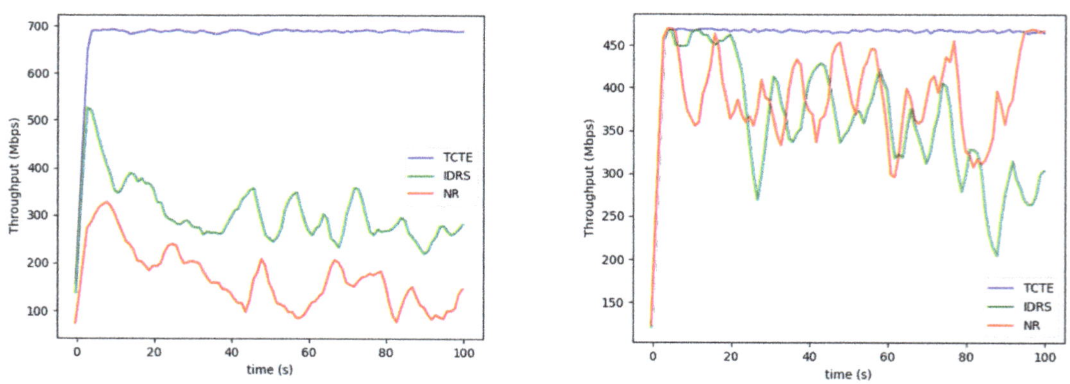

(**a**) Experiment Scenario 1. (**b**) Experiment Scenario 2.

Figure 8. Throughput of edge nodes.

Figure 8b shows the variation in throughput over time in Experimental Scenario 2. The highest throughput is still the TCTE algorithm, while IDRS and NR perform poorly. The performance of NR becomes better because the nearer replica node in this scenario has a higher probability of being an intermediate node between the farther replica node and the edge node. The bottleneck bandwidth between the edge nodes and the intermediate nodes is not lower than that between the edge nodes and the farther nodes. In addition, in both scenarios, TCTE has a very stable throughput, while the throughput of the other two algorithms has large fluctuations. Therefore, TCTE not only effectively improves the throughput of edge nodes, but also keeps the throughput stable.

In addition, in order to illustrate the reliability of the experimental results, we repeated the experiment 10 times, and took the average edge node throughput of each experiment for analysis. Figure 9 shows the mean and confidence interval (95% confidence level) for edge node throughput.

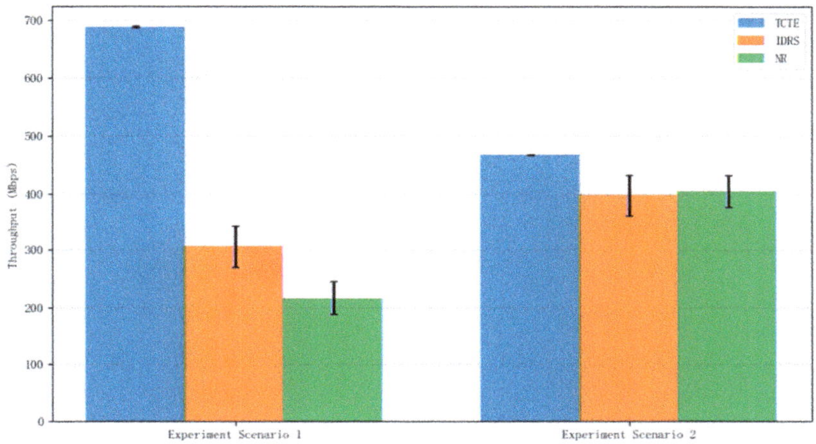

Figure 9. Mean and confidence intervals for edge node throughput.

5.2.3. Fairness

We use Jain's Fairness Index (JFI) to evaluate the fairness of the algorithm. JFI was proposed in [40], and its expression is:

$$F(x) = \frac{(\sum x_i)^2}{n(\sum x_i^2)} \tag{6}$$

where x_i is the transmission rate of user i. n is the number of users. The larger the $F(x)$ value, the better the fairness.

Figure 10 shows the variation in JFI over time in two experimental scenarios. TCTE always has the best fairness performance, while IDRS and NR have poor fairness performance. NR and IDRS perform similarly in Experimental Scenario 1, and NR outperforms IDRS in Experimental Scenario 2. IDRS and NR have high fairness at the beginning of the experiment, and then decline and stabilize at a small value. Unlike IDRS and NR, TCTE has a low JFI for a period of time at the beginning of the experiment, and then stabilizes at a value close to 1. As there were fewer data at the beginning of the experiment, TCTE could not perceive the network condition well.

In addition, in order to illustrate the reliability of the experimental results, we repeated the experiment 10 times, and took the JFI for analysis. Figure 11 shows the mean and confidence interval (95% confidence level) for JFI.

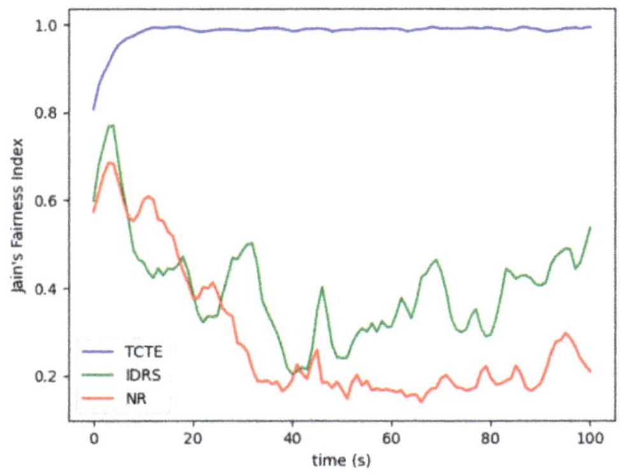

(**a**) Experiment Scenario 1.

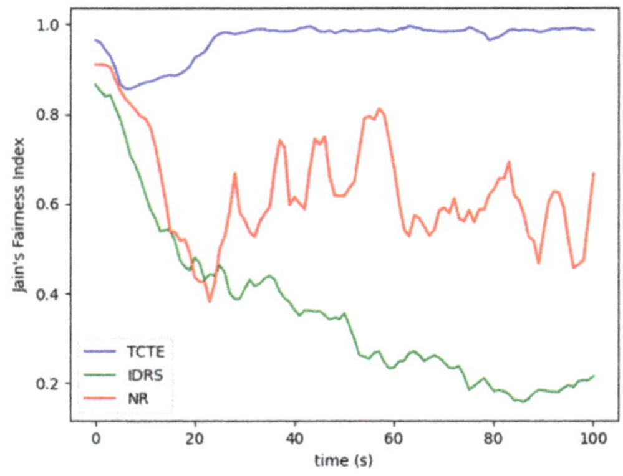

(**b**) Experiment Scenario 2.

Figure 10. Variation in Jain's Fairness Index over time.

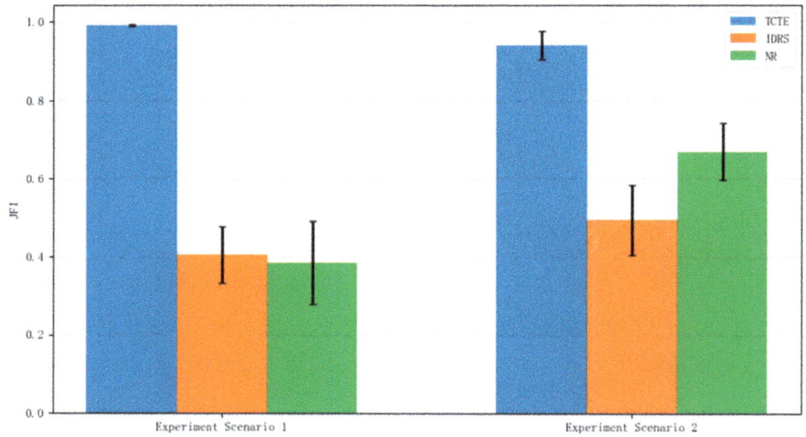

Figure 11. Mean and confidence intervals for JFI.

5.2.4. NDC Download Delay

In order to observe the relationship between NDC download delay and the number of users, we modified the number of users under the edge nodes in Experimental Scenario 1 and Experimental Scenario 2 for experiments. The number of users was 30, 40, 50, 60, 70, 80, 90, and 100. In addition, in order to make the experimental results universal, each experiment was repeated 10 times and the average and standard deviation of the average download delay of 10 experiments were extracted.

Figure 12 is the result of Experimental Scenario 1. The download delay of the TCTE algorithm is the lowest, the IDRS algorithm is second, and the NR algorithm is the worst. With the increase in the number of users, the download delay of the three algorithms continued to increase, and the gap became larger and larger. Figure 12b is a change in the standard deviation. It can be seen that TCTE has the most stable performance, and the performance of the other two algorithms has relatively large fluctuations.

Figure 13 is the result of Experimental Scenario 2. The download delay of the TCTE algorithm is the lowest, and the NR and DRS algorithm download delay is large. With the increase in the number of users, the download delay of the three algorithms continued to increase. Figure 13b is a change in the standard deviation. It can be seen that TCTE has the most stable performance, and the performance of the other two algorithms has relatively large fluctuations. Different from Experimental Scene 2 is that the NR algorithm is better than IDRS, and the gap between NR and TCTE download delay does not significantly expand.

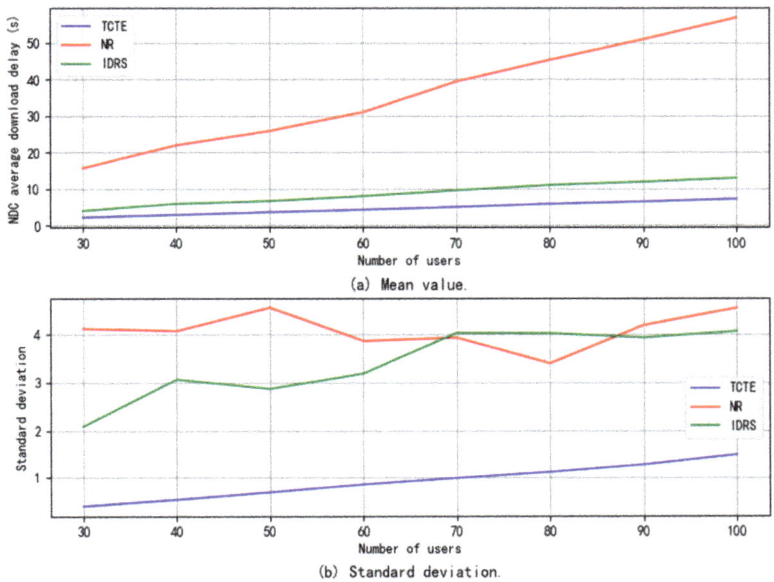

Figure 12. Mean and standard deviation of the NDC average download delay in Experimental Scenario 1.

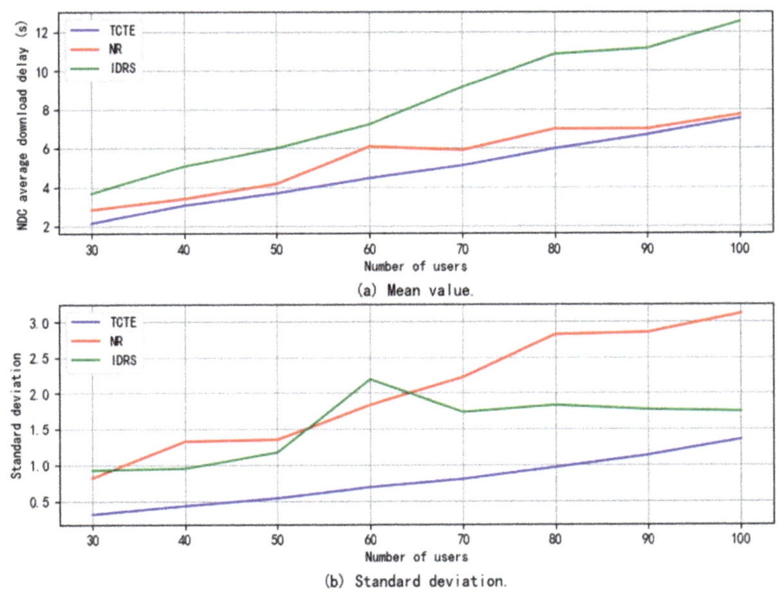

Figure 13. Mean and standard deviation of the NDC average download delay in Experimental Scenario 2.

Based on the results of the two experimental scenarios, TCTE always has the smallest NDC average download delay average and standard deviation, while NR and IDRS always have a large average and standard deviation. This means that NR and IDRS can only perform well in very few experiments, but TCTE has the best performance in most cases.

To illustrate the effectiveness of the experimental results and analysis in Sections 5.2.1–5.2.3, we conducted a t-test on the NDC download delay when the number of users is 60. Table 3 shows the results, and the p-value retains four significant figures. It can be seen that the maximum p-value in experimental scenario 1 is 4.512×10^{-3}, so the experimental results are significantly different at the 99.55% confidence level. In experimental scenario 2, the maximum p-value is 3.952×10^{-2}, so the experimental results are significantly different at the 96.05% confidence level. Therefore, the experimental results are statistically significant and were not obtained by chance.

Table 3. p-values of the t-test for the average NDC download delay with 60 users.

Experimental Scenario	TCTE-IDRS	IDRS-NR	NR-TCTE
Experimental Scenario 1	8.771×10^{-6}	4.512×10^{-3}	4.994×10^{-10}
Experimental Scenario 2	6.637×10^{-4}	2.350×10^{-2}	3.952×10^{-2}

6. Conclusions

In this paper, we propose an innovative approach based on transmission completion time estimation (TCTE) for the replica-selection problem in ICNs to ensure user experience and improve the throughput of the network. TCTE maintains the replica node information in the ENRS domain through passive measurement, and then estimates the transmission completion time and selects the replica node with the smallest transmission completion time. It should be noted that our algorithm only affects the distribution of traffic in the ENRS domain, and will not increase the traffic of the core network. Experiments show that TCTE not only effectively improves the user's download rate and edge node throughput, reduces download rate fluctuations, reduces user download delay, and improves fairness, but also has universal applicability.

Author Contributions: Methodology, Z.W., H.N. and R.H.; software, Z.W.; writing—original draft preparation, Z.W.; writing—review and editing, Z.W., H.N. and R.H.; supervision, R.H.; project administration, R.H.; funding acquisition, H.N. All authors have read and agreed to the published version of the manuscript.

Funding: This work was funded by the Strategic Leadership Project of Chinese Academy of Sciences: SEANET Technology Standardization Research System Development (Project No. XDC02070100).

Data Availability Statement: Not applicable.

Acknowledgments: We would like to express our gratitude to Jinlin Wang for their meaningful support in this work.

Conflicts of Interest: The authors declare no conflict of interest.

References

1. Nour, B.; Sharif, K.; Li, F.; Yang, S.; Moungla, H.; Wang, Y. ICN publisher-subscriber models: Challenges and group-based communication. *IEEE Netw.* **2019**, *33*, 156–163. [CrossRef]
2. Jacobson, V.; Smetters, D.K.; Thornton, J.D.; Plass, M.F.; Briggs, N.H.; Braynard, R.L. Networking named content. In Proceedings of the 5th International Conference on Emerging Networking Experiments and Technologies, Rome, Italy, 1–4 December 2009; pp. 1–12.
3. NDN. The Named Data Networking Project. Available online: http://www.named-data.net/ (accessed on 9 January 2023).
4. Raychaudhuri, D.; Nagaraja, K.; Venkataramani, A. Mobilityfirst: A robust and trustworthy mobility-centric architecture for the future internet. *ACM Sigmob. Mob. Comput. Commun. Rev.* **2012**, *16*, 2–13. [CrossRef]
5. Dannewitz, C.; Kutscher, D.; Ohlman, B.; Farrell, S.; Ahlgren, B.; Karl, H. Network of information (netinf)–an information-centric networking architecture. *Comput. Commun.* **2013**, *36*, 721–735. [CrossRef]
6. Koponen, T.; Chawla, M.; Chun, B.-G.; Ermolinskiy, A.; Kim, K.H.; Shenker, S.; Stoica, I. A data-oriented (and beyond) network architecture. In Proceedings of the 2007 Conference on Applications, Technologies, Architectures, and Protocols for Computer Communications, Kyoto, Japan, 27–31 August 2007; pp. 181–192.
7. Wang, J.; Chen, G.; You, J.; Sun, P. SEANet: Architecture and Technologies of an On-site, Elastic, Autonomous Network. *J. Netw. New Media* **2020**, *6*, 1–8.

8. Saadeh, H.; Almobaideen, W.; Sabri, K.E.; Saadeh, M. Hybrid SDN-ICN architecture design for the internet of things. In Proceedings of the 2019 Sixth International Conference on Software Defined Systems (SDS), Rome, Italy, 10–13 June 2019; pp. 96–101.
9. Mehmood, K.; Kralevska, K.; Palma, D.J. Intent-driven autonomous network and service management in future networks: A structured literature review. *arXiv* **2021**, arXiv:2108.04560. [CrossRef]
10. Xylomenos, G.; Ververidis, C.N.; Siris, V.A.; Fotiou, N.; Tsilopoulos, C.; Vasilakos, X.; Katsaros, K.V.; Polyzos, G.C. A survey of information-centric networking research. *IEEE Commun. Surv. Tutor.* **2013**, *16*, 1024–1049. [CrossRef]
11. Ioannou, A.; Weber, S. A Survey of Caching Policies and Forwarding Mechanisms in Information-Centric Networking. *IEEE Commun. Surv. Tutor.* **2016**, *18*, 2847–2886. [CrossRef]
12. Song, Y.; Ni, H.; Zhu, X. An Enhanced Replica Selection Approach Based on Distance Constraint in ICN. *Electronics* **2021**, *10*, 490. [CrossRef]
13. Liao, Y.; Sheng, Y.; Wang, J. A deterministic latency name resolution framework using network partitioning for 5G-ICN integration. *Int. J. Innov. Comput. Inf. Control* **2019**, *15*, 1865–1880.
14. Dong, L.; Wang, G. A Hybrid Approach for Name Resolution and Producer Selection in Information Centric Network. In Proceedings of the 2018 International Conference on Computing, Networking and Communications (ICNC), Maui, HI, USA, 5–8 March 2018; pp. 574–580.
15. Wang, L.; Hoque, A.; Yi, C.; Alyyan, A.; Zhang, B. *OSPFN: An OSPF Based Routing Protocol for Named Data Networking*. Available online: https://named-data.net/techreport/TR003-OSPFN.pdf (accessed on 5 February 2023).
16. Hoque, A.M.; Amin, S.O.; Alyyan, A.; Zhang, B.; Zhang, L.; Wang, L. NLSR: Named-data link state routing protocol. In Proceedings of the 3rd ACM SIGCOMM Workshop on Information-Centric Networking, Kyoto, Japan, 26–28 September 2016; pp. 15–20.
17. Yan, H.; Gao, D.; Su, W.; Foh, C.H.; Zhang, H.; Vasilakos, A.V. Caching Strategy Based on Hierarchical Cluster for Named Data Networking. *IEEE Access* **2017**, *5*, 8433–8443. [CrossRef]
18. Hasan, K.; Jeong, S.H. Efficient Caching for Delivery of Multimedia Information with Low Latency in ICN. In Proceedings of the 2019 Eleventh International Conference on Ubiquitous and Future Networks (ICUFN), Split, Croatia, 2–5 July 2019; pp. 745–747.
19. Rossini, G.; Rossi, D. Coupling caching and forwarding: Benefits, analysis, and implementation. In Proceedings of the 1st ACM Conference on Information-Centric Networking, Osaka, Japan, 19–21 September 2022; pp. 127–136.
20. Badov, M.; Seetharam, A.; Kurose, J.; Firoiu, V.; Nanda, S. Congestion-aware caching and search in information-centric networks. In Proceedings of the 1st ACM Conference on Information-Centric Networking, Osaka, Japan, 19–21 September 2022; pp. 37–46.
21. Chiocchetti, R.; Perino, D.; Carofiglio, G.; Rossi, D.; Rossini, G. Inform: A dynamic interest forwarding mechanism for information centric networking. In Proceedings of the 3rd ACM SIGCOMM Workshop on Information-Centric Networking, Kyoto, Japan, 26–28 September 2016; pp. 9–14.
22. Watkins, C.J.; Daya, P. Technical Note: Q-Learning. *Mach. Learn.* **1992**, *8*, 279–292. [CrossRef]
23. Yi, C.; Afanasyev, A.; Moiseenko, I.; Wang, L.; Zhang, B.; Zhang, L. A case for stateful forwarding plane. *Comput. Commun.* **2013**, *36*, 779–791. [CrossRef]
24. Lee, M.; Cho, K.; Park, K.; Kwon, T.; Choi, Y. SCAN: Scalable content routing for content-aware networking. In Proceedings of the 2011 IEEE International Conference on Communications (ICC), Kyoto, Japan, 5–9 June 2011; pp. 1–5.
25. Bloom, B.H. Space/time trade-offs in hash coding with allowable errors. *Commun. ACM* **1970**, *13*, 422–426. [CrossRef]
26. Domingues, G.d.M.B.; Leão, R.M.M.; Menasché, D.S. Enabling information centric networks through opportunistic search, routing and caching. *arXiv* **2013**, arXiv:1310.8258.
27. Sevilla, S.; Mahadevan, P.; Garcia-Luna-Aceves, J. iDNS: Enabling information centric networking through The DNS. In Proceedings of the 2014 IEEE Conference on Computer Communications Workshops (INFOCOM WKSHPS), Toronto, ON, Canada, 27 April–2 May 2014; pp. 476–481.
28. Fuller, V.; Farinacci, D. *RFC 6833: Locator/ID Separation Protocol (LISP) Map-Server Interface*; RFC, Ed.; ACM Digital Library: New York City, NY, USA, 2013.
29. Bogdanov, K.; Peón-Quirós, M.; Maguire, G.Q., Jr.; Kostić, D. The nearest replica can be farther than you think. In Proceedings of the Sixth ACM Symposium on Cloud Computing, Kohala Coast, HI, USA, 27–29 August 2015; pp. 16–29.
30. Carter, R.L.; Crovella, M.E. Server selection using dynamic path characterization in wide-area networks. In Proceedings of the INFOCOM'97, Kobe, Japan, 7–12 April 1997; pp. 1014–1021.
31. Hanna, K.M.; Natarajan, N.; Levine, B.N. Evaluation of a novel two-step server selection metric. In Proceedings of the Ninth International Conference on Network Protocols ICNP 2001, Riverside, CA, USA, 11–14 November 2001; pp. 290–300.
32. Mitzenmacher, M. The power of two choices in randomized load balancing. *IEEE Trans. Parallel Distrib. Syst.* **2001**, *12*, 1094–1104. [CrossRef]
33. Zeng, L.; Ni, H.; Han, R. An incrementally deployable IP-compatible-information-centric networking hierarchical cache system. *Appl. Sci.* **2020**, *10*, 6228. [CrossRef]
34. Salsano, S.; Detti, A.; Cancellieri, M.; Pomposini, M.; Blefari-Melazzi, N. Transport-layer issues in information centric networks. In Proceedings of the Second Edition of the ICN Workshop on Information-Centric Networking, Helsinki, Finland, 17 August 2012; pp. 19–24.

35. Song, Y.; Ni, H.; Zhu, X. Analytical Modeling of Optimal Chunk Size for Efficient Transmission in Information-Centric Networking. *J Int. J. Innov. Comput. Inf. Control* **2020**, *16*, 1511–1525.
36. Wang, Z.; Ni, H.; Han, R. Copa-ICN: Improving Copa as a Congestion Control Algorithm in Information-Centric Networking. *Electronics* **2022**, *11*, 1710. [CrossRef]
37. Saino, L.; Psaras, I.; Pavlou, G. Hash-routing schemes for information centric networking. In Proceedings of the 3rd ACM SIGCOMM Workshop on Information-Centric Networking, Kyoto, Japan, 26–28 September 2016; pp. 27–32.
38. NS-3 Project. Available online: https://www.nsnam.org (accessed on 9 January 2023).
39. Breslau, L.; Cao, P.; Fan, L.; Phillips, G.; Shenker, S. Web caching and Zipf-like distributions: Evidence and implications. In *The Future Is Now (Cat. No. 99CH36320), Proceedings of the IEEE INFOCOM'99 Conference on Computer Communications, Eighteenth Annual Joint Conference of the IEEE Computer and Communications Societies, New York, NY, USA, 21–25 March 1999*; IEEE: Piscataway, NJ, USA, 1999; pp. 126–134.
40. Chiu, D.-M.; Jain, R. Analysis of the increase and decrease algorithms for congestion avoidance in computer networks. *Comput. Netw. ISDN Syst.* **1989**, *17*, 1–14. [CrossRef]

future internet

MDPI

Article

NDN-BDA: A Blockchain-Based Decentralized Data Authentication Mechanism for Vehicular Named Data Networking

Ahmed Benmoussa [1,*], Chaker Abdelaziz Kerrache [2], Carlos T. Calafate [3,*] and Nasreddine Lagraa [2]

1 Department of Computer Science, University Center of Aflou, Laghouat 03000, Algeria
2 Laboratoire d'Informatique et de Mathématiques, Université Amar Telidji de Laghouat, Laghouat 03000, Algeria; ch.kerrache@lagh-univ.dz (C.A.K.); n.lagraa@lagh-univ.dz (N.L.)
3 Computer Engineering Department (DISCA), Universitat Politècnica de València, 46022 Valencia, Spain
* Correspondence: a.benmoussa@cu-aflou.edu.dz (A.B.); calafate@disca.upv.es (C.T.C.)

Abstract: Named Data Networking (NDN) is an implementation of Information-Centric Networking (ICN) that has emerged as a promising candidate for the Future Internet Architecture (FIA). In contrast to traditional networking protocols, NDN's focus is on content, rather than the source of the content. NDN enables name-based routing and location-independent data retrieval, which gives NDN the ability to support the highly dynamic nature of mobile networks. Among other important features, NDN integrates security mechanisms and prioritizes protecting content over communication channels through cryptographic signatures. However, the data verification process that NDN employs may cause significant delays, especially in mobile networks and vehicular networks. This aspect makes it unsuitable for time-critical and sensitive applications such as the sharing of safety messages. Therefore, in this work, we propose NDN-BDA, a blockchain-based decentralized mechanism that provides a faster and more efficient data authenticity mechanism for NDN-based vehicular networks.

Keywords: blockchain; data authentication; Named Data Networking (NDN); Future Internet Architecture (FIA); vehicular networks

Citation: Benmoussa, A.; Kerrache, C.A.; Calafate, C.T.; Lagraa, N. NDN-BDA: A Blockchain-Based Decentralized Data Authentication Mechanism for Vehicular Named Data Networking. *Future Internet* **2023**, *15*, 167. https://doi.org/10.3390/fi15050167

Academic Editor: Paolo Bellavista

Received: 31 March 2023
Revised: 21 April 2023
Accepted: 28 April 2023
Published: 29 April 2023

1. Introduction

Vehicular networks encounter many challenges, such as high mobility, heterogeneous communication, and low latency, which need to be addressed to ensure long-distance and reliable connections. However, the TCP/IP model that the actual Internet relies on was not designed for such networks as it does not adequately address these prerequisites. Furthermore, the current Internet architecture was not designed to handle the massive number of connected devices, which is expected to reach almost 30 billion in 2023 [1]. Additionally, people's use of the Internet has changed, with their focus shifting from the origin of the content to the content they want to access. To tackle these limitations, the research community has explored alternatives to support these requirements. One such alternative is Information-Centric Networking (ICN) [2], which is considered one of the most-attractive Future Internet Architectures (FIAs) [3]. Among the different ICN architectures proposed, such as DONA [4] and PUSUIT [5], Named Data Networking (NDN) [6] seems to be the most-promising contender for the upcoming Internet architecture. NDN finds its roots in the Content-Centric Networking (CCN) project [7], which emphasizes a content-oriented communication approach and prioritizes the data over their owner. Instead of relying on IP addresses, NDN uses data names for packet forwarding. In addition, NDN offers in-network caching capabilities, built-in multicast forwarding, mobility support, and security mechanisms. Unlike traditional networking protocols, NDN concentrates on securing content rather than communication channels.

Data signatures are required in NDN, which enables users to access and retrieve any content as long as its signature can be verified, regardless of its origin. NDN employs a mechanism that involves signing content by its producer and the authority who provided the certificate to the producer. To validate content, a user checks the authenticity of all the certificates involved within the certificate chain till it encounters a trusted or a self-signed network entity (usually a trusted authority). This process helps to enhance the trustworthiness of the data by verifying their authenticity. However, vehicular networks are prone to issues related to their highly dynamic nature. The data verification process employed by NDN may result in significant delays, which is unsuitable for sensitive applications, such as sharing safety messages. In traditional centralized authentication systems, data authentication is often associated with high overheads due to the need to communicate with a central authority. To this end, we propose an approach that enables local data verification while providing a secure authentication mechanism for producer vehicles. Our approach involves leveraging blockchain technology to decentralize the authentication process, making it more efficient for vehicular environments. By using blockchain, we can avoid the need for centralized authorities, thereby reducing the associated overheads. To the best of our knowledge, this study is the first attempt to use blockchain for decentralized data authentication in vehicular NDN. The validation process that we conducted showed that our proposal is secure against attackers.

The paper is structured as follows: Section 2 provides an overview of the NDN architecture. Then, Section 3 examines the related work. The threat model and assumptions for our study are outlined in Section 4. In Section 5, we describe our proposed solution, NDN-BDA. Our work is evaluated and validated in Section 6, and we conclude the paper and discuss future work in Section 7.

2. Named Data Networking: Nuts and Bolts

NDN is an Internet architecture that prioritizes data and intends to substitute the present TCP/IP-centric architecture of the Internet. In particular, it attempts to address the escalating need for communication that is centered on content [6]. NDN identifies two entities: *producers*, who are responsible for creating and offering content, and *consumers*, who seek and retrieve data. In order to obtain content, consumers transmit the name of the desired data via a specific packet. NDN relies on a hierarchical namespace to differentiate content, for example the present paper could be named: "`com/mdpi/future-Internet/2023/ndn-bda`".

NDN relies on two packet types: *Interest* packets and *Data* packets. Consumers use *Interest* packets to request content that producers include within *Data* packets. The most-recent NDN packet specifications [8] mandate that every *Interest* packet must comprise a combination of mandatory and optional parameters. A `Name` must be included in each *Interest* packet, representing the content being sought by the consumer. The `Nonce` field, consisting of a randomly generated string of four octets, is leveraged to uniquely identify *Interest* packets, and thus prevent looping in the network. *Interest* packets may include optional parameters: `CanBePrefix` for the *Interest* packet's name, `MustBeFresh` for the requested content, `InterestLifeTime` representing how long an NDN router will maintain the state for this *Interest*, being `HopLimit` and `ForwardingHint` utilized in forwarding. In addition, *Interest* packets may also include application-specific parameters in the `ApplicationParameters` field. Furthermore, *Interest* packets can be signed if required [9].

Once a consumer sends an *Interest* packet to request particular data, the corresponding response is transmitted through a *Data* packet. In NDN, the producer should sign every *Data* packet. Fundamentally, a *Data* packet comprises three components: the `Name`, the `Content`, which denotes the payload of the *Data* packet, and a `Signature`. Supplementary details such as `ContentType`, `FreshnessPeriod`, and `FinalBlockId` are also encompassed within the *Data* packet.

Each node with the NDN stack maintains three data structures [10]. The Pending Interest Table (PIT) stores all the not-yet-satisfied *Interest* packets, which remain in the PIT until a *Data* packet returns or times out. The PIT entry contains fields such as the name of the requested *Data*, incoming and outgoing interfaces, and an expiry timer. The Content Store (CS) enables NDN to offer in-network caching, where each NDN node can store passing *Data* packets in a local cache. To maintain its size, each CS needs to implement a caching policy, such as First In First Out (FIFO), Least Recently Used (LRU), and Least Frequently Used (LFU) [11]. The Forwarding Information Base (FIB) is used to forward incoming *Interest* packets to upstream nodes. FIB entries are indexed with name prefixes instead of IP addresses, and each FIB entry consists of a name prefix and a list of next hops. Routers can forward *Interest* packets to one or multiple hops, enabling multi-path forwarding based on their strategy.

2.1. Security Principles in NDN

In a named data network, the primary objective is to locate and retrieve data objects within a specified context by their names. The security features of the NDN design are intrinsic and require signatures to be used in *Data* packets [12]. Producers are responsible for digitally signing all *Data* they create, which enables consumers to validate the authenticity of the received information. Additionally, routers and repositories can store *Data*, and consumers can receive *Data* packets from any source. The NDN architecture relies on several security components to ensure data security [13].

2.1.1. Digital Signatures

Digital signatures in NDN represent the basic building blocks of object security. They enable securing data regardless of their location or transmission channel. The signature field is present in every *Data* packet and serves to bind the content of the packet to its name [14]. It consists of two components, namely `SignatureInfo` and `SignatureValue`. The former contains the producer's public key name and the cryptographic algorithm used to sign the *Data*, while the latter represents the signature's generated bits. The `SignatureInfo` field may also include `KeyLocator`, which includes information regarding the location of parent certificate(s). The signed hash of a *Data* packet covers all its fields except the `SignatureValue`, as illustrated in Figure 1.

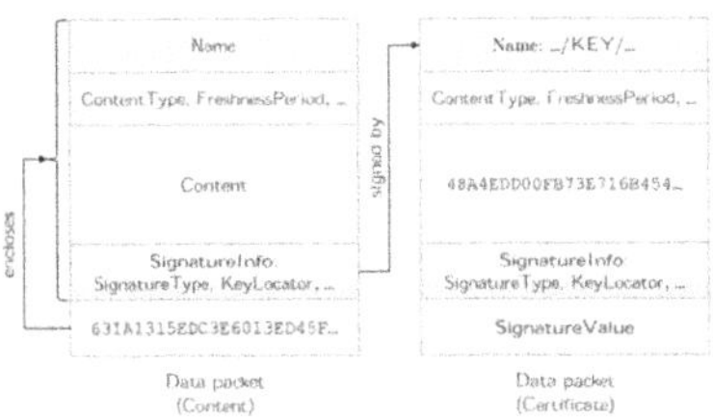

Figure 1. Example of a *Data* packet and its signing certificate.

While NDN does not mandate the use of signatures in *Interest* packets, they may be required in some scenarios where their authenticity is necessary, such as sending a new route announcement or a command packet to an IoT device. Compared to the signature field in *Data* packets, the *Interest* packet's signature field includes additional components, such as `SignatureNonce`, `SignatureTime`, and `SignatureSeqNum`, to add uniqueness to the signature.

2.1.2. NDN Certificates

To ensure secure communication in NDN, data producers need to possess at least one cryptographic key pair. Producers use their private keys to sign *Data* packets, and consumers verify them using public keys. Each public key is encoded in an X509 format digital certificate signed by an issuer [15]. The NDN certificate's name follows a specific naming convention: `/<IdentityName>/KEY/<KeyId>/<IssuerId>/<Version>`. `IdentityName` represents the namespace in which the key can operate, followed by the keyword `KEY`. The `KeyId` identifies an instance of the public key. Its value can be an 8-byte-long random value, an SHA-256 digest of the public key, a timestamp, or a numerical identifier. The `IssuerId` part identifies the issuer of the certificate. Its value is identical to the `KeyId`. Finally, the version of the certificate is included. In addition to its name, a certificate *Data* packet contains other fields: the `FreshnessPeriod` and the `Content`, which embeds the bits of the DER-encoded public key. Like any NDN Data packet, certificates contain a signature field, which is the signature of the issuer.

2.1.3. Trust Model

The *Data* packet's signature validates the authenticity of the received content and its origin. However, it does not indicate whether the signer has the authorization to produce the received content [16]. To verify whether a producer is authorized to generate data under a given namespace, consumers require a mechanism to check if the producer's key has the right to sign a *Data* packet under that namespace [17]. Application-based trust policies define which keys have the authorization to sign which *Data*, as illustrated in Figure 2. Furthermore, applications can use access control policies to protect the *Data* packet's content through encryption and permit only authorized nodes to decrypt it [18].

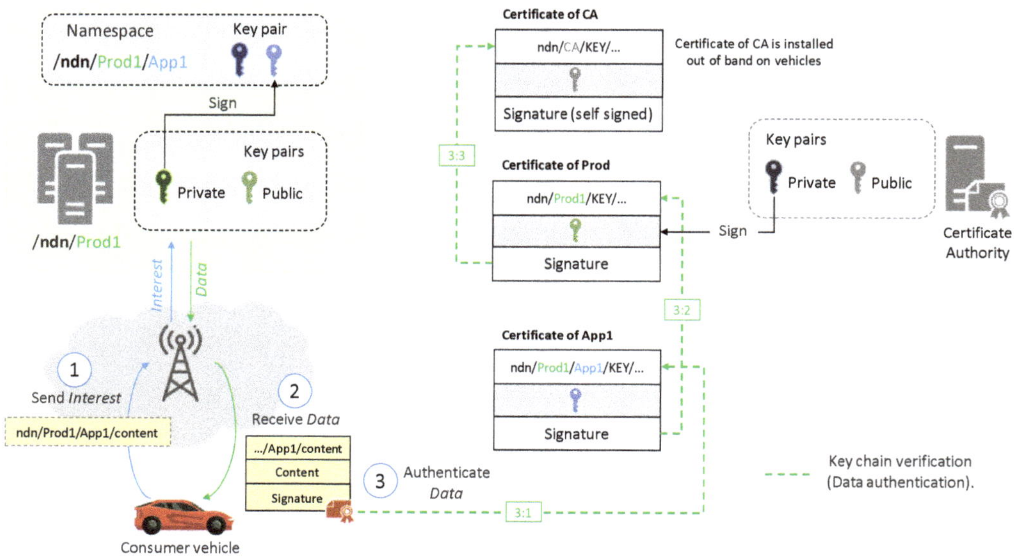

Figure 2. Example of the data authentication process.

2.2. *Packet Forwarding*

Instead of using IP addresses to forward packets, NDN routers utilize application namespaces [19]. NDN routers employ stateful forwarding, which implies that routers store details of received requests until they are satisfied or timed out. Upon the arrival of an *Interest* packet, the router checks whether it holds a copy of the *Data* in the CS. The NDN node forwards the *Data* downstream if it finds a copy of it. Otherwise, it checks the PIT for similar requests. If a pending request already exists, the router updates the incoming interfaces field. If no pending *Interest* exists, a new entry is created. The NDN node then forwards the *Interest* packet to the upstream neighbor(s) using a forwarding strategy and the FIB table [20]. The NDN node may respond with a NACK or drop the *Interest* packet if no route is found.

Once a *Data* packet is received by a router from a producer or in-network cache, the router checks for a matching pending *Interest* by examining the name of the received *Data* packet. If there is no match in the PIT, the router drops the received *Data* packet. However, if a match is found, the router caches (or not, depending on its caching strategy) the received *Data* and forwards them downstream to all the interfaces associated with the pending *Interest*. As a result, NDN routers provide a built-in multicast mechanism. Figure 3 illustrates a scenario of packet forwarding in NDN.

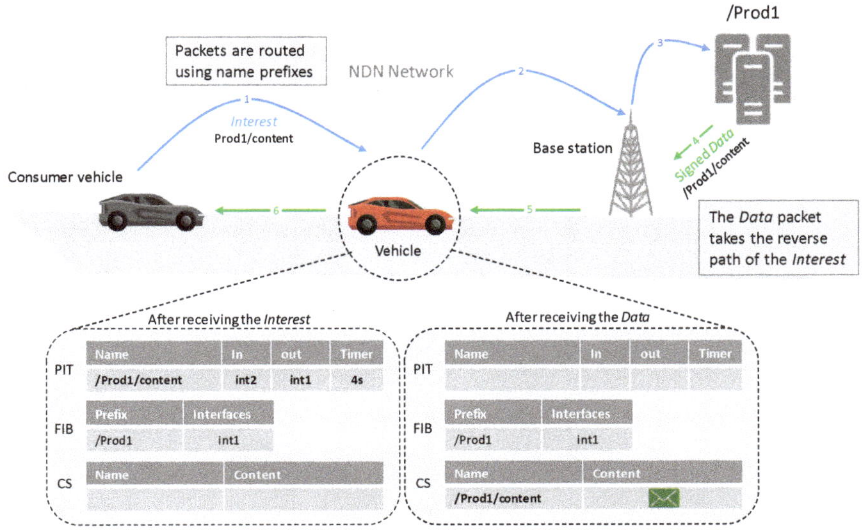

Figure 3. Example of packet forwarding in vehicular NDN.

2.3. Mobility Support in NDN

The packet forwarding process in mobile networks can be challenging due to the constantly changing network topology [21]. However, the NDN paradigm overcomes this issue by breaking the point-to-point nature of IP networks and enabling ubiquitous data retrieval. Unlike IP networks, NDN allows mobile nodes to retrieve data from anywhere in the network, instead of delivering data to a specific mobile node [22]. Furthermore, NDN's implementation allows it to operate over any transport protocol, which is a significant advantage. Additionally, the location-independent data retrieval enabled by NDN's data-centric security is another key advantage to mobility support.

In NDN, mobile consumers are not identified by addresses, which avoids the overhead associated with address management and connection sessions that TCP/IP requires. Instead, when a consumer requests content, the *Interest* packet creates reverse path entries in the PIT as it travels to the data producer. This enables a mobility-friendly forwarding process, as *Data* packets can follow the reverse path to reach the consumer. If the consumer moves to a new location, resending the *Interest* packet is sufficient to update the reverse route to the new location [23].

The simplest way of retrieving the content of a mobile producer consists of broadcasting *Interest* packets throughout the network until they reach the intended mobile producer or the network cache holding the content. However, this approach consumes significant network resources, making it unsuitable for large networks. Another approach involves using a DNS-like node to map produced data names to the actual location of a mobile producer [24]. A consumer sends a query to the *rendezvous* point to ask for the position of the producer. The *rendezvous* point can also act as a relay, which forwards requests to the mobile producer's position and sends back data to the requesting consumer. Data depots are another approach that leverages the built-in caching capabilities of NDN, allowing consumers to send *Interest* packets to a data depot instead of directly retrieving content from mobile producers [25]. The data depot can then send the requested data if it holds them or fetch them using one of the other approaches. To retrieve data associated with a specific region, a geographic-based forwarding strategy is used where any mobile node in that region can act as a producer. For example, a consumer can request traffic information for a particular road by sending an *Interest* packet to the relevant region using geographic-based routing. When a node in that region receives the request, it either sends the requested data (if it already has them in its cache) or generates them before sending them back to the consumer.

3. Related Work

Several authentication mechanisms have been proposed in recent years. The work in [26] presented a privacy mechanism for NDN-based vehicular networks. The proposal uses pseudonyms to identify vehicles instead of real names. The mechanism involves a vehicle broadcasting an *Interest* to retrieve a key from a nearby proxy. The vehicle then generates a pseudonym and a new key. Following that, it encrypts the new information and its actual key using the manufacturer's public key. It also encrypts the manufacturer's name using the proxy's key. The receiving proxy sends an *Interest* to the manufacturer for key issuance, which creates a certificate for the new key, and encrypts it using the actual vehicle's key. The vehicle decrypts and stores the new certificate and the manufacturer's pseudonym to use for future pseudonyms' generation. Similarly, the work in [27] proposed an anonymity-enabled consumer authentication mechanism. The proposed scheme aims at reducing the computational overhead associated with data authentication by verifying multiple signatures in a batch. When a consumer attempts to retrieve content for the first time, its edge node acts as a temporary content provider. It checks that the consumer has not been corrupted or impersonated before sending the requested content. The edge node and the consumer agree and create a shared key based on the randomly chosen pseudo-identity of each consumer.

The authors in [28] provided a solution for the long-living Data, which outlasts the lifetime of their signatures. The authentication framework, named DeLorean, uses a publicly auditable bookkeeping service that logs the fingerprints of signatures in a Merkle tree. By providing DeLorean proofs for data packets and certificates, the signature validity at the time of data creation can be checked, allowing authentication regardless of the signatures' expiration. The framework includes a data "chronicle", which consists of a sequence of volumes, each containing fingerprints of the witnessed signatures within a fixed timeslot. DeLorean requires auditors to continuously check the consistency of the chronicle to ensure correct and truthful operations. If an auditor detects any modification, it can share it publicly, serving as a deterrent for the service's misbehavior. To overcome the problem related to multiparty authentication, the authors of [29] presented a security support tool for applications named NDN-MPS. It consists of three components, a multi-signature trust schema to enable multiparty authentication policies, an extended key locator data object to keep data consistent across signers, and a signature collection protocol for multi-signature verification. The producer starts the signing process with the multi-signature trust schema, identifies the list of signers, and coordinates the delivery of the data object to be signed. Then, it collects signature values from signers. The producer uses another protocol such as NDN sync [30] to collect signatures from signers. The producer ensures consistency among signers by filling the key locator field with a placeholder name called the late-binding key locator. In the case of failure or the unavailability of signers, the producer can try reaching alternative signers permitted by the trust schema to provide a degree of resiliency.

Another authentication mechanism was proposed in [31]. The authors presented a secure authentication scheme for big-data-enabled ICN to prevent fake data caching. The proposal aimed at establishing trust among unpredictable entities during data acquisition. The mechanism involves three phases: initial trust establishment, data-centric certificate management, and forwarding-integrated authenticable data retrieval. In the first phase, certificates are issued to a group of highly trusted nodes called Intermediate Physical Entities (IPEs), which are used to cache big data. In the second phase, IPEs store the certificates of their neighboring nodes and other highly trusted entities in a local repository. The third phase consists of constructing a hop-by-hop trust chain to authenticate *Interest* and *Data* packets during forwarding. It utilizes highly trusted IPEs to replace highly trusted certificates of the chain. Additionally, it shortens the certificate chain into one certificate.

With the objective of reducing the overhead associated with signature verification, the work in [32] presented a lightweight communication strategy for ICN-based UAV networks. The mechanism utilizes the trust score of the node to determine whether data authenticity needs to be checked. In the case that the source UAV is considered trustworthy, the process of *Data* validation is postponed for a certain period, which is proportional to the UAV's trust score.

Blockchain technology is extensively employed in conjunction with the NDN architecture. Some authors have used it to provide solutions against security attacks, such as [33,34], while others have used it for securing content dissemination [35–37]. The authors of [38] proposed a blockchain-based mechanism to authenticate mobile producers in ICN. The protocol aims to authenticate producers' prefixes and ensure genuine routing updates. The architecture identifies two domains: the core network and the clusters. The clusters consist of both access gateways and a cluster head. Each cluster uses a private ledger managed by the cluster head. The core network includes a subset of core routers that manage the global blockchain, which stores the transactions generated to and from various clusters. The mechanism involves utilizing the mobile device's SIM card to securely distribute and register key pairs. When a producer connects to the network for the first time, the authorization server verifies the prefix announced by the producer by validating its digital signature. Then, it records the producer's public key and digital signature into the blockchain.

In a different context, the work in [39] presented a lightweight consensus mechanism for the IoT. The authors proposed a model named Proof of Evolutionary Model (PoEM) to enhance the consensus efficiency of Blockchain-as-a-Service (BaaS) in the IoT. The model relies on machine learning to achieve consensus. It also includes a mechanism to handle the dynamic nature of IoT environments by managing nodes joining and exiting in real-time.

4. Threat Model and Assumptions

The present paper aimed to address the challenges associated with the certificate chain verification process, which is a critical aspect of ensuring data authenticity and integrity. The proposed solution focuses on enabling consumers to authenticate and verify the content they receive, regardless of their location, while avoiding the overhead associated with validating each certificate involved in the received content. Unlike existing solutions, our approach does not require consumers to have prior knowledge of the trust schema used for data authentication. Instead, we assumed that consumers must possess the blockchain before requesting and verifying data, and each consumer in the network can request and receive the blockchain.

To account for potential adversarial nodes, we made several assumptions. Firstly, we assumed that attackers have the capability to initiate multiple protocol sessions simultaneously. Secondly, we assumed that attackers can access all public data related to the protocol, which gives them the ability to read, store, and block any sent messages. Furthermore, attackers can intercept, construct, and resend messages to their desired destinations. We also assumed that no key breaches occur. To ensure the effectiveness of our proposed mechanism, we needed to develop robust strategies to mitigate the impact of such attacks and enhance the overall security of the system. To this end, we investigated several potential security threats to the system, including impersonation attacks, blockchain tampering attacks, man-in-the-middle attacks, and (D)DoS attacks. In an impersonation attack, the adversary tries to masquerade as a legitimate producer. The blockchain tampering attack, on the other hand, involves the insertion of false or modified data into the blockchain by a malicious node. The man-in-the-middle attack, as the name implies, occurs when an attacker intercepts and alters the communication between two parties, often for the purpose of stealing or manipulating sensitive information. Finally, the (D)DoS attack involves flooding the system with *Interest* packets in an attempt to overwhelm it, and thereby prevent legitimate requests from being satisfied.

5. NDN-BDA

This section outlines the design of our proposed mechanism, NDN-BDA. It begins with an overview of the mechanism, followed by a detailed description of each step involved in its functioning. Table 1 lists the notations that are used in this paper.

Table 1. List of notations used in this paper.

Notation	Meaning
VRE	Vehicle Registration Entity
MS	Management Server
SCA	System Certification Authority
PV	Producer Vehicle
CV	Consumer Vehicle
si_i	The ith signed *Interest* packet
i_i	The ith *Interest* packet
d_i	The ith *Data* packet
K	Cryptographic Key
$\{m\}K$	Encrypted message m with K
K_{PV}	Public key of PV
K_{PV}^{-1}	Private key of PV
$decrypt(mK_{PV}^{-1}, K_{PV})$	Decrypts a message m encrypted with K_{PV}^{-1} using K_{PV}
BC	Blockchain
adv	Adversary node
$adv(PV) \hookrightarrow MS : m$	adv impersonates PV to send a message m to MS

5.1. Protocol Overview

The significant entities involved in the NDN-BDA protocol are: *Vehicle Registration Entity (VRE)*, *System Certificate Authority (SCA)*, *System Management Server (MS)*, and the blockchain. Figure 4 illustrates the different components of our proposed mechanism. The Vehicle Registration Entity represents the governmental body responsible for providing vehicle registration documents to a vehicle owner. It also acts as a certification authority, generating a personal certificate for the vehicle's owner. Our proposed mechanism involves the System Management Server as a central component responsible for receiving requests from producer vehicles and generating certificates for them using the System Certificate Authority. The Management Server is also tasked with maintaining and distributing the blockchain. It may also employ distribution servers that assist with blockchain distribution. The SCA is a local certification authority maintained by the MS. Finally, the blockchain contains the certificates of vehicle producers.

5.2. Blockchain

Blockchain technology is based on a decentralized and distributed database that is managed by a network of computers. Each block in the chain contains a record of transactions, and once a block is added to the chain, it cannot be altered or deleted. This immutability and transparency of the blockchain make it ideal for use cases where security and trust are critical. The use of blockchain involves utilizing a consensus algorithm, such as Proof of Work (PoW) and Proof of Stake (PoS). It consists of a mechanism that allows nodes to have an identical copy of the blockchain and to add transactions to the blockchain consistently and securely.

NDN-BDA relies on a centralized blockchain maintained by the Management Server. Each block header consists of several components, such as a unique identifier for the block, which is represented by the name of the certificate that it contains, a timestamp, a nonce, a hash pointer that directs to the previous block in the chain, and the hash of the block signed by the MS. The body of each block includes a producer's certificate. Figure 5 shows the structure of a block.

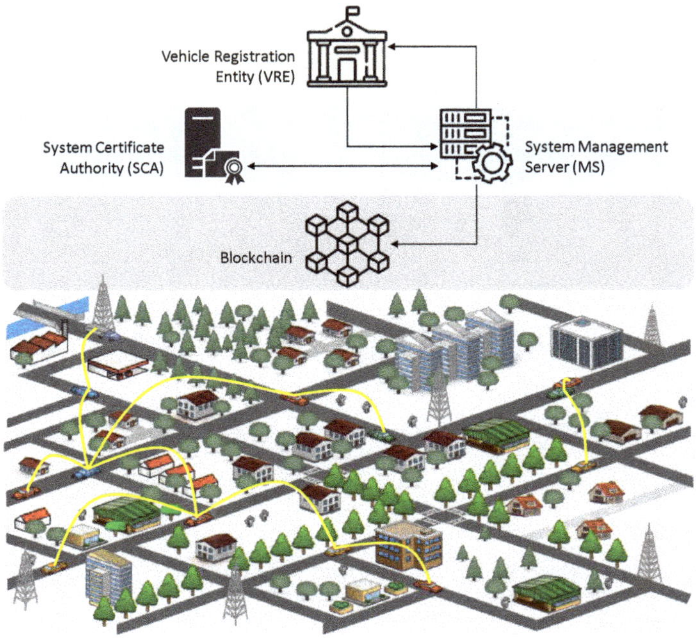

Figure 4. Overview of NDN-BDA.

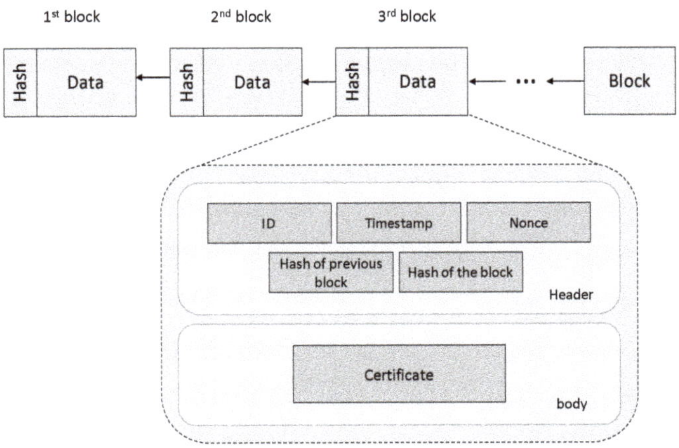

Figure 5. Overview of the structure of the blockchain used by NDN-BDA.

5.3. Trust Schema

The trust schema that NDN-BDA relies on is the Vehicle Registration Entity as the root node. The VRE grants authorization to the System Certification Authority and the Management Server on the name prefixes: "`/NDN-BDA/`" and "`/NDN-DBA/MS/`", respectively, enabling them to produce *Data* within these name prefixes. Additionally, the Vehicle Registration Entity generates a certificate for each citizen and assigns them a namespace in the format of "`/NDN-BDA/CID/`", where `CID` is a unique identifier for each citizen. Each registered citizen in the Vehicle Registration Entity possesses a car identified by its registration ID. Additionally, each producer vehicle in the network needs to have a certificate issued by the System Certification Authority. The namespace that each vehicle uses to produce *Data* is: "`/NDN-BDA/VID/`", where `VID` is the registration ID of the car. Figure 6 illustrates the trust schema that NDN-BDA uses.

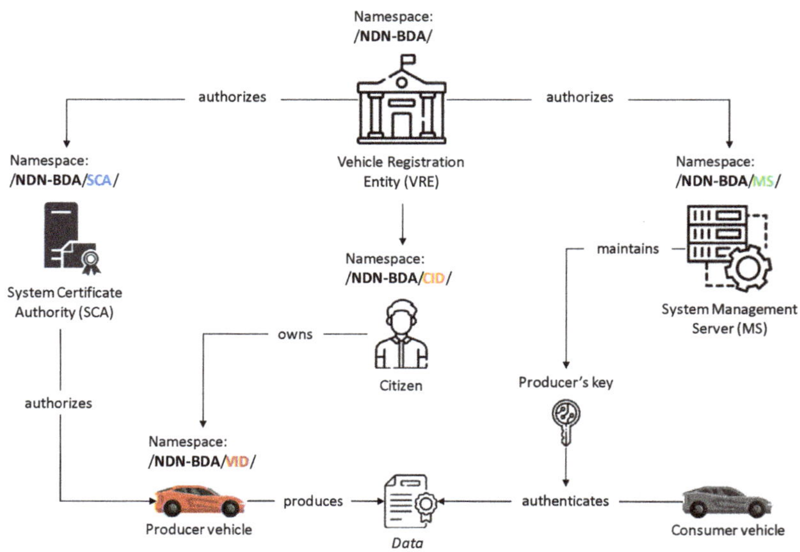

Figure 6. Overview of the trust schema used by NDN-BDA.

5.4. Producer Registration

The producer vehicle registration process in NDN-BDA involves multiple steps, which we discuss in detail below. Figure 7 provides a visual representation of the steps that are involved in the producer registration process.

Step 1: The producer vehicle *PV* sends an *Interest* packet si_1 with a name similar to: "`/NDN-BDA/MS/join/{CID/VID}`K_{MS}" and signs it with the vehicle owner's private key K_{CID}^{-1}. The name of the signed *Interest* starts with `/NDN-BDA/MS` to indicate the destination, which is, in this case, the *MS*. Keyword `join` indicates to the *MS* that this message is a request to join the network of producers. The Interest's name also includes the CID and VID information, encrypted with the MS's public key K_{MS}. This ensures that the information remains confidential during transmission, and only the *MS* can read it.

Step 2: Once the *MS* receives the signed *Interest* from the producer vehicle *PV*, it decrypts the associated CID and VID information using its private key: *decrypt*({CID/VID} K_{MS}, K_{MS}^{-1}). Following that, the *MS* sends a signed *Interest* si_2 with a name "`/NDN-BDA/MS/join/CID/VID`" to the *VRE* to verify whether the association CID and VID exist or not in the VRE's database. It checks whether the vehicle's owner, identified by its CID, possesses the vehicle identified by the received VID. Following the reception of the signed *Interest* si_2, the *VRE* checks the CID/VID association in its database. If it exists, the *VRE* responds with a *Data* packet d_1 containing the citizen's certificate. Otherwise, it notifies the *MS* that the received CID/VID association does not exist.

Step 3: After the reception of *Data* packet d_1, the *MS* stores the received certificate, if it exists, and sends a notification to *PV* regarding the outcome of its request, indicating whether it was accepted or not.

Step 4: The fourth step of the process involves *PV* generating an asymmetric key pair (k_{VP}, K_{VP}^{-1}). Once the key pair is generated, *PV* sends to the *MS* a signed *Interest* packet si_3 with the name "`/NDN-BDA/MS/sign/{VID/<key>}`K_{MS}". The name contains the keyword `sign`, which indicates that the *PV* requests a certificate for its generated public key. The name of si_3 embeds the value of the public key encrypted with the *MS*'s key K_{MS} to ensure information integrity and secrecy.

Step 5: When the `sign` request is received, the *MS* first verifies the authenticity of si_3 using the certificate that it stored in *Step 3*. Then, the *MS* decrypts the information of the key that the *PV* intends to sign, which is performed by using the *MS*'s private key: *decrypt*({VID/<key>}K_{MS},

K_{MS}^{-1}). Following that, the *MS* generates a signed *Interest* packet si_4 with a name that is identical to "`/NDN-BDA/SCA/sign/CID/VID/<key>`" and sends it to the System Certification Authority (*SCA*) to create a certificate for the public key K_{VID}.

Step 6: Upon the reception of si_4, the *SCA* generates a certificate for the received public key value in `<key>` and sends back to the *MS* a *Data* packet d_3 containing the generated certificate.

Step 7: When it receives the generated certificate from the *SCA*, the *MS* creates a new block in the blockchain *BC* containing the certificate of the *PV*. Afterward, the *MS* transmits the newly generated certificate to the *PV* within a *Data* packet d_4. With the reception of d_4, the process of the *PV* registration is complete, and the *PV* can now generate and send *Data* into the network.

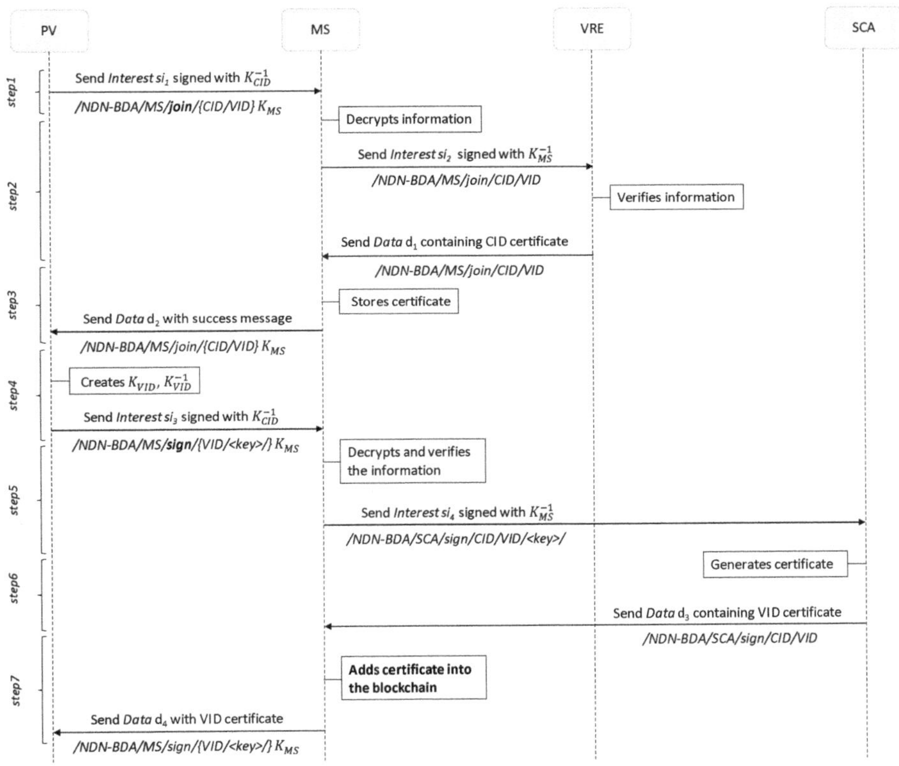

Figure 7. The steps involved in the producer vehicle registration process.

5.5. Data Authentication Process

The NDN-BDA employs a data authentication process that requires each consumer vehicle *CV* to send an *Interest* packet to request the associated blockchain of the network before it can request and receive the *Data*. To initiate this process, the *CV* sends an *Interest i* with the name "`/NDN-BDA/MS/BC/request`" towards the *MS*. Upon receiving the *Interest*, the *MS* responds with *Data* containing the requested blockchain. The *CV* will then use the received blockchain as an authentication tool for all the *Data* it receives from producer vehicles. In order to do so, the consumer vehicle needs to check the blockchain to find the certificate associated with the received *Data*. It checks if it can find the name included in the *Data's* `KeyLocator`. If the certificate is found, the *Data* are validated and accepted. If not, the *Data* are discarded. This process ensures the authenticity of the *Data* being transmitted within the network and provides a secure means of authentication for the *CV*.

6. Evaluation and Discussion

This section aims to validate our proposal, NDN-BDA. We began by performing a security analysis to test its resilience against the attacks mentioned in Section 4. Following that, we evaluated the effectiveness of NDN-BDA using the validation tool AVISPA [40]. Automated Validation of Internet Security Protocols and Applications (AVISPA) is a tool that analyses and validates cryptographic protocols. It also provides analysis techniques to detect possible vulnerabilities and flaws in protocols.

6.1. Security Analysis

The effectiveness of NDN-BDA against adversary nodes was tested in various scenarios in this section. It was assumed that the adversary *adv* can be any vehicle functioning as a producer or consumer in the network.

6.1.1. Impersonation Attack

In this attack scenario, an adversary *adv* attempts to impersonate another producer by sending an *Interest* packet with a fake or stolen producer CID and VID: $adv(PV) \hookrightarrow MS$: $si($`/NDN-BDA/MS/sign/{VID/<key>}`$K_{MS})$. If *adv* sends an *Interest* with the keyword `join` (*step1*), it will be discarded by the VRE as the `CID` included in the name does not match the key K_{adv}^{-1} that signed the *Interest*. Suppose that the *adv* fakes an *Interest* with the keyword `sign` (*step4*), willing to request a fraudulent certificate. This attempt will also fail, as the *Interest* packet will be rejected by the *MS* since it does not have the certificate that signed the *Interest*.

6.1.2. Blockchain Tempering Attack

In this attack scenario, an adversary *adv* attempts to insert fake blocks into the blockchain. Suppose that a legitimate consumer vehicle requests the blockchain so it can start verifying the received *Data*. It sends an *Interest* packet to *MS* asking for the blockchain: $CV \hookrightarrow MS$: $i($`/NDN-BDA/MS/BC/request`$)$. Since NDN supports in-network caching, a network cache may satisfy the consumer's request, providing the *adv* with the ability to fulfill the request with a falsified blockchain. The Consumer Vehicle (CV) can detect a fake blockchain using the *Data*'s certificate by relying on the *MS*'s key K_{MS}. The CV can also detect a manipulated block when it receives *Data*. For instance, when the CV receives a *Data* packet from the PV with a particular VID, it searches for the associated certificate in the blockchain. Upon finding it, it verifies the validity of the block by authenticating its hash, checking if it was signed by the *MS*'s key K_{MS}^{-1} or not. If the verification fails, the CV will consider the blockchain as invalid and then request another version from the *MS*.

6.1.3. Man-in-the-Middle Attack

In this attack scenario, the adversary (*adv*) tries to substitute the key value of a legitimate producer with its own in order to obtain a certificate. For instance, the *adv* intercepts the message: $PV \hookrightarrow MS$: $si($`/NDN-BDA/MS/sign/{VID/<key>}`$K_{MS})$ and then changes the values of `/VID/<key>/` with its own. When it receives the request, the *MS* will detect that the *Interest*'s information has been altered since the signature value does not match the tampered *Interest*'s hash. The adversary (*adv*) may also try to inject fake information into the message: $MS \hookrightarrow PV$: $si($`/NDN-BDA/MS/sign/{VID/<key>}`$K_{MS})$ containing the generated certificate for the PV. However, the PV will still be able to detect it when it authenticates the received *Data* with the *MS*'s key K_{MS}, thereby making NDN-BDA resilient against such attacks.

6.1.4. Denial of Service Attack

In NDN, Denial of Service (DoS) attacks are mainly associated with an *Interest* Flooding Attack (IFA) [41]. Such an attack involves sending a large number of *Interest* packets to a target to overwhelm it, making it unable to accept legitimate requests. NDN-BDA, and specifically the Management Server, is also susceptible to DoS attacks. The attack is characterized by the adversary sending an excessive number of `join` and `sign` *Interest* packets with random CID and VID values to overwhelm the MS. The aim is to fill the MS's PIT with these fake requests, making it impossible to add legitimate requests. However, the MS can limit the impact of the attack by discarding successive *Interest* packets with the same signature. The adversary, *adv*, is also capable of launching an IFA using *BC* `request` *Interest* packets. Nevertheless, the impact of this attack is considered less severe since the blockchain could be obtained from network caches, thereby satisfying the *Interest* packets before they reach the *MS*. Despite this, the design of NDN-BDA does not prevent DoS attacks, and the MS should use an IFA mechanism such as the one proposed in [42] to mitigate the attack.

6.2. Protocol Validation

In this section, we evaluated our proposal using AVISPA. AVISPA uses a formal language named High-level Protocol Specification Language (HLPSL) to verify the correctness and security of communication protocols [43]. To ensure the security of our proposed protocol, we identified four roles, namely `role_PV`, `role_MS`, `role_VRE`, and `role_SCA`, which represent the *PV*, *MS*, *VRE*, and *SCA*, respectively. We then established sessions between these entities to simulate the exchange of protocol-based messages. In accordance with the diagram presented in Figure 7, these sessions are as follows: (1) the `PV` and `MS` establish two sessions, one for the `join` message and another one for the `sign` message; (2) the *MS* and *VRE* establish a session to verify the information of the `join` message received from the *PV*; (3) the *MS* establishes a session with the *SCA* to send a certificate request message. The validation process aims to ensure the secrecy of the information sent within two messages:

- $PV \hookrightarrow MS$: *si*(`/NDN-BDA/MS/join/{CID/VID}`K_{MS}).
- $PV \hookrightarrow MS$: *si*(`/NDN-BDA/MS/sign/VID/<key>/`K_{MS}).

Additionally, it aims to authenticate several messages:

- $PV \hookrightarrow MS$: *si*(`/NDN-BDA/MS/join/{CID/VID}`K_{MS}).
- $MS \hookrightarrow VRE$: *si*(`/NDN-BDA/MS/join/CID/VID`).
- $VRE \hookrightarrow MS$: *d*(`/NDN-BDA/MS/join/CID/VID`).
- $MS \hookrightarrow PV$: *d*(`/NDN-BDA/MS/join/{CID/VID}`K_{MS}).
- $PV \hookrightarrow MS$: *si*(`/NDN-BDA/MS/sign/VID/<key>/`K_{MS}.
- $MS \hookrightarrow SCA$: *si*(`/NDN-BDA/SCA/sign/CID/VID/<key>`).
- $SCA \hookrightarrow MS$: *d*(`/NDN-BDA/SCA/sign/CID/VID`).
- $MS \hookrightarrow PV$: *d*(`/NDN-BDA/MS/sign/VID/<key>/`K_{MS}.

The HLPSL role specification of the system nodes, namely the *VP*, *MS*, *VRE*, and *SCA*, is presented in Figure 8. The definition code of the *MS* node is represented from *Line 3* to *Line 36*. The *MS* interacts with all the other entities in the system. Therefore, we need to add all the system nodes as agents within the definition code of *MS*, as shown in *Line 3*. The role specification of the *MS* utilizes the public keys belonging to the *MS*, *VP*, *VRE*, and *SCA*, denoted as `Kms`, `Kpv`, `Kvre`, and `Ksca`, respectively. It also employs two communication channels, `SND`, used to send messages, and `RCV`, to receive messages. It also uses local variables to store/create incoming/outgoing messages (*Line 7*). The *MS* role specification includes a set of transitions, which outline all the actions that the *MS* performs during communication sessions (*Lines 11–35*). The HLPSL specification defines a set of actions. Along with `SND` and `RCV`, an action may also be `secret`, which verifies message encryption, `witness` to enable message authentication, and `wrequest` to verify the message's authenticity. For example, in *Line 16*, the *MS* requests the message that it sends to the *VRE* to be authenticated. Similarly, in *Line 13*, the *MS* verifies the authenticity of the message that it received from *PV*.

The *PV* role specification starts from *Line 39* and ends in *Line 59*. *PV* employs only two agents: the `VP` and `MS`, as the *PV* communicates only with the *MS* (*Line 39*). The `role_PV` definition also defines the actions performed by the *PV*, which includes sending secret messages (*Lines 48 and 53*).

Similarly, the *VRE* specification ranges from *Line 62* to *Line 77*. It also defines two agents: the `VRE` and `MS`, with whom it communicates. The *VRE* performs two actions. It verifies the received message from the *MS* (*Lines 70–72*) and sends a data packet to *MS* (*Lines 75–77*). The *SCA* also defines two agents, and it performs two actions: receiving a message from the *MS* (*Line s89–90*) and sending the data to the *MS* (*Lines 92–94*).

```
1. %% Definition of the MS role
2. %% ========================
3. Role
   role_MS(MS:agent,PV:agent,VRE:agent,SCA:agent,S2,D4:text,m
   s,Kpv,Kvre,Ksca:public_key,SND,RCV:channel(dy))
4. played_by MS
5. def=
6. local
7.    State:nat,S,D,D3:text
8. init
9.    State := 0
10. transition
11. 1. State=0 /\ RCV({PV.{S'}_inv(Kpv)}_Kms)  =|>  State':=1
12. %% MS checks the autheticity of PV->MS:si1
13. /\ wrequest(MS,PV,auth_1,S')
14. %% Sending the message MS->VRE:si2
15. /\ SND({MS.{S2}_inv(Kms)}_Kvre)
16. /\ witness(MS,VRE,auth_2,S2)
17. %% Receiving the message VRE->MS:d1
18. /\ D':=new()
19. /\ RCV({VRE.{D'}_inv(Kvre)}_Kms)
20. /\ wrequest(MS,VRE,auth_3,D')
21. 2. State=1 /\ RCV({PV.{S'}_inv(Kpv)}_Kms)  =|>  State':=2
22. %% MS checks the autheticity of PV->MS:si1
23. /\ wrequest(MS,PV,auth_4,S')
24. %% Sending the message MS->SCA:si4
25. /\ S2':=new()
26. /\ SND({MS.{S2'}_inv(Kms)}_Ksca)
27. /\ witness(MS,SCA,auth_5,S2')
28. 3. State=2
29. %% Receiving the message SCA->MS:d3
30. /\ RCV({SCA.{D3'}_inv(Ksca)}_Kms)  =|>  State':=3
31. /\ wrequest(MS,SCA,auth_6,D3')
32. %% Sending the message MS->PV:d4
33. %/\ S2':=new()
34. /\ SND({MS.{D4}_inv(Kms)}_Kpv)
35. /\ witness(MS,PV,auth_7,D4)
36. end role

37. %% Definition of the PV role
38. %% ========================
39. role
   role_PV(PV:agent,MS:agent,S:text,Kms,Kpv:public_key,SND,RC
   V:channel(dy))
40. played_by PV
41. def=
42. local
43.    State:nat,D4:text
44. init
45.    State := 0
46. transition
47. 1. State=0 /\ RCV(start) =|> State':=1 /\
   SND({PV.{S}_inv(Kpv)}_Kms)
48. /\ secret(S,sec_1,{PV,MS})
49. /\ witness(PV,MS,auth_1,S)
50. %% Sending the message PV->MS:si3
51. /\ S':=new()
52. /\ SND({PV.{S'}_inv(Kpv)}_Kms)
53. /\ secret(S,sec_2,{PV,MS})
54. /\ witness(PV,MS,auth_4,S')
55. 2. State=1
56. %% Receiving the message MS->PV:d4
57. /\ RCV({MS.{D4'}_inv(Kms)}_Kpv)  =|> State':=2
58. /\ wrequest(PV,MS,auth_7,D4')
59. end role

60. %% Definition of the VRE role
61. %% ========================
62. role
   role_VRE(VRE:agent,MS:agent,D1:text,Kms,Kvre:public_key,SND,
   RCV:channel(dy))
63. played_by VRE
64. def=
65. local
66.    State:nat,S:text
67. init
68.    State := 0
69. transition
70. 1. State=0 /\ RCV({MS.{S'}_inv(Kms)}_Kvre)  =|>  State':=1
71. %% VRE checks the autheticity of MS->VRE:si2
72. /\ wrequest(VRE,MS,auth_2,S')
73. 2. State=1
74. %% VRE sends the message VRE->MS:d1
75. /\ SND({VRE.{D1}_inv(Kvre)}_Kms) =|>  State':=2
76. /\ witness(VRE,MS,auth_3,D1)
77. end role

78. %% Definition of the SCA role
79. %% ========================
80. role
   role_SCA(SCA:agent,MS:agent,D3:text,Kms,Ksca:public_key,SND,
   RCV:channel(dy))
81. played_by SCA
82. def=
83. local
84.    State:nat,S:text
85. init
86.    State := 0
87. transition
88. %% Receiving the message MS->SCA:si4
89. 1. State=0 /\ RCV({MS.{S'}_inv(Kms)}_Ksca)  =|>  State':=1
90. /\ wrequest(SCA,MS,auth_5,S')
91. %% Sending the message SCA->MS:d3
92. /\ SND({SCA.{D3}_inv(Ksca)}_Kms)
93. /\ witness(SCA,MS,auth_6,D3)
94. end role
```

Figure 8. HLPSL specification code used for the definition of the NDN-BDA roles.

Figure 9 shows the HLPSL specification code for the communication sessions (*Lines 97–104*), the simulation environment (*Lines 107–120*), and the security goals of the system (*Lines 123–127*). A session consists of the roles `PV`, `MS`, `VRE`, and `SCA`, respectively. The knowledge assumptions regarding the adversary node are specified using `intruder_knowledge` inside the environment role (*Line 115*). It assumes that the adversary knows all the agents in addition to their public keys. Finally, the security objectives are defined in the goal specification (*Lines 123–127*). The `secrecy_of` function in *Line 124* verifies the secrecy of messages `sec_1` and `sec_2` that *PV* sends, while the `weak_authentication` function was used to validate the authenticity of all exchanged messages.

```
%% Definition of the session
%% ==========================
role
 session(MS:agent,PV:agent,VRE:agent,SCA:agent,S:text,Kms,K
 pv,Kvre,Ksca:public_key)
def=
local
    SND1,RCV1,SND2,RCV2,SND3,RCV3,SND4,RCV4:channel(dy)
 composition
 role_PV(MS,PV,S,Kms,Kpv,SND2,RCV2)
 /\role_MS(MS,PV,VRE,SCA,S,S,Kms,Kpv,Kvre,Ksca,SND1,RCV1)
 /\role_VRE(VRE,MS,S,Kms,Kvre,SND3,RCV3)
 /\role_SCA(SCA,MS,S,Kms,Ksca,SND4,RCV4)
 end role

 %% Definition of the environment
 %% ==============================
 role environment()
 def=
 const
   pv,ms,vre,sca:agent,
   s1,s2,s3:text,
   kms,kpv,kvre,ksca:public_key,
   sec_1, sec_2, auth_1, auth_2, auth_3, auth_4, auth_5,
 auth_6, auth_7:protocol_id

 %% Knowledge of the adversary
 intruder_knowledge = {pv,ms,vre,kms,kpv,kvre,ksca}
 composition
   session(pv,ms,vre,sca,s1,kms,kpv,kvre,ksca)
   /\session(ms,vre,i,i,s2,kms,kpv,kvre,ksca)
   /\session(ms,pv,i,i,s3,kms,kpv,kvre,ksca)
 end role

 %% Definition of the objectives
 %% ============================
 goal
   secrecy_of sec_1, sec_2
   weak_authentication_on auth_1, auth_2, auth_3, auth_4,
 auth_5, auth_6, auth_7
 end goal

 environment()
```

Figure 9. HLPSL specification code used to define the simulation environment and communication sessions.

The specified scenario was executed in AVISPA using the ATtack SEarcher (ATSE) backend, and it was animated using a tool called Security Protocol Animator (SPAN). Our proposed mechanism was deemed safe against intruders and met the security requirements of the threat model implemented in AVISPA.

7. Conclusions and Future Works

In the context of vehicular networks, the process of verifying data in NDN can lead to significant delays due to their highly dynamic nature. In this paper, we introduced NDN-BDA, a decentralized data authentication mechanism based on blockchain technology, which is both efficient and straightforward. Our proposal was thoroughly evaluated for efficiency against various attacks, and its functionality was validated using the widely used AVISPA tool.

As future work, we aim to enhance our solution by including a certificate revocation process and expanding it to a distributed mechanism that utilizes multiple Management Servers maintaining a global blockchain.

Author Contributions: The authors equally contributed to this work in all aspects with more implementation-related efforts from A.B. All authors have read and agreed to the published version of the manuscript.

Funding: This work was derived from the R&D project PID2021-122580NB-I00, funded by MCIN/AEI/10.13039/501100011033 and "ERDF A way of making Europe".

Data Availability Statement: All implementation details, sources, and data will be delivered upon request to the corresponding author Carlos T. Calafate.

Conflicts of Interest: The authors declare no conflict of interest.

References

1. Cisco Annual Internet Report (2018–2023) White Paper. Available online: https://www.cisco.com/c/en/us/solutions/collateral/executive-perspectives/annual-Internet-report/white-paper-c11-741490.html (accessed on 7 March 2023).
2. Xylomenos, G.; Ververidis, C.N.; Siris, V.A.; Fotiou, N.; Tsilopoulos, C.; Vasilakos, X.; Katsaros, K.V.; Polyzos, G.C. A survey of information-centric networking research. *IEEE Commun. Surv. Tutor.* **2013**, *16*, 1024–1049. [CrossRef]
3. Pan, J.; Paul, S.; Jain, R. A survey of the research on future Internet architectures. *IEEE Commun. Mag.* **2011**, *49*, 26–36. [CrossRef]
4. Koponen, T.; Chawla, M.; Chun, B.G.; Ermolinskiy, A.; Kim, K.H.; Shenker, S.; Stoica, I. A data-oriented (and beyond) network architecture. In Proceedings of the 2007 Conference on Applications, Technologies, Architectures, and Protocols for Computer Communications, Kyoto, Japan, 27–31 August 2007; pp. 181–192.

5. Fotiou, N.; Nikander, P.; Trossen, D.; Polyzos, G.C. Developing information networking further: From PSIRP to PURSUIT. In *Proceedings of the Broadband Communications, Networks, and Systems, 7th International ICST Conference, BROADNETS 2010, Athens, Greece, 25–27 October 2010;* Revised Selected Papers 7; Springer: Berlin/Heidelberg, Germany, 2012; pp. 1–13.
6. Zhang, L.; Afanasyev, A.; Burke, J.; Jacobson, V.; Claffy, K.C.; Crowley, P.; Papadopoulos, C.; Wang, L.; Zhang, B. Named data networking. *ACM SIGCOMM Comput. Commun. Rev.* **2014**, *44*, 66–73. [CrossRef]
7. Jacobson, V.; Mosko, M.; Smetters, D.; Garcia-Luna-Aceves, J. *Content-Centric Networking;* Whitepaper; Palo Alto Research Center: Palo Alto, CA, USA , 2007; pp. 2–4.
8. NDN Packet Format Specification Version 0.3. Available online: https://named-data.net/doc/NDN-packet-spec/current (accessed on 3 March 2023).
9. Signed Interest. NDN Packet Format Specification Version 0.3. Available online: https://named-data.net/doc/NDN-packet-spec/current/signed-interest.html (accessed on 3 March 2023).
10. Zhang, H.; Li, Y.; Zhang, Z.; Afanasyev, A.; Zhang, L. Ndn host model. *ACM SIGCOMM Comput. Commun. Rev.* **2018**, *48*, 35–41. [CrossRef]
11. Zhang, M.; Luo, H.; Zhang, H. A survey of caching mechanisms in information-centric networking. *IEEE Commun. Surv. Tutor.* **2015**, *17*, 1473–1499. [CrossRef]
12. Zhang, Z.; Yu, Y.; Zhang, H.; Newberry, E.; Mastorakis, S.; Li, Y.; Afanasyev, A.; Zhang, L. An overview of security support in named data networking. *IEEE Commun. Mag.* **2018**, *56*, 62–68. [CrossRef]
13. Tehrani, P.F.; Osterweil, E.; Schmidt, T.C.; Wählisch, M. SoK: Public key and namespace management in NDN. In Proceedings of the 9th ACM Conference on Information-Centric Networking, Osaka, Japan, 19–21 September 2022; pp. 67–79.
14. Li, Y.; Zhang, Z.; Wang, X.; Lu, E.; Zhang, D.; Zhang, L. A secure sign-on protocol for smart homes over named data networking. *IEEE Commun. Mag.* **2019**, *57*, 62–68. [CrossRef]
15. Zhang, Z.; Yu, Y.; Afanasyev, A.; Zhang, L. *NDN Certificate Management Protocol (NDNCERT);* Technical Report NDN-0050; NDN: Helensvale, Australia, 2017.
16. Zhang, Z.; Yu, Y.; Ramani, S.K.; Afanasyev, A.; Zhang, L. Nac: Automating access control via named data. In Proceedings of the MILCOM 2018—2018 IEEE Military Communications Conference (MILCOM), Los Angeles, CA, USA, 29–31 October 2018; pp. 626–633.
17. Nour, B.; Khelifi, H.; Hussain, R.; Mastorakis, S.; Moungla, H. Access control mechanisms in named data networks: A comprehensive survey. *ACM Comput. Surv.* **2021**, *54*, 61. [CrossRef]
18. Lee, C.A.; Zhang, Z.; Tu, Y.; Afanasyev, A.; Zhang, L. Supporting virtual organizations using attribute-based encryption in named data networking. In Proceedings of the 2018 IEEE 4th International Conference on Collaboration and Internet Computing (CIC), Philadelphia, PA, USA, 18–20 October 2018; pp. 188–196.
19. Li, Z.; Xu, Y.; Zhang, B.; Yan, L.; Liu, K. Packet forwarding in named data networking requirements and survey of solutions. *IEEE Commun. Surv. Tutor.* **2018**, *21*, 1950–1987. [CrossRef]
20. Voitalov, I.; Aldecoa, R.; Wang, L.; Krioukov, D. Geohyperbolic routing and addressing schemes. *ACM SIGCOMM Comput. Commun. Rev.* **2017**, *47*, 11–18. [CrossRef]
21. Zhang, Y.; Afanasyev, A.; Burke, J.; Zhang, L. A survey of mobility support in named data networking. In Proceedings of the 2016 IEEE Conference on Computer Communications Workshops (INFOCOM WKSHPS), San Francisco, CA, USA, 10–14 April 2016; pp. 83–88.
22. Khelifi, H.; Luo, S.; Nour, B.; Moungla, H.; Faheem, Y.; Hussain, R.; Ksentini, A. Named data networking in vehicular ad hoc networks: State-of-the-art and challenges. *IEEE Commun. Surv. Tutor.* **2019**, *22*, 320–351. [CrossRef]
23. Zhu, Z.; Afanasyev, A.; Zhang, L. A New Perspective on Mobility Support. NDN, Technical Report NDN-0013. 2013. Available online: https://citeseerx.ist.psu.edu/document?repid=rep1&type=pdf&doi=a6fc488646c27d13191691f7145745ba828cff9b (accessed on 30 March 2023) .
24. Afanasyev, A.; Jiang, X.; Yu, Y.; Tan, J.; Xia, Y.; Mankin, A.; Zhang, L. NDNS: A DNS-like name service for NDN. In Proceedings of the 2017 26th International Conference on Computer Communication and Networks (ICCCN), Vancouver, BC, Canada, 31 July–3 August 2017; pp. 1–9.
25. Zhang, Y.; Xia, Z.; Mastorakis, S.; Zhang, L. Kite: Producer mobility support in named data networking. In Proceedings of the 5th ACM Conference on Information-Centric Networking, Boston, MA, USA, 21–23 September 2018; pp. 125–136.
26. Chowdhury, M.; Gawande, A.; Wang, L. Anonymous authentication and pseudonym-renewal for VANET in NDN. In Proceedings of the 4th ACM Conference on Information-Centric Networking, Berlin, Germany, 26–28 September 2017; pp. 222–223.
27. Lu, Y.; Wang, C.; Yue, M.; Wu, Z. Consumer-source authentication with conditional anonymity in information-centric networking. *Inf. Sci.* **2023**, *624*, 378–394. [CrossRef]
28. Yu, Y.; Afanasyev, A.; Seedorf, J.; Zhang, Z.; Zhang, L. NDN DeLorean: An authentication system for data archives in named data networking. In Proceedings of the 4th ACM Conference on Information-Centric Networking, Berlin, Germany, 26–28 September 2017; pp. 11–21.
29. Zhang, Z.; Liu, S.; King, R.; Zhang, L. Ndn-mps: Supporting multiparty authentication over named data networking. In Proceedings of the 8th ACM Conference on Information-Centric Networking, Paris, France, 22–24 September 2021; pp. 83–94.

30. Li, T.; Shang, W.; Afanasyev, A.; Wang, L.; Zhang, L. A brief introduction to ndn dataset synchronization (ndn sync). In Proceedings of the MILCOM 2018—2018 IEEE Military Communications Conference (MILCOM), Los Angeles, CA, USA, 29–31 October 2018; pp. 612–618.
31. Li, R.; Asaeda, H.; Wu, J. DCAuth: Data-centric authentication for secure in-network big-data retrieval. *IEEE Trans. Netw. Sci. Eng.* **2018**, *7*, 15–27. [CrossRef]
32. Barka, E.; Kerrache, C.A.; Hussain, R.; Lagraa, N.; Lakas, A.; Bouk, S.H. A trusted lightweight communication strategy for flying named data networking. *Sensors* **2018**, *18*, 2683. [CrossRef]
33. Zhu, K.; Chen, Z.; Yan, W.; Zhang, L. Security attacks in named data networking of things and a blockchain solution. *IEEE Internet Things J.* **2018**, *6*, 4733–4741. [CrossRef]
34. Goyal, S.; Sharma, N.; Kaushik, I.; Bhushan, B. Blockchain as a solution for security attacks in named data networking of things. In *Security and Privacy Issues in IoT Devices and Sensor Networks*; Academic Press: Cambridge, MA, USA, 2021; pp. 211–243.
35. Chen, C.; Wang, C.; Qiu, T.; Lv, N.; Pei, Q. A secure content sharing scheme based on blockchain in vehicular named data networks. *IEEE Trans. Ind. Inform.* **2019**, *16*, 3278–3289. [CrossRef]
36. Rawat, D.B.; Doku, R.; Adebayo, A.; Bajracharya, C.; Kamhoua, C. Blockchain enabled named data networking for secure vehicle-to-everything communications. *IEEE Netw.* **2020**, *34*, 185–189. [CrossRef]
37. Khelifi, H.; Luo, S.; Nour, B.; Moungla, H.; Ahmed, S.H.; Guizani, M. A blockchain-based architecture for secure vehicular Named Data Networks. *Comput. Electr. Eng.* **2020**, *86*, 106715. [CrossRef]
38. Conti, M.; Hassan, M.; Lal, C. BlockAuth: BlockChain based distributed producer authentication in ICN. *Comput. Netw.* **2019**, *164*, 106888. [CrossRef]
39. Zhao, Y.; Qu, Y.; Xiang, Y.; Zhang, Y.; Gao, L. A Lightweight Model-Based Evolutionary Consensus Protocol in Blockchain as a Service for IoT. *IEEE Trans. Serv. Comput.* **2023**, *Early Access*.
40. Vigano, L. Automated security protocol analysis with the AVISPA tool. *Electron. Notes Theor. Comput. Sci.* **2006**, *155*, 61–86. [CrossRef]
41. Benmoussa, A.; Kerrache, C.A.; Lagraa, N.; Mastorakis, S.; Lakas, A.; Tahari, A.E.K. Interest Flooding Attacks in Named Data Networking: Survey of Existing Solutions, Open Issues, Requirements and Future Directions. *ACM Comput. Surv.* **2021**, *55*, 139. [CrossRef]
42. Benmoussa, A.; El Karim Tahari, A.; Kerrache, C.A.; Lagraa, N.; Lakas, A.; Hussain, R.; Ahmad, F. MSIDN: Mitigation of sophisticated interest flooding-based DDoS attacks in named data networking. *Future Gener. Comput. Syst.* **2020**, *107*, 293–306. [CrossRef]
43. Chevalier, Y.; Compagna, L.; Cuellar, J.; Drielsma, P.H.; Mantovani, J.; Mödersheim, S.; Vigneron, L. A high level protocol specification language for industrial security-sensitive protocols. In Proceedings of the Workshop on Specification and Automated Processing of Security Requirements-SAPS'2004, Linz, Austria, 20–25 September 2004; Austrian Computer Society: Vienna, Austria, 2004; p. 13.

future internet

Article

Intelligent Caching with Graph Neural Network-Based Deep Reinforcement Learning on SDN-Based ICN

Jiacheng Hou [1,*], Tianhao Tao [1], Haoye Lu [2] and Amiya Nayak [1]

1 School of Electrical Engineering and Computer Science, University of Ottawa, Ottawa, ON K1N 6N5, Canada; ttao061@uottawa.ca (T.T.); nayak@uottawa.ca (A.N.)
2 David R. Cheriton School of Computer Science, University of Waterloo, Waterloo, ON N2L 3G1, Canada; haoye.lu@uwaterloo.ca
* Correspondence: jhou013@uottawa.ca

Abstract: Information-centric networking (ICN) has gained significant attention due to its in-network caching and named-based routing capabilities. Caching plays a crucial role in managing the increasing network traffic and improving the content delivery efficiency. However, caching faces challenges as routers have limited cache space while the network hosts tens of thousands of items. This paper focuses on enhancing the cache performance by maximizing the cache hit ratio in the context of software-defined networking–ICN (SDN-ICN). We propose a statistical model that generates users' content preferences, incorporating key elements observed in real-world scenarios. Furthermore, we introduce a graph neural network–double deep Q-network (GNN-DDQN) agent to make caching decisions for each node based on the user request history. Simulation results demonstrate that our caching strategy achieves a cache hit ratio 34.42% higher than the state-of-the-art policy. We also establish the robustness of our approach, consistently outperforming various benchmark strategies.

Keywords: information-centric networking; software-defined networking; graph neural network; deep reinforcement learning; intelligent caching

Citation: Hou, J.; Tao, T.; Lu, H.; Nayak, A. Intelligent Caching with Graph Neural Network-Based Deep Reinforcement Learning on SDN-Based ICN. *Future Internet* **2023**, *15*, 251. https://doi.org/10.3390/fi15080251

Academic Editor: Danda B. Rawat

Received: 22 June 2023
Revised: 9 July 2023
Accepted: 24 July 2023
Published: 26 July 2023

1. Introduction

Information-centric networking (ICN) has received significant interest and attention in recent years as a promising paradigm for network communication. ICN introduces a shift in focus from the traditional host-centric model to a content-centric approach. Named-based routing and in-network caching are two key features of ICN that have contributed to its growing popularity and adoption in various domains.

Researchers have explored the potential applications of ICN in diverse areas such as the Internet of Things (IoTs), where the efficient dissemination and retrieval of data is crucial for IoT devices to interact and exchange information [1]. In the context of the Internet of Vehicles (IoVs), ICN can enable efficient content delivery, facilitate real-time communication, and support intelligent transportation systems [2]. Furthermore, in the realm of 5G networks, ICN has been investigated for its potential to enhance content delivery, reduce latency, and improve overall network performance [3]. ICN has also been explored in software-defined networking (SDN) environments, offering flexibility, scalability, and efficient resource management [4].

One of the fundamental challenges in ICN is the effective caching of content across the network. Caching reduces latency, minimizes network congestion, and improves content delivery efficiency. However, it is a complex task due to the limited cache space available at each router and the potentially vast number of items distributed throughout the network. Researchers have been actively investigating caching strategies and policies to optimize cache performance.

In recent years, deep reinforcement learning (DRL) has allowed significant advancements in decision making, particularly in caching decisions. Numerous studies (see [5,6])

have demonstrated the exceptional performance of DRL in solving caching problems. Researchers have adopted deep Q-learning network architectures, such as multi-layer perceptron (MLP) and convolutional neural networks (CNNs), to replace traditional Q-tables. However, MLP and CNN architectures struggle to effectively utilize the neighbourhood information in arbitrary graph data, such as network topologies and knowledge graphs. While CNNs have been extensively optimized for the processing of Euclidean space data such as images and grids, they face challenges when dealing with graph-structured data. This limitation hampers their ability to capture the relational information necessary for efficient caching decision making.

Graph neural networks (GNNs) offer distinct advantages over traditional MLP and CNN architectures, as they are purpose-built to handle graph-structured data and excel in non-Euclidean spaces. This unique capability has made GNNs a popular choice in a wide range of domains that involve data represented as arbitrary graphs [7]. Notably, GNNs have demonstrated remarkable success in network routing optimization [8], where the underlying graph structure captures the intricate relationships between network nodes and facilitates efficient path planning. Additionally, in the domain of traffic prediction [9], GNNs leverage the graph structures of road intersections and their connectivity to forecast traffic flow patterns accurately.

Moreover, recent research has highlighted the remarkable generalization capabilities of GNNs [10]. GNNs can generalize effectively over different network topologies, allowing them to adapt to various environments and scenarios. This has been substantiated by studies such as [11–13], which have showcased the impressive generalization performance of GNNs across diverse network architectures.

The inherent suitability of GNNs for graph-structured data and their exceptional generalization capabilities make them an ideal choice in tackling complex problems in network-related domains. In the context of our research, leveraging the power of GNNs allows us to capture the intricate relationships and dependencies present in network caching scenarios, ultimately enhancing the network caching performance.

This paper aims to enhance the caching performance in the SDN-ICN scenario by leveraging DRL and GNN. Specifically, we introduce a GNN–double deep Q-network [14] (GNN-DDQN) caching agent within the SDN controller. The SDN controller provides a real-time and comprehensive view of the traffic situation in the SDN-ICN environment, while the network nodes are equipped with caching capabilities. The GNN-DDQN agent determines optimal caching decisions for individual nodes by considering the traffic conditions at each time step. The controller then communicates these decisions to the respective nodes, enabling them to update their cache stores accordingly.

The contributions of this paper are as follows.

- We develop a statistical model to generate users' preferences. Initially, we employ matrix factorization based on the Neural Collaborative Filtering Model [15] to learn content and user embeddings using the real-world dataset MovieLens100K [16]. Next, we employ a Gaussian mixture model to cluster users and content based on their embeddings. Subsequently, we employ a statistical model to generate the request behaviour of each user group.
- We introduce a GNN-DDQN agent within the SDN-ICN scenario. Incorporating a GNN in DRL is advantageous as GNNs excel in modelling graph-structured data, enabling nodes to engage in cooperative caching and enhancing the overall caching performance. Additionally, with only a single forward pass through the neural network, the GNN-DDQN agent can make caching decisions for all nodes in the network at each time step.
- We extensively evaluate the proposed caching scheme through simulations across various scenarios. These scenarios include different numbers of items, cache sizes, and network topologies, such as GEANT [17], ROCKETFUEL [18], TISCALI [19], and GARR [19]. Notably, our proposed caching scheme outperforms the state-of-the-art DRL-based caching strategy. Furthermore, it exhibits a significant performance advan-

tage over several benchmark caching schemes (Leave Copy Down (LCD), Probabilistic Caching (PROB_CACHE), Cache Less for More (CL4M), and Leave Copy Everywhere (LCE) [20–23]). The evaluations demonstrate the robustness of our proposed strategy to simulation parameters and variations in network topology.

It is worth noting that GNN-DDQN has several advantages.

- Computational Efficiency: GNN-DDQN is computationally efficient, requiring only one DRL agent to make caching decisions for all network nodes in a single forward pass.
- Multi-Action Capability: GNN-DDQN enables the agent to take multiple actions for each network node at each time step, demonstrating strong performance even with the incorporation of multi-actions.
- Applicability: GNN-DDQN can be applied in various real-world scenarios.
  - Content Delivery Networks (CDNs): Our proposed caching scheme can be employed within CDNs to improve caching decisions at edge nodes.
  - Mobile Edge Computing (MEC): Our caching scheme can benefit MEC environments by strategically caching frequently accessed content at edge servers.
  - Internet Service Providers (ISPs): By deploying our scheme, ISPs can enhance their caching infrastructure, effectively reducing the bandwidth requirements for popular content and providing faster access to frequently accessed data for their subscribers.
  - Video Streaming Platforms: By caching popular videos at appropriate network nodes, our algorithm can reduce buffering times and enhance the overall streaming experience for users.

However, there are also limitations to consider.

- Scalability: GNN-DDQN may face challenges in terms of scalability when dealing with a large number of network nodes. With only one SDN controller monitoring the entire network traffic, it may experience high latency, impacting the overall performance of the caching algorithm.
- Overfitting and Underfitting: GNN-DDQN, as with other deep learning algorithms, may suffer from overfitting or underfitting, depending on various factors.

The rest of this paper is organized as follows. Section 2 overviews related work. Sections 3 and 4 present our system model and proposed methodology. Section 5 shows the experimental results. Section 6 concludes the paper.

2. Related Work

Classical caching placement algorithms commonly used in the literature include LCE, LCD, PROB_CACHE, and CL4M [20–23]. LCE involves copying content at any cache between the serving and receiving nodes, while LCD caches content in the immediate neighbourhood of the serving node in the receiver's direction. PROB_CACHE probabilistically caches content on a path, considering various factors. CL4M aims to place content in nodes with high graph-based centrality. In addition to placement algorithms, traditional caching replacement algorithms such as Least Recently Used (LRU), Least Frequently Used (LFU), and First-In-First-Out (FIFO) are commonly employed [24,25]. LRU discards the least recently accessed content, LFU replaces the least frequently used content first, and FIFO evicts the first item inserted in the cache. However, these traditional algorithms are often considered inefficient, yielding poorer performance than deep learning-based caching algorithms.

DRL-based caching algorithms have demonstrated remarkable achievements in recent years [26,27]. In [5], the authors developed a deep Q-network (DQN)-based caching algorithm designed explicitly for mobile edge networks. The application of DRL in the Internet of Vehicles (IoV) field has also gained substantial attention. As the demand for computation and entertainment in autonomous driving and vehicular scenarios increases, researchers have been actively advancing caching strategies to enhance the user experience.

In [28], the authors propose CoCaRL, a caching strategy leveraging DRL and a multi-level federated learning framework. They utilize a DDQN [14] to optimize the cache hit ratio of local roadside units (RSUs), neighbour RSUs, and cloud data centers in vehicular networks. Their approach also incorporates federated learning to enable decentralized model training. Another study [29] introduces a quality of experience (QoE)-driven RSU caching model based on DRL. Their caching algorithm addresses the growing demand for time-sensitive short videos in a 5G-based IoV scenario. The reward in their DRL model is defined as the ratio of the number of videos interesting to each user to the total number of videos stored in the RSU. Furthermore, in [6], the authors design a DQN-based strategy to optimize joint computing and edge caching in a three-layer IoV-ICN network architecture, encompassing vehicles, edges, and cloud layers. In [30], the authors proposed a social-aware vehicular edge computing architecture to efficiently deliver popular content to end-users in vehicular social networks. They introduce a social-aware graph pruning search algorithm to assign content consumer vehicles to the shortest path with the most relevant content providers. Additionally, they utilize a DRL method to optimize content distribution across the network. In [31], the authors develop an IoV-specific edge caching model that enables collaborative content caching among mobile vehicles and considers varying content popularity and channel conditions. Additionally, the framework empowers each vehicle agent to make caching decisions based on environmental observations autonomously. In [32], the authors proposed a spatial–temporal correlation approach to predict content popularity in the IoV. They introduce a DRL-based multi-agent caching strategy, where each RSU is an independent agent, to optimize caching decisions. In [33], the authors investigate joint computation offloading, data caching, transmission path selection, subchannel assignment, and caching management in the IoV-based environment. Dynamic online algorithms such as the Simulated Annealing Genetic Algorithm (SAGA) and DQN are adopted to minimize the content access latency.

In addition to DRL, GNNs have emerged as another effective approach in addressing caching problems. One notable application of GNNs in caching is presented in [34]. The authors introduce a GNN-based caching algorithm to optimize the cache hit ratio in a named data networking (NDN) context. Their approach involves two key steps. First, they utilize a 1D-CNN to predict the popularity of content in each node. Subsequently, a GNN is employed to propagate the content popularity predictions among neighbouring nodes. Finally, each node makes caching decisions based on node-level caching probability ranking. Leveraging the message-passing capabilities of GNNs, their caching approach outperforms the CNN-based caching algorithm, leading to improved caching performance in the NDN scenario. Moreover, in [35], the authors propose GNN-GM to enhance the caching performance in NDN. In this work, a GNN is utilized to predict users' ratings of unviewed movies within a bipartite graph representation. Leveraging the accurate rating predictions achieved by GNN, the proposed approach achieves a higher cache hit ratio compared to state-of-the-art caching schemes. The successful application of GNNs in these studies highlights their efficacy in addressing caching challenges. By leveraging the GNN's ability to capture complex dependencies and propagate information across nodes, these approaches demonstrate improved caching performance and provide valuable insights for the optimization of cache hit ratios in various network scenarios.

The integration of DRL and GNN has additionally emerged as a growing trend, delivering numerous benefits. The authors in [36] employ dynamic graph convolutional networks (GCNs) and RL for long-term traffic flow prediction. They represent traffic flow as a graph, where each station is a node and directed weighted edges are used to indicate traffic flow occurrence. A graph convolutional policy network (GCPN) model generates dynamic graphs at each time step, and the RL agent receives a reward if the generated graph closely resembles the target graph. The paper further utilizes a GCN and long short-term memory (LSTM) to extract spatial and temporal features from the generated dynamic graph sequences, enabling traffic flow prediction in future time steps. Another study [37] introduces the Inductive Heterogeneous Graph Multi-Agent Actor–Critic (IHG-MA) algorithm for traffic signal control. The traffic network is modelled as a heterogeneous graph, with each traffic signal controller considered an agent. An inductive GNN algorithm is applied to learn the embeddings of the agents and their neighbours. The learned representations are then fed into an actor–critic network to optimize traffic control. Additionally, [38] proposes an innovative approach using a GCN and DQN for the multi-agent cooperative control of connected autonomous vehicles (CAVs). Each CAV is treated as an agent, and a GCN is utilized to extract embeddings for each agent. These representations are then fed into a Q-network to determine the actions of each agent, facilitating effective cooperative control among the CAVs. Furthermore, in an SDN-based scenario, [13] presents a centralized agent that leverages DRL and GNNs to optimize routing strategies. They utilize a GNN to model the network and DRL to calculate the Q-value of an action. By embedding routing paths into node representations and feeding them into the Q-network, they evaluate various routing strategies and select the optimal one when a traffic demand is issued. In [39], the authors propose a method that combines prediction, caching, and offloading techniques to optimize computation in 6G-enabled IoV. The prediction method is based on a spatial–temporal graph neural network (STGNN), the caching decision method is realized using the simplex algorithm, and the offloading method is based on Twin Delayed Deterministic Policy Gradient (TD3).

These studies demonstrate the efficacy of combining DRL and GNNs in tackling various issues, including traffic prediction, traffic signal control, the cooperative control of autonomous vehicles, routing optimization in SDN scenarios, and caching in IoV environments. By leveraging the respective strengths of DRL and GNNs, these approaches enable intelligent decision making and enhance performance in intricate systems. We are confident that the amalgamation of DRL and GNNs can similarly bring advantages to caching in the SDN-ICN context.

3. System Model

In this section, we present the system architecture of our proposed caching scheme and provide a comprehensive overview of the key components. We also define the concept of content popularity. Furthermore, we develop a user preference model based on real-world data from the MovieLens100K dataset [16]. Important notations used throughout the paper are listed in Table 1.

Table 1. Important notations.

Notation	Definition
N	Number of network nodes
C	Number of content items
T	Number of time slots
$\mathcal{N} = \{n_1, \ldots, n_N\}$	Set of network nodes
$\mathcal{C} = \{c_1, \ldots, c_C\}$	Set of content
$\mathcal{U} = \{u_1, \ldots, u_U\}$	Set of users
$\mathcal{T} = \{t_0, t_1, \ldots, t_T\}$	Set of time steps
$b_{t_l}^{c_i,n_k}$	"Cache" or "not cache" content c_i at node n_k at time step t_l (i.e., availability of content c_i at node n_k's cache store during the time interval between t_l and t_{l+1})
$\mathcal{S}_{t_l} = \{s_{t_l}^{n_1}, \ldots, s_{t_l}^{n_N}\}$	Set of all nodes' states at time step t_l
$\mathcal{A}_{t_l} = \{a_{t_l}^{n_1}, \ldots, a_{t_l}^{n_N}\}$	Set of all nodes' caching actions at time step t_l
$a_{t_l}^{n_k} = \{b_{t_l}^{c_1,n_k}, \ldots, b_{t_l}^{c_C,n_k}\}$	Set of node n_k's caching action at time step t_l
$\mathcal{R}_{t_l} = \{r_{t_l}^{n_1}, \ldots, r_{t_l}^{n_N}\}$	Set of all nodes' rewards (i.e., cache hits) at time step t_l
$r_{t_l}^{n_k} = \{r_{t_l}^{c_1,n_k}, \ldots, r_{t_l}^{c_C,n_k}\}$	Set of node n_k's reward for each content item at time step t_l
z	Router's cache size
x_{c_i}	Content c_i's embedding
y_{u_j}	User u_j's embedding

3.1. System Architecture

We propose an intelligent caching strategy in an SDN-based ICN (SDN-ICN) architecture, as depicted in Figure 1. The SDN-ICN architecture separates the control plane from the data plane, and the OpenFlow protocol facilitates the data transfer between them. We introduce the GNN-DDQN [14,40] agent responsible for making caching decisions in the control plane. The data plane comprises network nodes that perform caching actions.

Figure 1 illustrates the control plane, consisting of two modules: (i) the GNN-DDQN agent module, which plays a crucial role in caching decisions for ICN nodes in the data plane; and (ii) the content caching management module, which handles the content caching of each ICN node. We assume that the data plane's network topology consists of N ICN nodes, denoted as $\mathcal{N} = \{n_1, n_2, \ldots, n_N\}$.

The data plane encompasses ICN nodes, each fulfilling specific roles: (i) source nodes responsible for content publication without caching capabilities, (ii) receiver nodes accountable for sending requests to source nodes, also without caching capabilities, and (iii) router nodes responsible for forwarding requests and data packets across the network. Instead of assuming that all router nodes have the caching capability, we consider only some of them to have this. These router nodes equipped with caching capabilities have a cache capacity of z, defined as the number of content items.

The network contains C distinct content items, represented by the set $\mathcal{C} = \{c_1, \ldots, c_C\}$. We assume that an experimental time round can be divided into T slots of equal duration, denoted by $\mathcal{T} = \{t_0, t_1, \ldots, t_T\}$. To indicate whether a node n_k caches content c_i at time step t_l, we employ a binary variable $\{b_{t_l}^{c_i,n_k}\}$, where $t_l \in \mathcal{T}$, $c_i \in \mathcal{C}$, and $n_k \in \mathcal{N}$. Specifically, $b_{t_l}^{c_i,n_k} = 1$ if and only if node n_k caches content c_i at time step t_l, implying that content c_i is available at node n_k during the time interval between t_l and t_{l+1}.

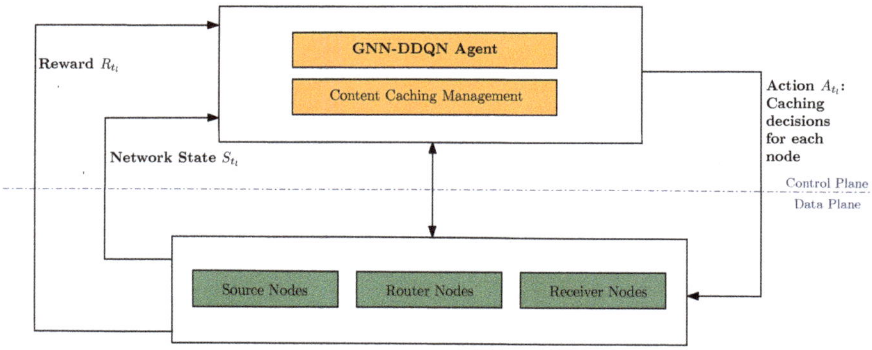

Figure 1. The SDN-ICN architecture. In the controller, the GNN-DDQN agent receives a network state S_{t_l} and generates an action A_{t_l} at each time step t_l. Subsequently, it receives a reward R_{t_l} at the next time step t_{l+1}.

In the SDN-ICN architecture, the controller can see the network's traffic. Therefore, at each time step t_l, the GNN-DDQN agent observes the network state $\mathcal{S}_{t_l}$, which encompasses information about the network's status during the time interval between t_{l-1} and t_l. Subsequently, the GNN-DDQN agent performs content caching predictions at the node level, denoted as $\mathcal{A}_{t_l}$, and communicates the recommended content to be cached to each router node with caching capabilities. When a router node with caching capabilities receives a request packet for content c_i within this period, it can fulfill the request directly if the requested content is cached or forward the request to the source node. At the subsequent time step t_{l+1}, a set of rewards $\mathcal{R}_{t_l}$ is sent to the GNN-DDQN agent. Specifically, $\mathcal{R}_{t_l} = r_{t_l}^{n_1}, r_{t_l}^{n_2}, \ldots, r_{t_l}^{n_N}$ represents the rewards of all nodes at time step t_l. Furthermore, $r_{t_l}^{n_k} = r_{t_l}^{c_1,n_k}, r_{t_l}^{c_2,n_k}, \ldots, r_{t_l}^{c_C,n_k}$ represents the set of rewards for each content item received by node n_k at time step t_l.

3.2. Content Popularity and User Preference

In a realistic computer network, users exhibit preferences for specific types of content, leading to varying request frequencies. We develop a statistical model that incorporates content popularity and user preferences to simulate this network traffic. In our network, receiver nodes correspond to users, and we denote the users as $\mathcal{U} = \{u_1, \ldots, u_U\}$. Our objective is to determine the probability distribution $P(c_i, u_j)$, which represents the likelihood of a request for the ith content by the jth user.

Content popularity refers to the probability distribution of requesting the ith content within the network, represented by $P(c_i)$. Research studies [41] have shown that the content popularity in a network can be modelled using a Zipfian distribution,

$$P(c_i) = \frac{1}{(i)^{\alpha} \sum_{k=1}^{C} (k)^{-\alpha}} \quad (1)$$

where α is a skewness factor with a value of 0.8.

User preference refers to the relationship between users and content items. In our study, we employ collaborative filtering [15,42] to capture user preferences. Collaborative filtering is a popular recommendation system technique that predicts a user's preference by identifying users with similar tastes based on their historical behaviours. Collaborative filtering has two primary approaches: the neighbourhood-based method and the latent factor method. The neighbourhood-based method identifies similar users or items based on their historical preferences and recommends items that similar users or items have liked. On the other hand, the latent factor method discovers latent factors that represent underlying characteristics of users and items and uses these factors to predict user preferences. In our case, we utilize matrix factorization, a latent factor method, to extract the latent factors of users and content items. By decomposing the user–item interaction matrix into lower-dimensional matrices, we can represent users and items in terms of these latent factors. Subsequently, we calculate user preferences by analyzing the relationships between users and content items derived from matrix factorization.

To capture the user–content relation, we construct a matrix M with dimensions $C \times U$, where C represents the number of content items and U represents the number of users. Each element $m_{i,j}$ in the matrix corresponds to the relationship between content c_i and user u_j. This relationship can be based on various factors, such as ratings given by the user, the time spent on the content, or any other relevant metric. We utilize trainable embedding layers to process the user–content matrix further to generate embedding vectors for each content item and user. Specifically, for each content item c_i, we apply an embedding layer that maps it to a continuous vector representation $x_{c_i} \in \mathbb{R}^e$, where e denotes the dimensionality of the embedding. Similarly, for each user u_j, we employ an embedding layer to obtain the embedding vector $y_{u_j} \in \mathbb{R}^e$.

Our research uses the well-known MovieLens100K dataset [16] as a real-world dataset for our experiments. This dataset consists of user ratings for movies and is widely used in evaluating recommendation systems. We focus on learning embedding vectors for 943 users and 1682 content items within this dataset.

To obtain user and content embeddings, we employ a matrix factorization technique combined with a neural network architecture inspired by the works [15,42]. Our model consists of two trainable embedding layers, one for users and another for content items. These layers enable the learning of dense and low-dimensional representations that capture users' and content's underlying characteristics and preferences. The next step in our model involves computing the element-wise product of the content and user embedding vectors. This element-wise product represents the interaction between a specific content item and a user. Subsequently, the resulting products are fed into a linear layer with an activation function. For a given content embedding x_{c_i} and a user embedding y_{u_j}, the output is computed as follows:

$$\hat{m}_{i,j} = \sigma(\mathbf{w}^T (x_{c_i} \odot y_{u_j})) \quad (2)$$

in this equation, σ denotes the sigmoid activation function, **w** represents a trainable matrix, and $\odot$ signifies the element-wise product. To train the matrix factorization model, we minimize the binary cross-entropy (BCE) loss between the ground truth values $m_{i,j}$ and the predicted values $\hat{m}_{i,j}$. It is worth mentioning that we label $m_{i,j}$ as 1 if the jth user has provided a rating for the ith content item, and 0 otherwise. The key training parameters for the Neural Collaborative Filtering (NCF) model are summarized in Table 2.

Table 2. Key NCF model training parameters.

Parameters	Values
Epoch	100
Learning rate	0.001
Batch size	256
Embedding dimension e	8
Optimizer	Adam

In order to fit the number of content items and users in our network, all users and content items in the dataset are divided into groups. If the content embeddings are close to each other, we cluster them into groups, and users are grouped in the same way. We utilize a Gaussian mixture model (GMM) to cluster the embedding vectors, and then compute a representative embedding for each group by taking the element-wise mean.

Since the inner product ${x_{c_i}}^T y_{u_j}$ captures the correlation between content c_i and a user u_j, we apply the softmax function on the inner products to obtain the probability $P(u_j|c_i)$ for given content c_i. This probability represents the preference of user u_j for content c_i. Inspired by the works [43,44], we calculate the joint probability $P(c_i, u_j)$ of content c_i being requested by user u_j as follows:

$$
\begin{aligned}
P(c_i, u_j) &= P(c_i)P(u_j|c_i) \\
&= P(c_i)\frac{\exp({x_{c_i}}^T y_{u_j})}{\sum_{j=1}^{U} \exp({x_{c_i}}^T y_{u_j})}
\end{aligned}
\tag{3}
$$

where $P(c_i)$ represents the content popularity, while $P(u_j|c_i)$ reflects the preference of user u_j for content c_i. Combining these probabilities establishes a link between the user preference and content popularity.

It is important to note that our approach differs from the methods proposed in [43,44], as we obtain user and content embeddings from a real-world dataset. Furthermore, we consider the inner products of the learned user embeddings and content embeddings to measure their associations, enabling us to capture the relationships between users and content meaningfully.

4. Proposed Methodology

This section presents our GNN-DDQN agent, which incorporates a GNN as the Q-network within the DDQN framework [14]. DDQN improves upon the original DQN algorithm [40] by mitigating Q-value overestimation and enhancing the overall performance.

The GNN-DDQN agent predicts Q-values based on the observed state $\mathcal{S}_{t_l}$ and the chosen action $\mathcal{A}_{t_l}$ at each time step t_l. The predicted Q-value is denoted as $Q(\mathcal{S}_{t_l}, \mathcal{A}_{t_l})$. The objective of the agent is to learn an optimized policy that maximizes the expected Q-value $Q^*(\mathcal{S}_{t_l}, \mathcal{A}_{t_l})$. This section describes the state space, action space, and reward function used in our DDQN. Additionally, we explain the GNN architecture employed to map network states to action rewards for each node. Finally, we provide an overview of the GNN-DDQN agent, including its key components and functionality.

4.1. State Space

The network state $\mathcal{S}_{t_l} = \{s_{t_l}^{n_1}, \ldots, s_{t_l}^{n_N}\}$ captures the state of each network node at time step t_l. Each node's state feature vector $s_{t_l}^{n_k}$ at time step t_l is represented by $s_{t_l}^{n_k} \in \mathbb{R}^{C\times 3}$, where C is the total number of items in the network.

The state $s_{t_l}^{n_k}$ of a network node n_k at time step t_l consists of three components:

- 1st component: The number of requests for each content item c_i that have traversed the node during the previous time interval (t_{l-1} to t_l). This count is stored only for the requested content in receiver nodes, cached content in router nodes, and published content in source nodes.
- 2nd component: The cache storage of the node, represented by a binary variable for each content item c_i. A value of 1 indicates that the node caches the content during the previous time interval, while a value of 0 is used otherwise.
- 3rd component: The content publication of the node is also represented by a binary variable for each content item c_i. A value of 1 indicates that the node has published the content during the previous time interval, while a value of 0 is used otherwise.

4.2. Action Space

At a time step t_l, each node n_k can choose z out of C content items to cache. We record its cache scheme in a binary tuple,

$$a_{t_l}^{n_k} = \{b_{t_l}^{c_1,n_k}, \ldots, b_{t_l}^{c_C,n_k}\}, \tag{4}$$

where 1 means to 'cache' and 0 means to 'not cache', and the sum of all entries cannot exceed the assumed router's cache size z. We also use $\mathcal{A}_{t_l}$ to denote the cache scheme of all nodes such that,

$$\mathcal{A}_{t_l} = \{a_{t_l}^{n_1}, \ldots, a_{t_l}^{n_N}\} \tag{5}$$

and refer to it as the agent's action at the time step t_l. When the agent takes action $\mathcal{A}_{t_l}$ at time step t_l, the node n_k caches content according to $a_{t_l}^{n_k}$, which can be used to satisfy the request in the future.

4.3. Reward Function

Our objective is to maximize the cache hit ratio. Thus, we use cache hits as the agent's reward, denoted as $\mathcal{R}_{t_l} = \{r_{t_l}^{n_1}, \ldots, r_{t_l}^{n_N}\}$, which includes the cache hits of each node. For a node n_k at time step t_l, its cache hits for each content item are $r_{t_l}^{n_k} = \{r_{t_l}^{c_1,n_k}, \ldots, r_{t_l}^{c_C,n_k}\}$. Let us assume that a node n_k's reward sum for all content at time step t_l is $cacheHits_{t_l}^{n_k} = r_{t_l}^{c_1,n_k} + r_{t_l}^{c_2,n_k} + \cdots + r_{t_l}^{c_C,n_k}$; then, the objective function can be formulated as follows:

$$\begin{aligned} &\max \sum_{n_k\in\mathcal{N}} \sum_{t_l\in\mathcal{T}} cacheHits_{t_l}^{n_k} \\ &\text{s.t.} \\ &\sum_{c_i\in C} b_{t_l}^{c_i,n_k} \leq \begin{cases} z, & \text{node } n_k \text{ has the caching capability} \\ 0, & \text{otherwise} \end{cases}, \forall t_l \in \mathcal{T}, \forall n_k \in \mathcal{N} \end{aligned} \tag{6}$$

this objective aims to maximize the cache hit ratio for the stored content while ensuring that the number of content items stored in a node does not exceed z if it is a router node with caching capability and zero otherwise. We apply this constraint because only router nodes with caching capability can cache content, while other nodes, such as source and receiver nodes, can only distribute or receive content.

4.4. GNN Architecture

Our methodology utilizes a GNN for node-level Q-value predictions. The model architecture, depicted in Figure 2, operates on a network graph consisting of node embeddings and an adjacency matrix.

The GNN takes as input the graph structure data $G = (\mathcal{N}, \mathcal{E}, \mathcal{S})$, where $\mathcal{N}$ represents the set of nodes, $\mathcal{E}$ denotes the set of edges, and $\mathcal{S}$ represents the network state. With this input, the GNN model generates Q-value predictions for each action of every node, allowing us to estimate the outcome of each action through a single forward propagation of the GNN model.

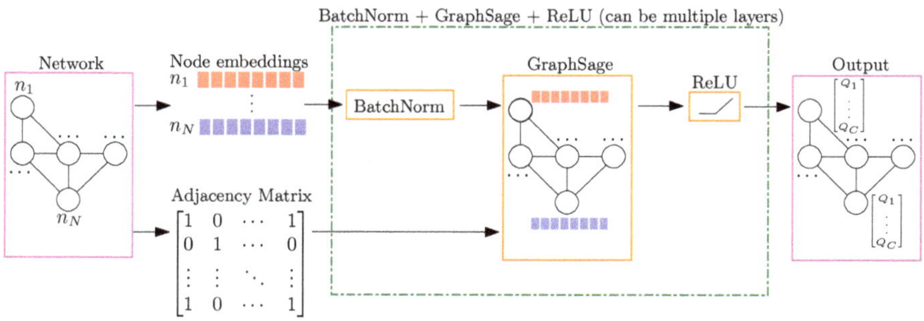

Figure 2. Model architecture.

For the GNN architecture, our approach utilizes four GraphSage layers [45]. Each layer has different hidden embedding dimensions, specifically 1024, 512, 256, and C. GraphSage is an inductive framework that leverages sampling and aggregation techniques to generate node embeddings. It allows for efficient embedding generation even for previously unseen data. By incorporating four GraphSage layers, we can aggregate information from up to four-hop neighbouring nodes at each step. This enables the GNN to capture the network structure and traffic patterns, resulting in more informative node embeddings for Q-value prediction.

The aggregation process in GraphSage is described by the equation

$$h^k_{N(v)} = AGG_k(h^{k-1}_u, \forall u \in N(v)), \tag{7}$$

where $N(v)$ represents the one-hop neighbours of node v, and h^{k-1}_u is the embedding of node u at the previous $(k-1)^{\text{th}}$ step. In each step, the GNN aggregates the embeddings of the one-hop neighbours of a node v from the previous step to obtain $h^k_{N(v)}$. The aggregation function AGG is typically permutation-invariant, meaning that it is not affected by the ordering of the aggregated embeddings. In our approach, we use a mean aggregator, which calculates the element-wise mean of the vectors $h^{k-1}_u, \forall u \in N(v)$.

After the aggregation step, the GNN performs concatenation by combining the embeddings of each central node from the previous $(k-1)^{\text{th}}$ step with the embeddings of its neighbouring nodes from the current k^{th} step. The concatenated embeddings are then fed into a fully connected layer with a nonlinear activation function:

$$h^k_v = \sigma\left[W^k \cdot \text{CONCAT}\left(h^{k-1}_v, h^k_{N(v)}\right)\right], \tag{8}$$

where W^k represents a learned matrix specific to the kth step, σ denotes the rectified linear activation function (ReLU), and h^k_v corresponds to the embedding of the central node v at the kth step. The CONCAT operation refers to the concatenation of the embeddings h^{k-1}_v and $h^k_{N(v)}$.

4.5. The GNN-DDQN Agent

The GNN-DDQN agent operates based on the procedure described in Algorithm 1. In the beginning, we initialize a replay buffer P, a Q-network (Q) implemented as a GNN with randomly generated parameters θ, and a target Q-network ($\hat{Q}$) with the same network architecture and parameters as Q. Each episode corresponds to a complete round of experimentation, and the time is divided into T slots. The GNN-DDQN agent takes actions at each time step, denoted as t_l (starting from t_1, as t_0 represents the initial point of the experimentation).

To balance exploration and exploitation, we utilize an ϵ-greedy exploration strategy [40]. This strategy involves randomly selecting actions with a probability of ϵ and selecting the action with the highest expected Q-value with a probability of $1-\epsilon$. The purpose is to encourage initial exploration and gradually decrease exploration over time. We employ an exponential decay strategy for ϵ, starting with an initial value of $\epsilon_s = 0.9$ and decaying to a minimum value of $\epsilon_e = 0.01$ with a decay rate of 0.01, denoted as $\epsilon_d = 100$.

Since each router with caching capability has a cache size of z, the GNN-DDQN agent selects z actions for each node at each time step. It chooses the top z actions with the highest Q-values for each node during greedy action selection. For random actions, it randomly selects z actions for each node. It is crucial to emphasize that the agent precisely chooses z actions for each node at every time step. However, it only executes these actions for router nodes with caching capabilities, excluding others.

Algorithm 1 GNN-DDQN Agent Operation

Input: number of episodes E, batch size B, target network update step K, replay buffer capacity R, epsilon start ϵ_s, epsilon end ϵ_e, epsilon decay ϵ_d, number of steps k, discount factor γ

Initialize replay buffer P with capacity R
Initialize Q-network with random weights θ
Initialize target $\hat{Q}$-network with weights $\hat{\theta} = \theta$
for *episode* $\in \{1, \ldots, E\}$ **do**
  for $t_l \in \{t_1, \ldots, t_T\}$ **do**
    Randomly pick $\epsilon' \in [0,1]$
    $\epsilon_t = \epsilon_e \cdot (\epsilon_s - \epsilon_e) \cdot \exp\left(\frac{-k}{\epsilon_d}\right)$
    $k = k + 1$
    for $n_k \in \{n_1, \ldots, n_N\}$ **do**
      if $\epsilon' < \epsilon_t$ **then**
        Randomly select z actions
      else
        Select z actions with the highest $Q(\mathcal{S}_{t_l}, \mathcal{A}_{t_l}|\theta)$
      end if
    end for
    Take action $\mathcal{A}_{t_l}$, get reward $\mathcal{R}_{t_l}$ and next state $\mathcal{S}_{t_{l+1}}$
    Store transition $(\mathcal{S}_{t_l}, \mathcal{A}_{t_l}, \mathcal{R}_{t_l}, \mathcal{S}_{t_{l+1}})$ into P
    Randomly sample B transitions $(\mathcal{S}_{b_j}, \mathcal{A}_{b_j}, \mathcal{R}_{b_j}, \mathcal{S}_{b_{j+1}})$ from P
    Use Equation (10) to compute $\mathcal{Y}_{b_j}$
    Perform a gradient descent step on $L(\theta)$ with respect to the network parameters θ, where $L(\theta)$ is computed in Equation (9)
    Update $\hat{\theta} = \theta$ every K steps
  end for
end for

At time step t_l, the GNN-DDQN agent interacts with the environment by taking action A_{t_l} and receiving a reward R_{t_l} and the subsequent state $S_{t_{l+1}}$ at time step t_{l+1}. The rewards, denoted by $\mathcal{R}_{t_l}$, are node-level rewards, where each node has C rewards corresponding to different actions. The newly generated transition $(\mathcal{S}_{t_l}, \mathcal{A}_{t_l}, \mathcal{R}_{t_l}, \mathcal{S}_{t_{l+1}})$ is then stored in the replay buffer P.

- We train the Q-network by randomly sampling a batch of transitions $(\mathcal{S}_{b_j}, \mathcal{A}_{b_j}, \mathcal{R}_{b_j}, \mathcal{S}_{b_{j+1}})$ from the replay buffer P. The Q-network is trained using gradient descent on a loss function $L(\theta)$, which measures the discrepancy between the predicted Q-values and the target Q-values. For the sampled transitions b_j, the loss function is defined as follows:

$$L(\theta) = \sqrt{\mathbb{E}\left[\sum_{b_j}((\mathcal{Y}_{b_j} - Q(\mathcal{S}_{b_j}, \mathcal{A}_{b_j}|\theta))^2 \cdot mask)\right]} \tag{9}$$

where $\mathcal{Y}_{b_j}$ is defined as follows:

$$\mathcal{Y}_{b_j} = \begin{cases} \mathcal{R}_{b_j}, & \text{if episode terminates at } b_{j+1} \\ \mathcal{R}_{b_j} + \frac{1}{z}\gamma \sum_{r \in \hat{Q}(\mathbf{S}_{b_{j+1}}, \arg\max_{\mathcal{A}'_{b_{j+1}}, |\mathcal{A}'_{b_{j+1}}|=z} Q(\mathbf{S}_{b_{j+1}}, \mathcal{A}'_{b_{j+1}}|\theta)|\hat{\theta})} r, & \text{otherwise} \end{cases} \tag{10}$$

and *mask* is defined as follows:

$$mask = \begin{cases} 1, & \text{if } n_k \text{ is a router with the caching capability} \\ 0, & \text{otherwise} \end{cases} \tag{11}$$

where $\mathcal{Y}_{b_j}$ represents the ground truth Q-values. If the episode terminates at transition b_{j+1}, $\mathcal{Y}_{b_j}$ is equal to $\mathcal{R}_{b_j}$. Otherwise, it is computed as the sum of $\mathcal{R}_{b_j}$ and the discounted expected reward of the next state. To estimate the expected future reward, the Q-network selects the top z greedy actions based on state $S_{b_{j+1}}$, and the corresponding Q-values are computed using the target Q-network $\hat{Q}$. The discount factor γ determines the importance of long-term rewards and is typically between 0 and 1. To ensure that each action taken at the next time step contributes equally, the sum of the expected long-term rewards is divided by z. To focus the loss contribution on routers with caching capabilities, a mask is applied in Equation (9). Nodes without caching capabilities are assigned a mask value of 0, while routers with caching capabilities have a mask value of 1.

To maintain the training stability, the parameters of the Q-network Q are periodically copied to the target Q-network $\hat{Q}$ every K steps. This helps to reduce the potential for the overestimation of the Q-values during training.

The key training parameters for the GNN-DDQN model are summarized in Table 3.

Table 3. Key GNN-DDQN model training parameters.

Parameters	Value
Number of episodes E	1000
Learning rate	0.001
Batch size B	32
Target network update step K	10
Replay buffer capacity R	1000
Epsilon start ϵ_s	0.9
Epsilon end ϵ_e	0.01
Epsilon decay ϵ_d	100
Discount factor γ	1
Optimizer	Adam

5. Experimentation and Results

In this section, we present simulation results to demonstrate the effectiveness of the proposed caching strategy in various network scenarios. To conduct these experiments, we utilized Icarus[46], a Python-based ICN caching simulator that comprehensively evaluates different caching strategies. Not bound to any specific architecture, such as content-centric networking (CCN) or named data networking (NDN), Icarus provides functionalities for more generalized ICN.

We employed the LRU strategy as the caching replacement policy for all our experiments. Moreover, the content popularity and user preference distributions mentioned in Section 3.2 were considered. Table 4 lists the key simulation parameters used. We followed the recommendations from a previous study [47] and set the internal and external link delays to 2 milliseconds (ms) and 34 ms, respectively, for all network topologies.

The experiments involved a set of distinct content, ranging from 600 to 1000, uniformly distributed among all source nodes in the network. The router's cache size varied from 1 to 4, denoting the number of content items that it could store. Each experiment consisted of a warm-up phase with 2000 requests, followed by 4000 requests that were measured to evaluate the performance of different caching schemes. User requests followed a Poisson distribution with a mean of 100 requests per second.

We divided each experiment into T segments, each representing 10 s. We conducted 600 experiments for each caching scenario and calculated the average evaluation metrics based on the results of the last 200 experiments.

Table 4. Key simulation parameters.

Parameters	Values
Network topology	GEANT [17], ROCKETFUEL [18], TISCALI [19], and GARR [19]
Internal link delay (all networks)	2 ms
External link delay (all networks)	34 ms
Number of distinct content items	Range: 600–1000 items
Content size	1500 bytes
Request size	150 bytes
Cache size	Range: 1–4 items
Number of warm-up requests	2000
Number of measured requests	4000
Request distribution	Poisson distribution with a mean of 100 requests per second
Time slot	10 s
Number of experimentations	600

The evaluation of different caching strategies relied on four key metrics.

- Cache Hit Ratio (CHR): The cache hit ratio represents the percentage of requests that can be fulfilled by retrieving data packets from the cache in the router nodes,

$$CHR = \frac{cacheHits}{cacheHits + cacheMiss}, \tag{12}$$

  where $cacheHits$ refers to the count of $Interest$ packets that are successfully satisfied by retrieving the corresponding $Data$ packet from the router's cache. On the other hand, $cacheMiss$ represents the count of $Interest$ packets that cannot be fulfilled by the cache and require fetching from external sources. An $Interest$ packet carries the name of the requested content and is transmitted from the receiver node, while a $Data$ packet contains the requested content itself and can serve as a response to the corresponding $Interest$ packet.
- Average Latency Time (ALT): The average latency time represents the average delay between the moment that a user sends an $Interest$ packet and the moment that it receives the corresponding $Data$ packet,

$$ALT = \frac{\sum_{i=1}^{I}(i_t + i_r)}{I}, \tag{13}$$

where I denotes the total number of user requests. i_t represents the travel time of the *Interest* packet from the receiver node to the node that fulfills the request, while i_r denotes the travel time of the responding *Data* packet.

- Average Path Stretch (APS): The average path stretch measures the average increase in path length for each user request,

$$APS = \sum_{i=1}^{I} \frac{path_{i,n_u,n_r}}{Path_{i,n_u,n_s}}, \quad (14)$$

where I represents the total number of user requests. n_u denotes the receiver node that sends the request, n_r refers to the node that responds to the request, and n_s represents the source node that publishes the requested content. $path_{i,n_u,n_r}$ denotes the number of hops travelled by the ith request, while $Path_{i,n_u,n_s}$ represents the shortest path from the receiver to the source.

- Average Link Load (ALL): The average link load represents the average ratio of the total link load to the total number of links in the network,

$$ALL = \frac{\sum_{l=1}^{L} \mathcal{L}_l}{L}, \quad (15)$$

where L denotes the total number of links in the network, and $\mathcal{L}_l$ represents the link load of the specific link l.

The caching performance of our proposed GNN-DDQN scheme was evaluated and compared with the state-of-the-art caching scheme MLP-DDQN. MLP-DDQN, which has been extensively studied in various research works [5,6], was used as a baseline for comparison. We adapted the MLP-DQN framework to incorporate the DDQN technique to ensure a fair comparison. The MLP-DDQN agent consisted of four linear layers with dimensions of 1024, 512, 256, and C.

There are some differences between the state representations of the MLP-DDQN agent and our proposed GNN-DDQN approach. In the MLP-DDQN agent, the first component of the state representation includes the number of requests for each content item c_i passed through each node, covering all types of nodes (receivers, routers, and sources). This provides more general traffic-related information to assist the MLP agent in making predictions, as it lacks the ability to gather neighbouring information as in the GNN approach.

Additionally, we compared our caching strategy with classical caching algorithms, including LCD, PROB_CACHE, LCE, and CL4M. These algorithms served as additional baselines to assess the performance of our proposed approach.

5.1. *Effect of Content Item Number*

This section examines the impact of the number of content items on caching performance. The number of content items ranged from 600 to 1000. Figure 3 illustrates how the caching performance varied with the number of items in the GEANT [17] network, where routers with caching capability had a uniform cache size of one item. The GEANT network is a well-known real-world topology comprising 53 nodes and 74 edges. Within the network are 13 source nodes responsible for content production, 32 router nodes, and 8 receiver nodes that initiate requests. However, it is worth noting that only router nodes with a degree higher than 2 have cache capabilities, which amounts to 19 nodes in this case.

Figure 3 demonstrates that GNN-DDQN consistently outperformed all other caching strategies across different numbers of distinct content items. GNN-DDQN achieved a maximum improvement of 34.42% in CHR, 4.76% in ALT, 3.77% in APS, and 5.21% in ALL compared to MLP-DDQN. On average, GNN-DDQN surpassed LCD and PROB_CACHE by 41.33% and 103.92% in CHR, respectively. It also achieved significantly lower ALT, APS, and ALL than LCD and PROB_CACHE. Furthermore, the performance gap between GNN-DDQN and LCE and CL4M was even more pronounced regarding all evaluation metrics.

Overall, GNN-DDQN consistently exhibited exceptional caching performance regardless of the number of content items. Its superiority over MLP-DDQN stemmed from its ability to facilitate cooperative caching among neighbouring router nodes. By efficiently utilizing the caching space of all router nodes, GNN-DDQN enhanced the network performance. Additionally, GNN-DDQN outperformed traditional caching algorithms by quickly capturing user preferences and proactively placing popular content on appropriate router nodes. Consequently, the cache hit ratio improved, alleviating network traffic congestion.

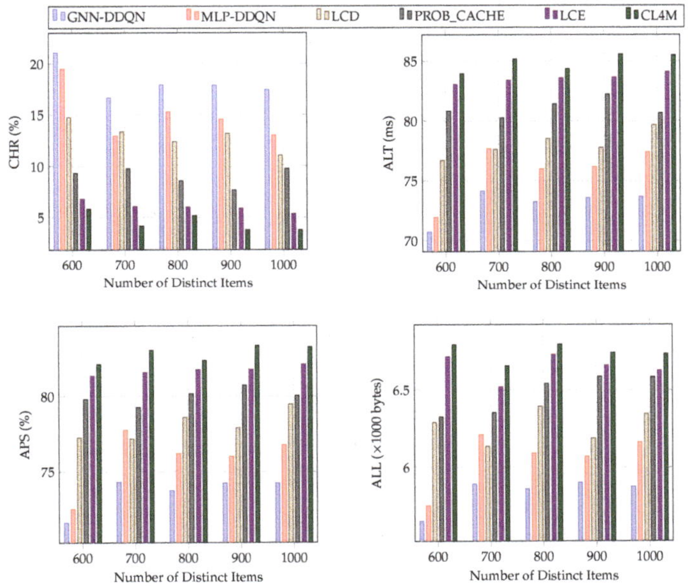

Figure 3. The cache performance of GNN-DDQN, MLP-DDQN, LCD, PROB_CACHE, LCE, and CL4M varied with the number of content items in the GEANT network.

5.2. Effect of Cache Size

This section investigates the performance of different caching schemes across various router cache sizes, defined as the number of content items. Figure 4 presents the caching performance of GNN-DDQN, MLP-DDQN, LCD, PROB_CACHE, LCE, and CL4M under different caching scenarios. The router cache sizes ranged from 1 to 4, while the number of content items was fixed at 1000.

GNN-DDQN exhibited a substantial performance advantage over MLP-DDQN when the cache size was limited to one item. For cache sizes of two and four, GNN-DDQN and MLP-DDQN performed similarly. However, when the cache size was set to three items, GNN-DDQN outperformed MLP-DDQN by achieving an 11.87% higher CHR, 3.57% lower ALT, 1.54% APS, and 2.20% lower ALL.

Significantly, regardless of the router cache size, GNN-DDQN consistently reduced the latency time by at least 14.96%, 29.88%, 92.20%, and 76.37% compared to LCD, PROB_CACHE, LCE, and CL4M, respectively. The advantages of GNN-DDQN stem from its ability to predict popular content in advance and proactively cache them.

5.3. Effect of Network Topology

To further evaluate the effectiveness of the proposed caching scheme, we conducted experiments on different network topologies, namely ROCKETFUEL [18], TISCALI [19], and GARR [19]. The aim was to assess the robustness of the caching scheme in diverse network environments.

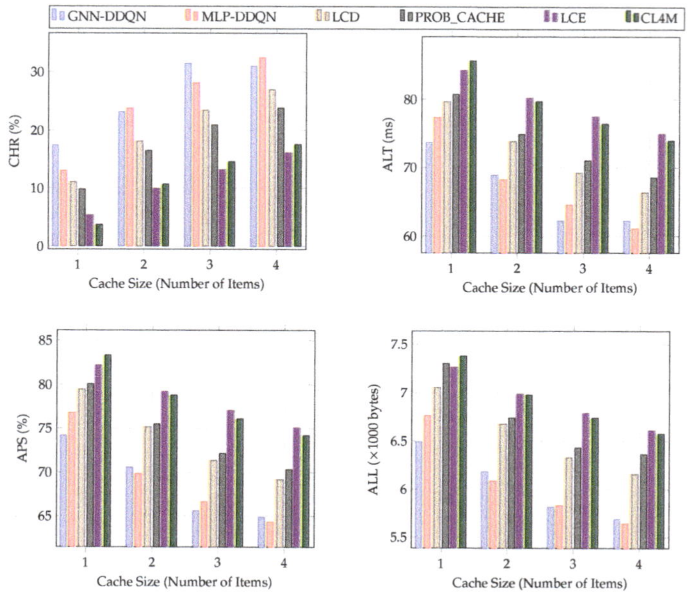

Figure 4. The cache performance of GNN-DDQN, MLP-DDQN, LCD, PROB_CACHE, LCE, and CL4M varied with the router's cache size in the GEANT network.

Table 5 presents the distribution of each network topology's source, router, and receiver nodes. It is important to note that, in the TISCALI network, only router nodes with a degree higher than 6 possess caching capabilities, resulting in 36 router nodes equipped with cache functionality.

Table 5. The number of source, router, and receiver nodes for different network topologies.

Topologies	Source Nodes	Router Nodes	Receiver Nodes
ROCKETFUEL [18]	10	104	104
TISCALI [19]	44	160	36
GARR [19]	13	27	21

This section evaluates the caching performance of different strategies in the ROCKETFUEL, TISCALI, and GARR network topologies. The experiments were conducted with an item number of 1000, and all routers with caching capabilities had a uniform cache size of one item. The results are summarized in Table 6.

Across all network topologies, GNN-DDQN consistently outperformed the other strategies. Specifically, in ROCKETFUEL, GNN-DDQN achieved a 2.89% higher CHR than MLP-DDQN. In TISCALI, the margin became even more significant, with GNN-DDQN achieving a 25.72% higher CHR than MLP-DDQN. These results highlight the superior caching performance of GNN-DDQN, particularly in large networks such as ROCKETFUEL and TISCALI.

Furthermore, GNN-DDQN demonstrated a significant margin over MLP-DDQN and other traditional caching schemes in the GARR network. This further emphasizes the robustness and effectiveness of GNN-DDQN across various network topologies.

Table 6. The caching performance of GNN-DDQN, MLP-DDQN, LCD, PROB_CACHE, LCE and CL4M in ROCKETFUEL, TISCALI, and GARR.

Strategy \ Topology	ROCKETFUEL			
	CHR	ALT	APS	ALL
GNN-DDQN	18.41%	75.05 ms	73.78%	4081.90 bytes
MLP-DDQN	17.89%	75.27 ms	74.00%	4104.82 bytes
LCD	13.00%	80.12 ms	78.61%	4410.05 bytes
PROB_CACHE	9.32%	82.67 ms	78.78%	4425.06 bytes
LCE	8.35%	83.16 ms	79.36%	4598.08 bytes
CL4M	10.20%	82.13 ms	79.31%	4530.41 bytes
Strategy \ Topology	**TISCALI**			
	CHR	**ALT**	**APS**	**ALL**
GNN-DDQN	15.57%	82.19 ms	82.76%	2498.73 bytes
MLP-DDQN	12.38%	84.41 ms	83.49%	2525.98 bytes
LCD	12.07%	85.14 ms	84.75%	2626.72 bytes
PROB_CACHE	7.90%	88.11 ms	85.55%	2656.99 bytes
LCE	7.22%	88.71 ms	86.03%	2692.65 bytes
CL4M	3.67%	91.22 ms	86.61%	2727.35 bytes
Strategy \ Topology	**GARR**			
	CHR	**ALT**	**APS**	**ALL**
GNN-DDQN	12.18%	71.96 ms	74.48%	5559.79 bytes
MLP-DDQN	5.65%	76.31 ms	76.38%	5762.81 bytes
LCD	8.12%	74.77 ms	76.26%	5665.82 bytes
PROB_CACHE	4.02%	77.73 ms	77.48%	5749.57 bytes
LCE	3.67%	77.91 ms	77.76%	5832.152 bytes
CL4M	4.32%	77.42 ms	77.34%	5801.03 bytes

6. Conclusions

In this paper, we introduced GNN-DDQN, an intelligent caching scheme designed for the SDN-ICN scenario. GNNs have gained significant attention recently for their ability to handle graph-structured data. Leveraging this capability, we applied GNNs to process network topologies, enabling cooperative caching among nodes and promoting a wider variety of cached content. By integrating GNNs into DRL, our proposed approach empowered the DRL agent to make caching decisions for all nodes in the network with only one forward pass through the neural network. This integration not only streamlined the caching decision-making process but also harnessed the power of GNN-DRL synergy in optimizing the caching strategies.

Firstly, we generated user preferences for content based on a real-world dataset. This step ensured that the evaluation reflected realistic user behaviour and content demand patterns. Next, we developed a GNN-DDQN agent within the SDN controller, enabling the agent to make intelligent caching decisions for all router nodes equipped with caching capabilities in the ICN network. Finally, we compared the performance of our proposed GNN-DDQN caching scheme with the state-of-the-art MLP-DDQN strategy and several classical benchmark caching schemes, including LCD, PROB_CACHE, CL4M, and LCE. The extensive evaluation revealed that GNN-DDQN consistently outperformed MLP-DDQN in most scenarios. Notably, in the best-case scenario, GNN-DDQN achieved a remarkable 34.42% higher CHR, a 4.76% lower ALT, a 3.77% lower APS, and a 5.21% lower ALL compared to MLP-DDQN. Furthermore, GNN-DDQN demonstrated superior performance compared to classical caching schemes. To assess the robustness of our proposed scheme, we conducted experiments on benchmark network topologies, including GEANT, ROCKET-FUEL, TISCALI, and GARR. GNN-DDQN consistently delivered outstanding performance across these diverse network topologies, reinforcing its reliability and applicability in real-world scenarios.

Some potential directions for future research include the following.

- Latency Consideration: Investigating the latency of the SDN controller and exploring techniques to mitigate the latency issue when dealing with a large number of network nodes.
- IoV-Based Environment: Integrating the proposed caching strategy in an IoV environment. This may involve studying the unique characteristics of vehicular networks and exploring how the methodology can be adapted to optimize content caching and delivery in such dynamic and mobile scenarios.

Author Contributions: Conceptualization, J.H., T.T., H.L. and A.N.; Methodology, J.H., T.T. and H.L.; Software, J.H. and T.T.; Supervision, A.N.; Validation, T.T. and H.L.; Visualization, J.H.; Writing—original draft, J.H. and T.T.; Writing—review and editing, H.L. and A.N. All authors have read and agreed to the published version of the manuscript.

Funding: This research received no external funding.

Data Availability Statement: The data will be available on request from the corresponding author.

Conflicts of Interest: The authors declare no conflict of interest.

References

1. Zhang, Z.; Lung, C.H.; Wei, X.; Chen, M.; Chatterjee, S.; Zhang, Z. In-network Caching for ICN-based IoT (ICN-IoT): A Comprehensive Survey. *IEEE Internet Things J.* **2023**. [CrossRef]
2. Musa, S.S.; Zennaro, M.; Libsie, M.; Pietrosemoli, E. Convergence of Information-Centric Networks and Edge Intelligence for IoV: Challenges and Future Directions. *Future Internet* **2022**, *14*, 192. [CrossRef]
3. Gür, G.; Kalla, A.; de Alwis, C.; Pham, Q.V.; Ngo, K.H.; Liyanage, M.; Porambage, P. Integration of ICN and MEC in 5G and Beyond Networks: Mutual Benefits, Use Cases, Challenges, Standardization, and Future Research. *IEEE Open J. Commun. Soc.* **2022**, *3*, 1382–1412. [CrossRef]
4. Aldaoud, M.; Al-Abri, D.; Awadalla, M.; Kausar, F. Leveraging ICN and SDN for Future Internet Architecture: A Survey. *Electronics* **2023**, *12*, 1723. [CrossRef]
5. Sun, S.; Zhou, J.; Wen, J.; Wei, Y.; Wang, X. A DQN-based cache strategy for mobile edge networks. *Comput. Mater. Contin.* **2022**, *71*, 3277–3291. [CrossRef]
6. Li, J.; Tang, J.; Li, J.; Zou, F. Deep reinforcement learning for intelligent computing and content edge service in ICN-based IoV. In Proceedings of the 2021 IEEE International Conference on Communications Workshops (ICC Workshops), Montreal, QC, Canada, 14–23 June 2021; pp. 1–7.
7. Wu, L.; Cui, P.; Pei, J.; Zhao, L.; Song, L. *Graph Neural Networks*; Springer: Berlin/Heidelberg, Germany, 2022.
8. Almasan, P.; Suárez-Varela, J.; Rusek, K.; Barlet-Ros, P.; Cabellos-Aparicio, A. Deep reinforcement learning meets graph neural networks: Exploring a routing optimization use case. *Comput. Commun.* **2022**, *196*, 184–194. [CrossRef]
9. Jiang, W.; Luo, J. Graph neural network for traffic forecasting: A survey. *Expert Syst. Appl.* **2022**, *207* , 117921. [CrossRef]
10. Fan, S.; Wang, X.; Shi, C.; Cui, P.; Wang, B. Generalizing Graph Neural Networks on Out-Of-Distribution Graphs. *arXiv* **2021**, arXiv:2111.10657.
11. Rusek, K.; Suárez-Varela, J.; Almasan, P.; Barlet-Ros, P.; Cabellos-Aparicio, A. RouteNet: Leveraging Graph Neural Networks for network modeling and optimization in SDN. *IEEE J. Sel. Areas Commun.* **2020**, *38*, 2260–2270. [CrossRef]
12. Suárez-Varela, J.; Carol-Bosch, S.; Rusek, K.; Almasan, P.; Arias, M.; Barlet-Ros, P.; Cabellos-Aparicio, A. Challenging the generalization capabilities of Graph Neural Networks for network modeling. In Proceedings of the ACM SIGCOMM 2019 Conference Posters and Demos, Beijing, China, 19–23 August 2019; pp. 114–115.
13. Almasan, P.; Suárez-Varela, J.; Badia-Sampera, A.; Rusek, K.; Barlet-Ros, P.; Cabellos-Aparicio, A. Deep reinforcement learning meets graph neural networks: Exploring a routing optimization use case. *arXiv* **2019**, arXiv:1910.07421.
14. Van Hasselt, H.; Guez, A.; Silver, D. Deep reinforcement learning with double q-learning. In Proceedings of the AAAI Conference on Artificial Intelligence, Phoenix, AZ, USA, 12–17 February 2016; Volume 30.
15. He, X.; Liao, L.; Zhang, H.; Nie, L.; Hu, X.; Chua, T.S. Neural collaborative filtering. In Proceedings of the 26th International Conference on World Wide Web, Perth, Australia, 3–7 April 2017; pp. 173–182.

16. Harper, F.M.; Konstan, J.A. The movielens datasets: History and context. *Acm Trans. Interact. Intell. Syst. TIIS* **2015**, *5*, 1–19. [CrossRef]
17. Géant Homepage. 2020. Available online: https://geant3plus.archive.geant.net/Pages/home.html (accessed on 18 August 2022).
18. Spring, N.; Mahajan, R.; Wetherall, D. Measuring ISP topologies with Rocketfuel. *ACM SIGCOMM Comput. Commun. Rev.* **2002**, *32*, 133–145. [CrossRef]
19. Knight, S.; Nguyen, H.X.; Falkner, N.; Bowden, R.; Roughan, M. The internet topology zoo. *IEEE J. Sel. Areas Commun.* **2011**, *29*, 1765–1775. [CrossRef]
20. Zhang, L.; Estrin, D.; Burke, J.; Jacobson, V.; Thornton, J.D.; Smetters, D.K.; Zhang, B.; Tsudik, G.; Massey, D.; Papadopoulos, C.; et al. Named data networking (ndn) project. *Relat. Téc. NDN-0001 Xerox Palo Alto Res. Cent.-PARC* **2010**, *157*, 158.
21. Laoutaris, N.; Che, H.; Stavrakakis, I. The LCD interconnection of LRU caches and its analysis. *Perform. Eval.* **2006**, *63*, 609–634. [CrossRef]
22. Psaras, I.; Chai, W.K.; Pavlou, G. Probabilistic in-network caching for information-centric networks. In Proceedings of the Second Edition of the ICN Workshop on Information-Centric Networking, Helsinki, Finland, 13–17 August 2012; pp. 55–60.
23. Chai, W.K.; He, D.; Psaras, I.; Pavlou, G. Cache "less for more" in information-centric networks. In Proceedings of the International Conference on Research in Networking, Chennai, India, 1–3 February 2012; Springer: Berlin/Heidelberg, Germany, 2012; pp. 27–40.
24. Li, Z.; Simon, G.; Gravey, A. Caching policies for in-network caching. In Proceedings of the 2012 21st International Conference on Computer Communications and Networks (ICCCN), Munich, Germany, 30 July–2 August 2012; pp. 1–7.
25. Shailendra, S.; Sengottuvelan, S.; Rath, H.K.; Panigrahi, B.; Simha, A. Performance evaluation of caching policies in ndn-an icn architecture. In Proceedings of the 2016 IEEE Region 10 Conference (TENCON), Singapore, 22–25 November 2016; pp. 1117–1121.
26. Munikoti, S.; Agarwal, D.; Das, L.; Halappanavar, M.; Natarajan, B. Challenges and opportunities in deep reinforcement learning with graph neural networks: A comprehensive review of algorithms and applications. *arXiv* **2022**, arXiv:2206.07922.
27. Nomikos, N.; Zoupanos, S.; Charalambous, T.; Krikidis, I. A Survey on Reinforcement Learning-Aided Caching in Heterogeneous Mobile Edge Networks. *IEEE Access* **2022**, *10* , 4380–4413. [CrossRef]
28. Zhao, L.; Ran, Y.; Wang, H.; Wang, J.; Luo, J. Towards Cooperative Caching for Vehicular Networks with Multi-level Federated Reinforcement Learning. In Proceedings of the ICC 2021-IEEE International Conference on Communications, Montreal, QC, Canada, 14–23 June 2021; pp. 1–6.
29. Song, C.; Xu, W.; Wu, T.; Yu, S.; Zeng, P.; Zhang, N. QoE-driven edge caching in vehicle networks based on deep reinforcement learning. *IEEE Trans. Veh. Technol.* **2021**, *70*, 5286–5295. [CrossRef]
30. Aung, N.; Dhelim, S.; Chen, L.; Lakas, A.; Zhang, W.; Ning, H.; Chaib, S.; Kechadi, M.T. VeSoNet: Traffic-Aware Content Caching for Vehicular Social Networks Using Deep Reinforcement Learning. *IEEE Trans. Intell. Transp. Syst.* **2023** . [CrossRef]
31. Zhang, D.; Wang, W.; Zhang, J.; Zhang, T.; Du, J.; Yang, C. Novel edge caching approach based on multi-agent deep reinforcement learning for Internet of vehicles. *IEEE Trans. Intell. Transp. Syst.* **2023** . [CrossRef]
32. He, P.; Cao, L.; Cui, Y.; Wang, R.; Wu, D. Multi-Agent Caching Strategy for Spatial-Temporal Popularity in IoV. *IEEE Trans. Veh. Technol.* **2023** . [CrossRef]
33. Liu, L.; Yuan, X.; Zhang, N.; Chen, D.; Yu, K.; Taherkordi, A. Joint Computation Offloading and Data Caching in Multi-Access Edge Computing Enabled Internet of Vehicles. *IEEE Trans. Veh. Technol.* **2023** . [CrossRef]
34. Hou, J.; Xia, H.; Lu, H.; Nayak, A. A gnn-based approach to optimize cache hit ratio in ndn networks. In Proceedings of the 2021 IEEE Global Communications Conference (GLOBECOM), Madrid, Spain, 7–11 December 2021; pp. 1–6.
35. Hou, J.; Lu, H.; Nayak, A. GNN-GM: A Proactive Caching Scheme for Named Data Networking. In Proceedings of the 2022 IEEE International Conference on Communications Workshops (ICC Workshops), Seoul, Republic of Korea, 16–20 May 2022; pp. 1–6.
36. Peng, H.; Du, B.; Liu, M.; Liu, M.; Ji, S.; Wang, S.; Zhang, X.; He, L. Dynamic graph convolutional network for long-term traffic flow prediction with reinforcement learning. *Inf. Sci.* **2021**, *578*, 401–416. [CrossRef]
37. Yang, S.; Yang, B.; Kang, Z.; Deng, L. IHG-MA: Inductive heterogeneous graph multi-agent reinforcement learning for multi-intersection traffic signal control. *Neural Netw.* **2021**, *139*, 265–277. [CrossRef]
38. Chen, S.; Dong, J.; Ha, P.; Li, Y.; Labi, S. Graph neural network and reinforcement learning for multi-agent cooperative control of connected autonomous vehicles. *Comput.-Aided Civ. Infrastruct. Eng.* **2021**, *36*, 838–857. [CrossRef]
39. Zhou, X.; Bilal, M.; Dou, R.; Rodrigues, J.J.; Zhao, Q.; Dai, J.; Xu, X. Edge Computation Offloading with Content Caching in 6G-Enabled IoV. *IEEE Trans. Intell. Transp. Syst.* **2023** . [CrossRef]
40. Mnih, V.; Kavukcuoglu, K.; Silver, D.; Rusu, A.A.; Veness, J.; Bellemare, M.G.; Graves, A.; Riedmiller, M.; Fidjeland, A.K.; Ostrovski, G.; et al. Human-level control through deep reinforcement learning. *Nature* **2015**, *518*, 529–533. [CrossRef]
41. Breslau, L.; Cao, P.; Fan, L.; Phillips, G.; Shenker, S. Web caching and Zipf-like distributions: Evidence and implications. In Proceedings of the IEEE INFOCOM'99, Conference on Computer Communications, Proceedings, Eighteenth Annual Joint Conference of the IEEE Computer and Communications Societies, The Future is Now (Cat. No. 99CH36320), New York, NY, USA, 21–25 March 1999; Volume 1, pp. 126–134.
42. Rendle, S.; Krichene, W.; Zhang, L.; Anderson, J. Neural collaborative filtering vs. matrix factorization revisited. In Proceedings of the Fourteenth ACM Conference on Recommender Systems, Virtual, 22–26 September 2020; pp. 240–248.
43. Chen, B.; Yang, C. Caching policy optimization for D2D communications by learning user preference. In Proceedings of the 2017 IEEE 85th Vehicular Technology Conference (VTC Spring), Sydney, Australia, 4–7 June 2017; pp. 1–6.

44. Chen, B.; Yang, C. Caching policy for cache-enabled D2D communications by learning user preference. *IEEE Trans. Commun.* **2018**, *66*, 6586–6601. [CrossRef]
45. Hamilton, W.L.; Ying, R.; Leskovec, J. Inductive representation learning on large graphs. *arXiv* **2017**, arXiv:1706.02216.
46. Saino, L.; Psaras, I.; Pavlou, G. Icarus: A caching simulator for information centric networking (icn). In Proceedings of the SimuTools, ICST, Lisbon, Portugal, 17–19 March 2014; Volume 7, pp. 66–75.
47. Zhang, B.; Ng, T.E.; Nandi, A.; Riedi, R.; Druschel, P.; Wang, G. Measurement based analysis, modeling, and synthesis of the internet delay space. In Proceedings of the 6th ACM SIGCOMM Conference on Internet Measurement, Rio de Janeriro, Brazil, 25–27 October 2006; pp. 85–98.

MDPI
St. Alban-Anlage 66
4052 Basel
Switzerland
www.mdpi.com

Future Internet Editorial Office
E-mail: futureinternet@mdpi.com
www.mdpi.com/journal/futureinternet

www.ingramcontent.com/pod-product-compliance
Lightning Source LLC
Chambersburg PA
CBHW061515210926
51850CB00008B/1932
9783036598574